THE TABLE SAW BOOK

THE TABLE SAW BOOK

R.J. De Cristoforo

FIRST EDITION
SEVENTH PRINTING

Printed in the United States of America

Library of Congress Cataloging-in-Publication Data

De Cristoforo, R.J.
The table saw book / by R.J. De Cristoforo.
p. cm.
Includes index.
ISBN 0-8306-7789-5 ISBN 0-8306-2789-8 (pbk.)
1. Radial saws. 2. Woodworking. I. Title.
TT186.D44 1987 87-29046
684′.083—dc19 CIP

TAB BOOKS offers software for sale. For information and a catalog, please contact TAB Software Department, Blue Ridge Summit, PA 17294-0850.

Questions regarding the content of this book should be addressed to:

Reader Inquiry Branch
TAB BOOKS
Blue Ridge Summit, PA 17294-0214

Cover photograph courtesy of Delta International Machinery Corp.

Contents

Introduction

IF YOU HAVE EVER CUT A PIECE OF WOOD WITH A hand saw, you will immediately be awed by the advantages of a table saw. Much of the possibility of human error and all of the effort of hand-tool work are eliminated because of the power in a table saw and its built-in accuracy. For example, rip cuts and crosscuts, which are basic woodworking procedures no matter what the project is, will automatically be straight, and the sawed edges will be square to adjacent surfaces. To cut straight and square with hand saws requires considerable apprenticeship. I am not suggesting that you will be an expert with a table saw by tomorrow, but you will get there sooner than you might think.

What is very impressive is that the rudimentary applications of ripping and crosscutting are a drop in the ocean when the real potential of the table saw is realized. There are no secrets, however. You become knowledgeable, and that's what this book is all about. The magic of the table saw is there for everyone. The machine doesn't care who flicks the switch—amateur or professional, male or female, dedicated woodworker or part time user, the tool turns on for all. Knowing how to get the most out of the saw is simply a matter of following the procedures that get you there, and this holds true for all standard, and some not-so-standard, applications.

If all you did with a table saw was use its basic components—a miter gauge and rip fence—for routine wood-sizing chores, you would be using only a small percentage of the saw's myriad functions. By using just the saw blade and repeat-pass and two-pass techniques, you can form many classic joint designs. By using special techniques with a standard saw blade, you can shape an arch or cove, form a raised panel, even cut a circle, or do *kerfing*, which allows wood to be bent without steaming equipment. By adding a dadoing accessory, you can do joint-making, among other things, faster, easier, and more accurately. By adding a molding head, also an accessory, you can use the tool for many jobs that usually require a separate shaper.

One of the fascinating aspects of using a table saw is that you can add to its functions, or do routine work more safely and more accurately, by making special jigs and fixtures. Some of these make-it-yourself items are not available commercially. Others might have counterparts that can be purchased, but you can save money and custom-design the accessory for your use by making it yourself. The idea, of course, is to make the accessories as the need for them occurs. There is little point in stocking a shop with idle materials.

Work carefully when you do make an accessory. Since it will become a lifetime tool, it pays to take ten minutes to do a five-minute job. You can use these accessories for the projects in TAB Book No. 2964, *24 Table Saw Projects*, by Percy W. Blandford.

I'd like to stress the importance of safety in the shop. I have worked with power tools for better than 30 years and I'm happy to say that I am still whole. I attribute this fact, to a considerable extent, to my healthy respect for power equipment, and to a degree of fear that I have never lost. Millions of table saws are in use and injuries are a matter of statistics, but an interesting fact emerges. More professionals are hurt than amateurs. The reason? Expertise can lead to overconfidence, which makes the operator vulnerable. The expert often feels that he is good enough to bypass a safety requirement, to his regret. Don't ever allow "casual" to become part of "expertise." You are always the master, the creative partner of the tool, but being the boss doesn't change the machine's disinterest in what you present for it to cut.

Follow all procedures as they are outlined. Many times, especially when the application is strange to you, it makes sense to go through a dry run; that is, follow the instructions, but with the machine turned off. In this way, you can preview how the wood must move and where your hands must be positioned for safety before you flick the switch.

Table saws are supplied with guards, which include a blade-covering hood, a splitter, and anti-kickback fingers. Always be sure to follow the manufacturer's instructions so the guard will be correctly installed. You will see illustrations throughout this book that do not show the tool's guard in place. This is not the way to work. The guard was removed for clarity in the photographs. If it were left in place, you would not be able to see the cut being demonstrated. Always use the guard!

Remember to "measure twice, saw once" and to "think twice before sawing."

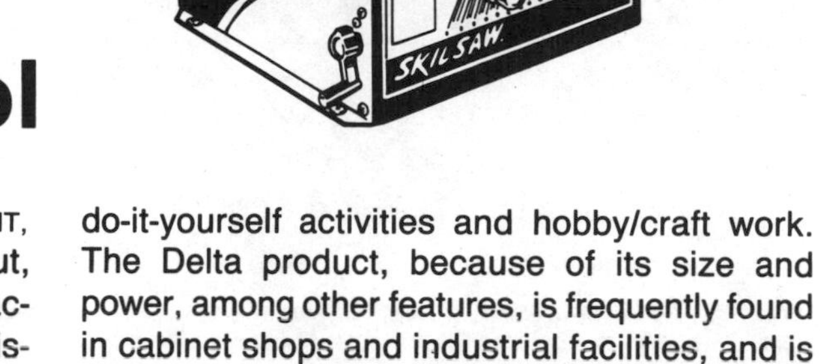

Chapter 1

The Tool

TABLE SAWS DIFFER IN PHYSICAL SIZE AND WEIGHT, in horesepower, in maximum depth of cut, in location of controls, and in some other factors, but they all have an essential characteristic: they drive a circular saw blade that does the cutting. Without the saw blade, you might as well increase the size of the tool's table with a plywood panel and use it as a workbench.

An example of extremes in sizes and capacities is illustrated by the units shown in Figs. 1-1 and 1-2. Both machines can crosscut, and allow ripping, mitering, and beveling, all of which are basic table saw functions, but the Delta Unisaw (Fig. 1-1) weighs about 400 pounds, drives a 10-inch blade with a 3-horsepower motor, and cuts as deep as 3 ⅛ inches. The Dremel product, which weighs less than 10 pounds (Fig. 1-2), works with a 4-inch blade and is limited to cutting through 1-inch stock at 90 degrees. The small tool warrants no apologies, so long as the user accepts its horsepower and capacity limitations. Many a homeowner is happy with this easily moved and stored tool, using it for do-it-yourself activities and hobby/craft work. The Delta product, because of its size and power, among other features, is frequently found in cabinet shops and industrial facilities, and is the dream tool of the amateur who longs to work with professional equipment.

The size of a saw is determined by the diameter of the blade it can drive. You might not see much difference in the overall dimensions of various machines. The 12-inch concept shown in Fig. 1-3 doesn't require much more floor space than a 9- or 10-inch unit, but it will provide greater depth of cut at both 45 and 90 degrees. It also will, or should, have the horsepower that is needed for the additional capacity. It simply requires more oomph to get through a 4-inch-thick beam than a 2-inch-thick one.

The major capacity criterion that is accepted today, logically, is: for general shop use, furniture making, woodworking, and home maintenance, the saw should be able to cut through a 2 × 4 at 45 and 90 degrees. Even the new compact saws, or benchtop saws as they are of-

Fig. 1-1. The Delta Unisaw, shown here equipped with an accessory sliding table, turns a 10-inch blade and uses, or should use, 2- to 3-horsepower, 220 V motor. Maximum depth of cut is 3⅛ inches. The distance from the front of the table to the saw blade at maximum elevation is 12¼ inches.

ten called (Figs. 1-4 through 1-6), are designed to meet this criterion. What the small tools might fall behind on is table area. The larger the table, the easier it is to establish control when sawing large panel material like sheets of plywood or particleboard. The table area of most saws can be increased through the use of side extensions, which are supplied as standard equipment or as extra-cost accessories.

TABLE SIZE

The size of the table also affects the distance from the blade to the front edge of the table. If this distance is 12 inches or more, then it is easy to establish good support when crosscutting a standard-sized board, which is 11¼ inches wide. When the distance is less, it is often necessary to start a crosscut with the head of the miter gauge off the table.

Another factor influenced by table size is the maximum distance that can be established between the saw blade and the rip fence. Some tools have bars or tubes for the rip fence that are longer than the table's width. Thus, the ripping capacity is increased without adding table area.

Many operators increase the support area of a small saw, sometimes even a big saw, by constructing a large table around it. One way to increase the support area without using additional floor space is to build the saw into a workbench (Fig. 1-7). This, in effect, increases the width of the saw's table, which helps when crosscutting long boards, and provides con-

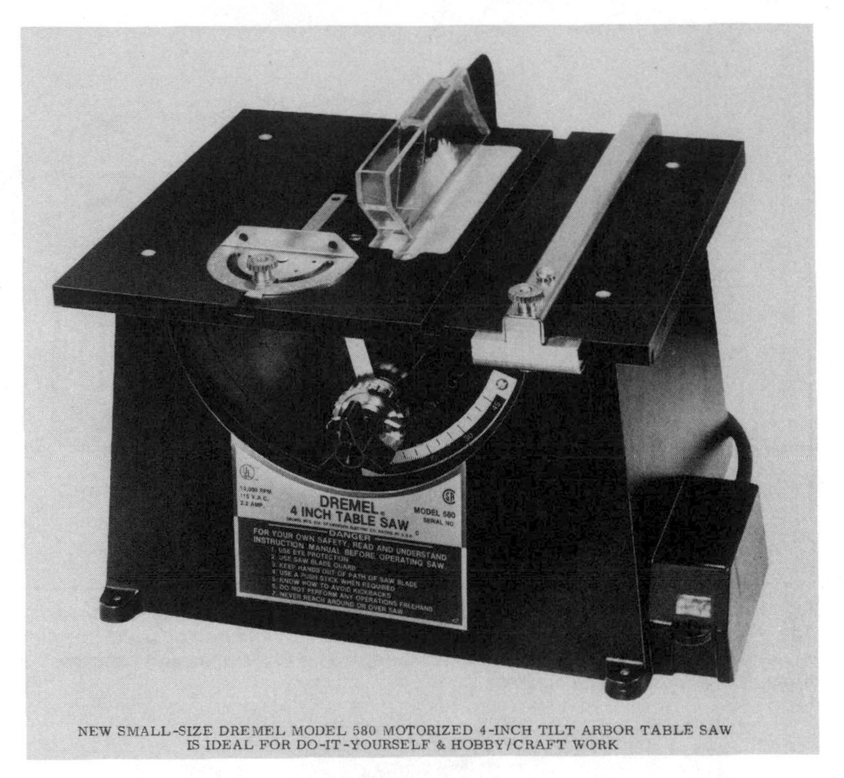

Fig. 1-2. Dremel's small saw has a 10- × -12-inch work table and drives a 4-inch-diameter blade. The tool can be used for perpendicular cuts up to 1 inch and, with its tilting arbor, can do mitering in stock up to ¾ inch thick. A single front control is used for both blade elevation and blade tilt.

Fig. 1-3. A 12-inch saw doesn't require much more floor space than a 9- or 10-inch version, but it can saw through thicker wood at both 90- and 45-degree settings. Note the built-in sliding extension at the left side of the table.

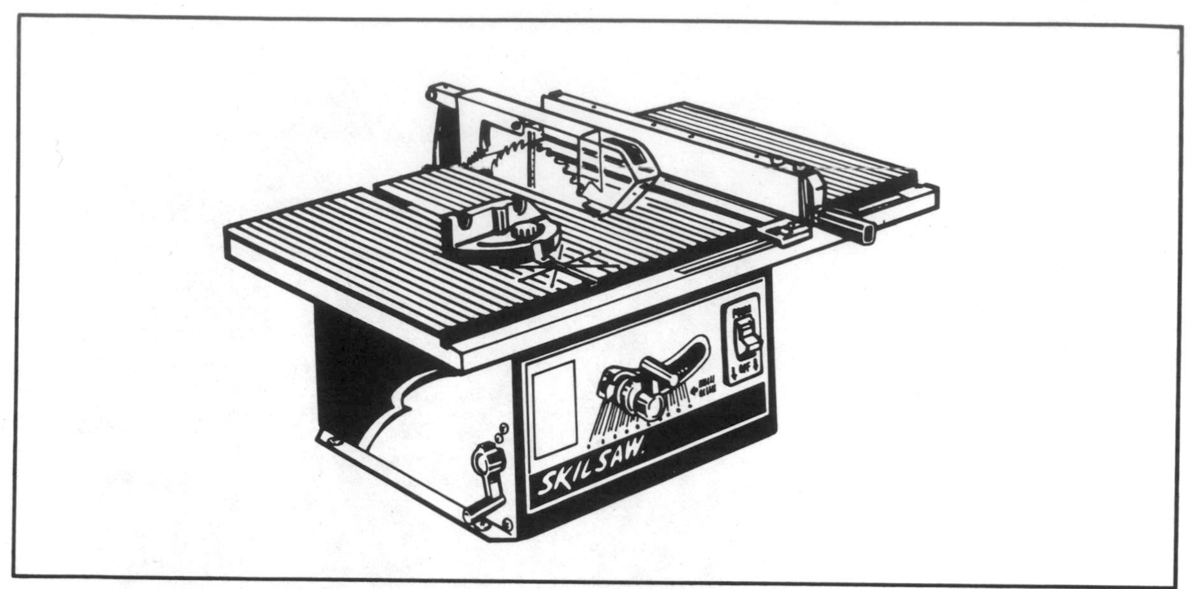

Fig. 1-4. Skil's compact saw can be used with blades from 5½ to 8¼ inches in diameter and is rated at 2 horsepower. The die-cast aluminum table measures 16 × 27 inches and can be used for rip cuts up to 12 inches wide without extensions. Maximum cuts are 2⅙ inches at 90 degrees and 2 inches at 45 degrees. The tool has separate controls for blade projection and bevel adjustments.

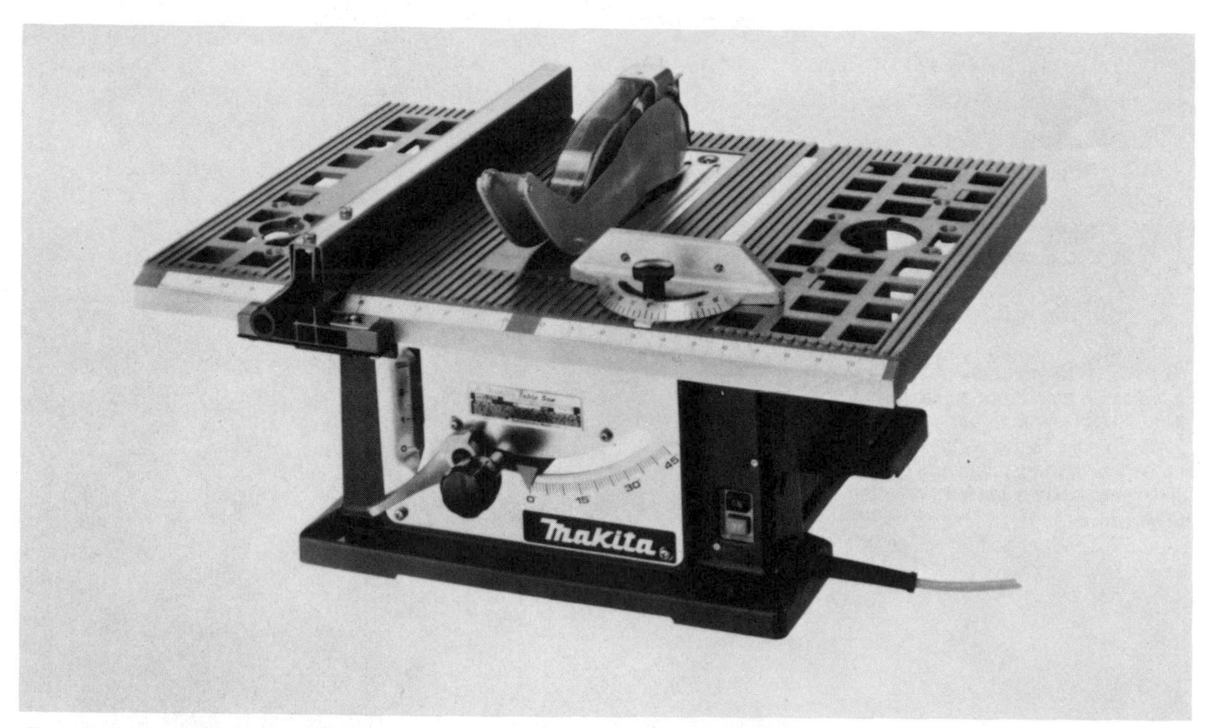

Fig. 1-5. Makita's 8¼-inch saw has a powerful 12-amp motor and a no-load speed of 4,500 rpm. The 37½-pound machine has a 26- × -18⅛-inch table and can handle 2 ½-inch stock at 90 degrees and 1⅝-inch stock at 45 degrees. It can be used with a 6-inch dadoing tool for cuts up to ½ inch wide. Machines of this type can be bolted to a benchtop or to a floor stand.

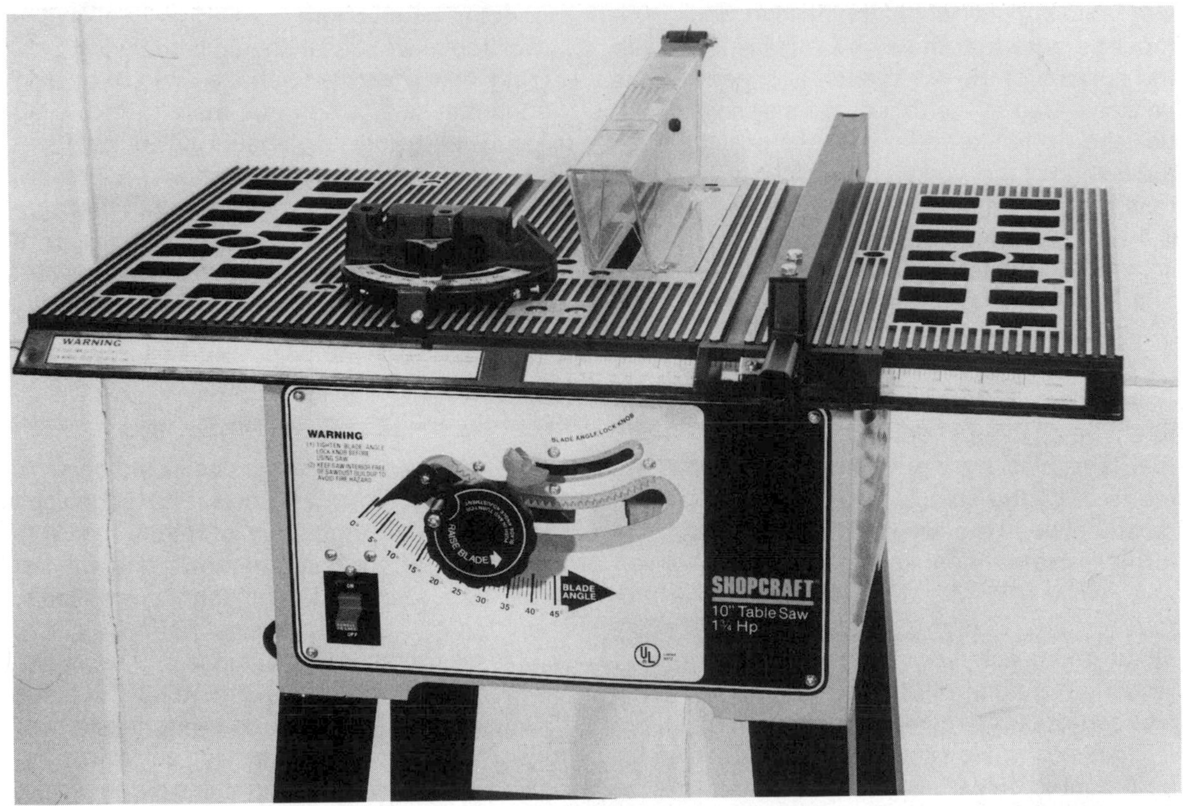

Fig. 1-6. The Shopcraft 1¾-horsepower saw is a motorized benchtop concept that drives a 10-inch blade. The table measures 19 × 28 inches and has a scale along its front edge for rip-fence settings. It easily cuts through a 2 × 4 at 90 degrees or 45 degrees, which is a basic capacity criterion for table saws.

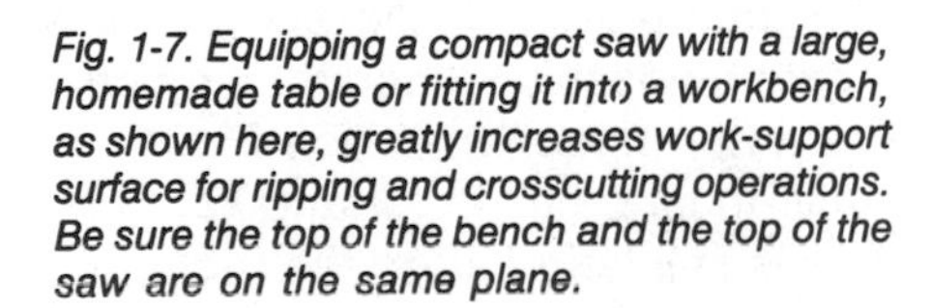

Fig. 1-7. Equipping a compact saw with a large, homemade table or fitting it into a workbench, as shown here, greatly increases work-support surface for ripping and crosscutting operations. Be sure the top of the bench and the top of the saw are on the same plane.

siderable support when ripping long stock. You must be sure that the saw's table is level with the surface of the workbench and that there is an open area at the rear of the saw so you can use the rip fence and mount the guard. When the design of the workbench or special table permits, design an opening into the shelf the saw will rest on. Then, you can use a container of some sort to catch the bulk of the sawdust (Fig. 1-8).

ARBOR DRIVES

Saw blades are secured on an arbor, which might be an independent, bearing-mounted unit that is rotated by a pulley and V-belt arrangement, or by the spindle of the motor. To distinguish between the two methods, a saw with the former arrangement is called a *motor-driven* saw, while one with the latter is called a *motorized saw*. Many available saws, especially small ones, are motorized. This arrangement lessens the number of necessary components and contributes to compactness of the machine.

A motor-driven saw requires that the motor be mounted on its own plate, usually hinged to provide correct belt tension. A V-belt connects between a pulley on the motor spindle and one on the arbor. The motor might be mounted under the saw's table or behind it. Often, more than one V-belt is used, on the assumption that multiple belts ensure maximum motor power, minimize vibration, and are less likely to slip. The Delta Unisaw, for example, drives the saw's arbor with three belts. In such cases, replacement belts must be purchased in matched sets.

A point that is made for motorized saws is that a direct drive eliminates the possibility of belt slippage and therefore ensures full power. The rebuttal is that should the saw blade be taxed or stalled for some reason, the motor would be abused, whereas in a similar situation with a belt drive, it's more likely that the belt would slip and the motor would not be harassed. It's really a moot point. Abuse situations like encountering a particularly hard grain area or a knot can happen, but the remedy is for the operator not to be stubborn. Easing up to allow the saw blade to work as it should is the way to go.

It's also possible to abuse a motor by asking the machine to do more than it is powered to do. That, of course, should never happen.

A unique drive system is shown in Fig. 1-9. The arbor is rotated by a flexible shaft that connects between it and the motor's spindle. Like the motorized concept, the idea is for the shaft to provide direct power to the blade. No belts, no possibility of slippage.

ELECTRONICS

It was inevitable that electronics would begin to affect the design of power tools. The technology has already become part of many portable power tools and it has appeared, and will become more evident, on stationary power tools. The 10-inch Craftsman product shown in Fig. 1-10, is microprocessor-equipped and has a digital information display panel. Settings for blade elevation and blade tilt are electronically controlled. The action is called an *automatic jog movement*. Press a button and the blade projection changes by 5/1000 inch. Blade tilt is controllable in increments of 1/10 degree.

TILTING ARBOR/TILTING TABLE

In order to make cuts across the grain or with the grain at an angle to adjacent surfaces, it is necessary to tilt the table or the saw blade, actions that depend on how the tool is designed. When the tool has a tilting arbor, so the blade is tilted rather than the table, the workpiece is held in a horizontal position when making cuts like the cross miter demonstrated in Fig. 1-11. When the table must be tilted for angular cuts, the workpiece is still held flat on the table, but on the same plane as the tilt, as shown in Figs. 1-12 and 1-13.

The question of which design is best is not as debatable as it once was. For one thing, you would need to search hard today, especially among saws offered to amateur woodworkers,

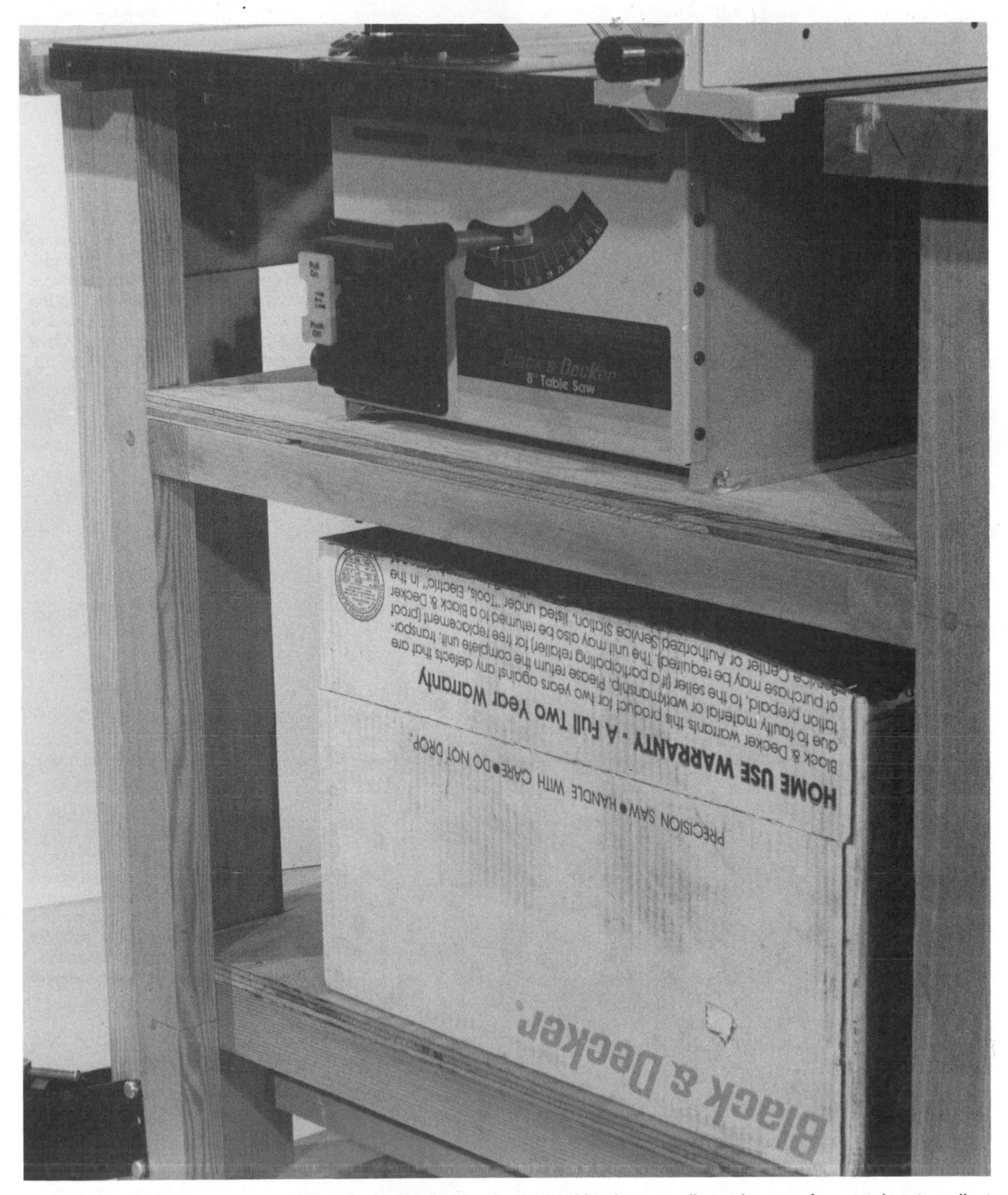

Fig. 1-8. Mounting the saw on a shelf with an opening that is spanned by the saw allows the use of a container to collect the bulk of the sawdust.

Fig. 1-9. The special drive system found on some Craftsman saws consists of a flexible shaft that connects between the motor's spindle and the saw's arbor. Systems like this are touted as providing more positive motor power to the saw blade.

Fig. 1-10. Blade elevation and blade tilt on this microprocessor-equipped table saw are electronically controlled. At a touch of a button on a control panel, the blade can be raised or tilted in specific increments. The control panel includes a digital information display.

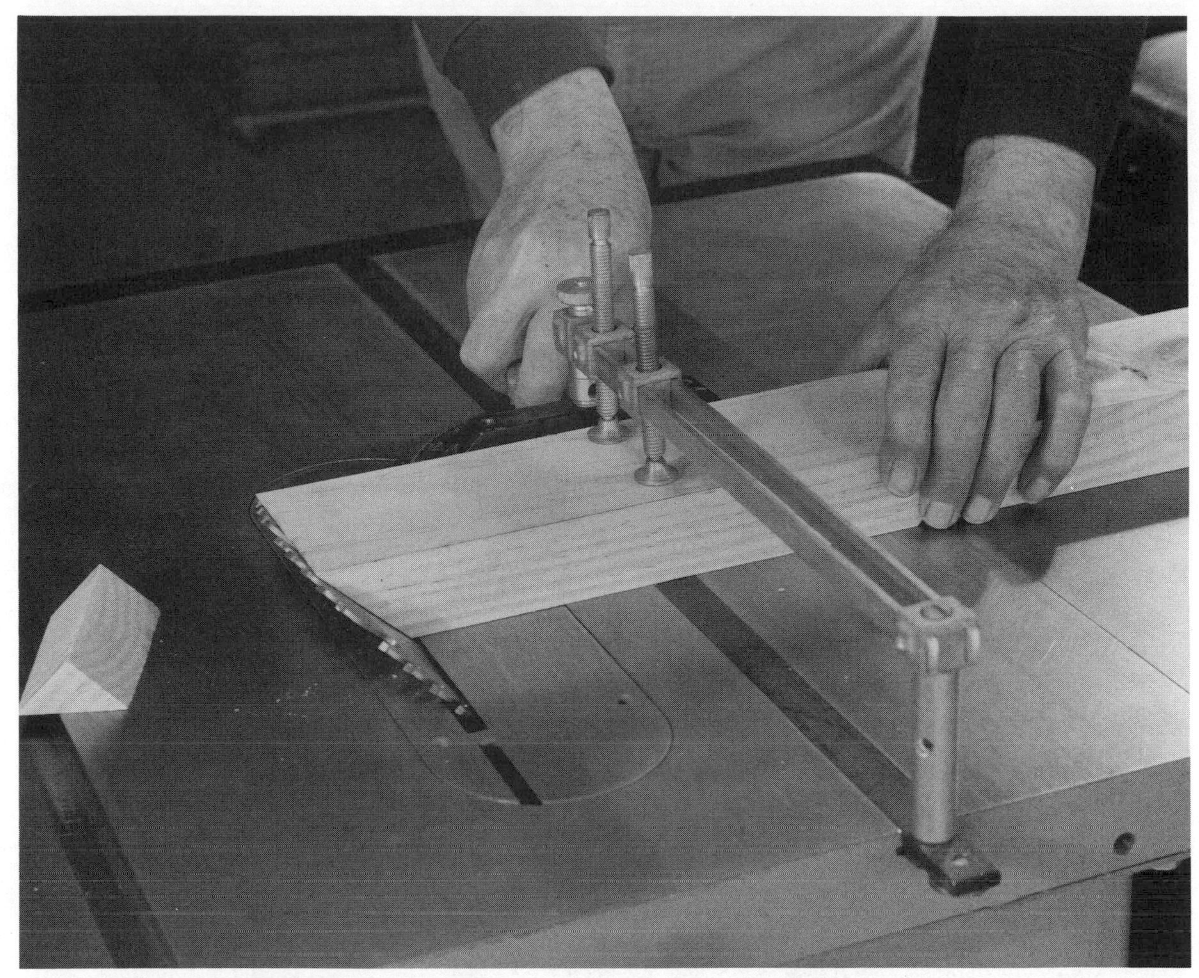

Fig. 1-11. A table saw designed with a tilting arbor can make cross miter cuts, like the one shown here, and bevel cuts—which, in essence, are angular rip cuts. Keeping the work flat on the table is convenient and aids accuracy.

for one that does not have a tilting arbor. Exceptions are found among the multipurpose tools like the Shopsmith shown in Fig. 1-14, where the concept does not permit incorporating a tilting arbor. Folks who buy this type of tool are interested in the entire concept, not just its sawing mode.

There is little doubt that a tilting arbor adds convenience to many sawing operations, but for either type, accuracy depends primarily on how carefully adjustments are made and how the wood is controlled during the pass.

OTHER BASIC FEATURES

A complete table saw includes a guard, a miter gauge, and a rip fence (Fig. 1-15). It's a rare offering that does not include the basic accessories, but there are considerable differences in design. For example, the unit shown in Fig. 1-16 employs a heavy U-shaped channel as a rip fence, and the fence is designed with an integral "pusher" that has a replaceable finger to move the workpiece on ripping operations so fingers can stay well away from the saw blade. Original designs for pushers that can be used on any

Fig. 1-12. For saws without a tilting arbor, the table must be tilted to provide the cut angle required. Such saws require a little more control to be precise. A hold-down, shown here on a Shopsmith miter gauge, can be a big help.

type of rip fence are detailed in Chapter 6.

The maximum distance available between the saw blade and the rip fence depends on the length of the bars or tubes on which the fence slides. Many times when the standard fence supports are not much longer than the table's width, you can replace them with accessory units that allow the fence to be safely moved off the table to increase maximum ripping capacity (Fig. 1-17).

How a rip fence is secured can differ among tools. A common design is shown in Fig. 1-18. A knob at the front end, by means of a rod inside the hollow fence, pulls a clamp tightly against the rear of the table. The base of the fence at the front end bears tightly against the front edge of the table so the fence locks parallel to the saw blade, as it should.

Some units, like the new Delta Unifence (Fig. 1-19), employ only a front locking device; in this case, a lever. When the mechanism is correctly adjusted, moving the lever downward (Fig. 1-20) ensures that the fence will hold the correct adjustment and that locking will be positive. Adjustable nylon screws in the base of the unit are used to control how the fence will slide and the angle between the face of the fence and the saw's table, which should be 90 degrees.

Some rip fence support units are fitted with scales so the fence can be positioned for the cut

Fig. 1-13. Bevel cuts also require that the table be tilted when the unit does not have a tilting arbor. This type of cut can be controlled more easily than a cross miter since the work is guided and supported by the rip fence.

without fussing too much with measuring between fence and blade (Fig. 1-21). Not all of them, however, are as accurate as you might wish, so it usually pays to double-check by actually measuring. I have found, though, that the Unifence, when correctly installed and maintained, can be accurate within 1⁄64 inch.

The Miter Gauge

The miter gauge (Fig. 1-22) is standard equipment for all table saws. The unit slides in slots that are milled into the table on both sides of, and parallel to, the saw blade. On some saws, the slots are somewhat T-shaped, and the bar of the gauge has a screw-attached circular plate

Fig. 1-14. Tilting tables are rarely found on modern table saws, but are common on multipurpose tools since the concept does not permit the use of a tilting arbor. Tools like this are purchased for everything they can do, not for just the saw function.

Fig. 1-15. Basic accessories for a table saw—and it's rare to find a unit that does not include them in the purchase price—are a miter gauge, a rip fence, and a guard. The guard will usually include a splitter and anti-kickback fingers.

Fig. 1-16. Designs of basic components differ among tools. This heavy, channel-type rip fence has a built-in pusher that is used in place of fingers to get thin work past the saw blade. Note the basket-type saw guard.

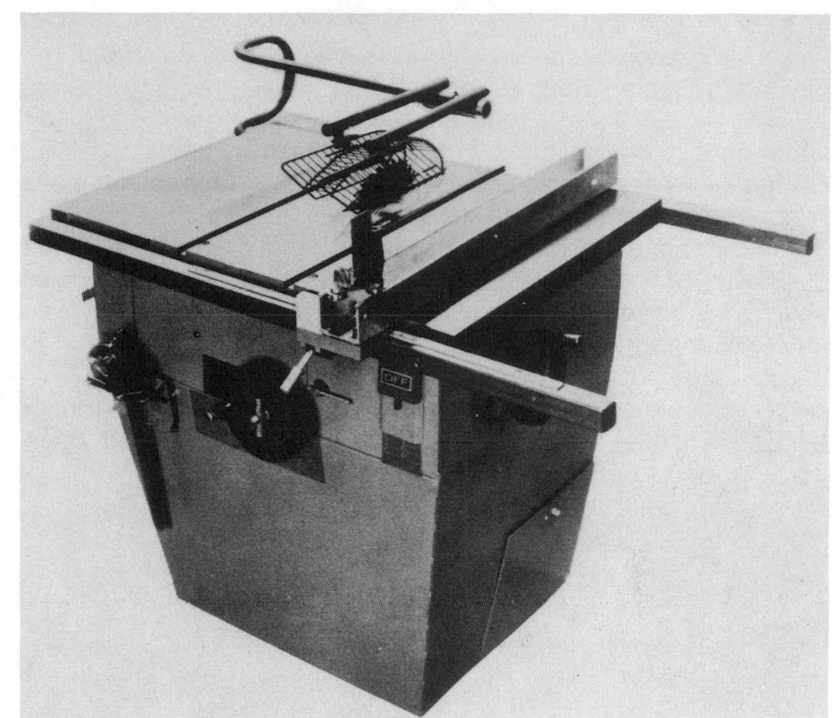

Fig. 1-17. Extra-long supports for the rip fence might be provided with the tool, but are usually offered as an accessory. The purpose of the units is to increase rip capacity. Often, the open space between the rails is filled with a plywood panel.

at its front end that engages in the slot and keeps the gauge from falling when it is retracted beyond the table's front edge.

The head of the miter gauge can be pivoted left or right so the tool can be used for angular sawing, as well as simple crosscuts (Fig. 1-23). A good miter gauge will be equipped with individually adjustable index stops at the 90-degree setting and at 45 degrees both left and right. The tool can, of course, be set and locked at angles that are not determined by the built-in stops.

A valuable accessory for a miter gauge is a *hold-down*, often simply called a clamping device, which might be designed along one of the lines shown in Figs. 1-24 through 1-26. Whatever the arrangement, the purpose of the hold-down is to hold the workpiece securely against the head of the miter gauge and flat on the table as you make the cut. The benefits are twofold. You gain in accuracy since it is less likely that the work will move as you cut, and you work more safely since you do not use your fingers to keep the wood correctly positioned.

Controls

In addition to the switch used to turn the tool on and off, there are mechanisms for adjusting

Fig. 1-18. A common rip fence design has a locking knob at the front that works through a connecting rod to pull a clamp tightly against the rear edge of the table. The fence locks securely at each end and will be parallel to the saw blade.

the elevation of the saw blade (projection above the table) and for setting the blade at an angle for a cross miter or bevel cut. Usually, the blade-elevating control is a handwheel located on the front of the machine (Fig. 1-27). Usually, the wheel is turned clockwise to raise the blade, counterclockwise to lower it. An additional control, in this case the triangular knob in the center of the handwheel in Fig. 1-27, is used to lock the blade's elevation where you have set it. Be sure to loosen this knob before you make an adjustment and to tighten it afterwards. Positive stops are provided so you can't raise or lower the blade beyond the extreme positions.

Blade-tilt settings are also controlled with a handwheel, which is usually located on a side of the saw (Fig. 1-28). You can determine the tilt of the blade by reading a scale located on the front of the saw, as shown in Fig. 1-27. It's okay to use the scale for approximate settings, but determine what the true setting must be by making the final adjustment with a T-bevel or other measuring device. The tilt mechanism will have adjustable auto stops at the 0- and 45-degree positions. Check these stops right off, and frequently afterwards, by using a square for the 0-degree stop (90 degrees between the side of the blade and the table) and an accurate triangle for the 45-degree stop (45 degrees between the side of the blade and the table). More information on these critical adjustment factors is provided in Chapter 3.

Table Inserts

Table inserts, sometimes called *throat plates*, fit in an opening that is in the saw's ta-

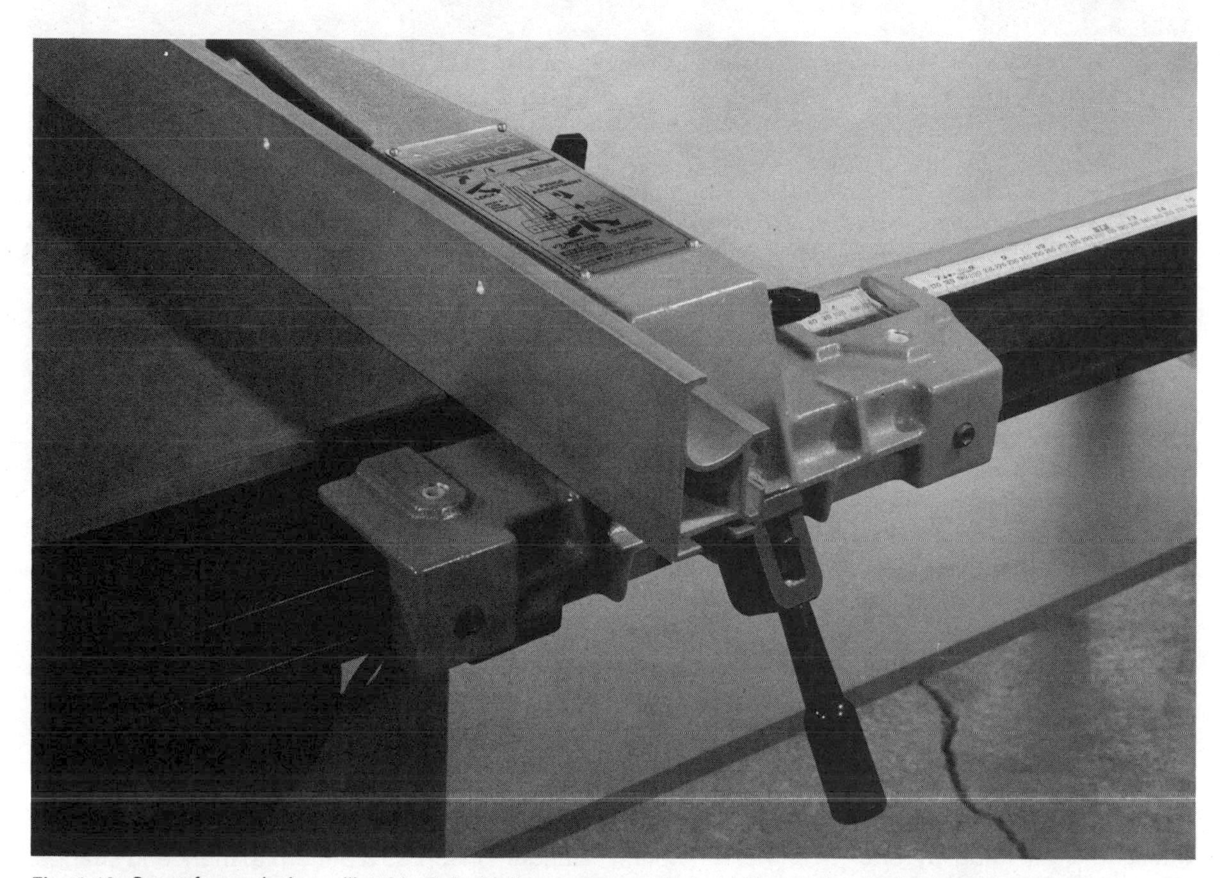

Fig. 1-19. Some fence designs, like this Delta Unifence, have a single locking device, in this case a lever.

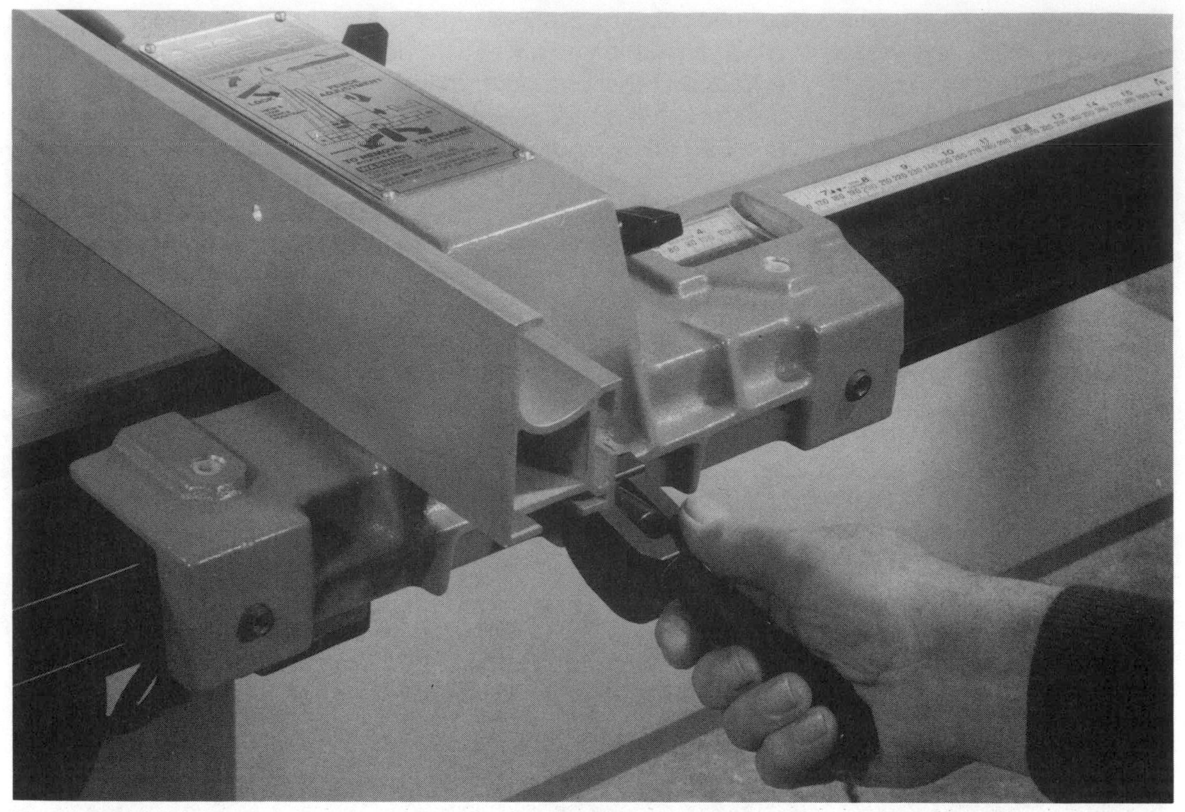

Fig. 1-20. When the mechanism is adjusted correctly, the single lever, moved downward, ensures a firm grip for the carriage that supports the rip fence. A clamp device at the rear of the fence is not needed.

ble and are removable for access to the arbor, arbor washers, and the arbor nut that secures the saw blade. These inserts allow the use of various cutting tools, while minimizing the open space around them. The inserts used for a particular saw will be of identical size and shape, but will have different internal cutouts to accommodate various cutting tools that can be mounted on the arbor (Fig. 1-29). The primary insert will have a slim, long slot so a conventional saw blade can poke through. One designed for use with a dadoing tool will have a shorter but wider slot, while one designed for use with a molding head will have an even shorter but wider slot.

The top surface of inserts must be level with the surface of the table. Some are adjustable and can be leveled by one means or another (Fig. 1-30). Others are designed to snap into position so they automatically achieve the necessary levelness. Some merely sit in place; others are secured with screws. What is important is to use the right one for the particular cutting tool. Actually, being right is the only option. For example, you can't possibly use the standard saw-blade insert with a molding head.

Many times, it is good practice to make special inserts. For example, when ripping very slim pieces of wood, even the small opening around the blade in a standard insert is wide enough for the action of the saw blade to pull the strips down into the table. The solution is to make an insert so the opening around the blade is nil. To custom-make inserts like those shown in Fig. 1-31, rip stock—plywood, solid lumber, or even plastic—to the necessary width, and use the standard insert as a pattern to mark end shapes. You can form the end shapes with a jig, band,

Fig. 1-21. The miter gauge is used for crosscutting and mitering. It can be used in either of the table slots that parallel the saw blade.

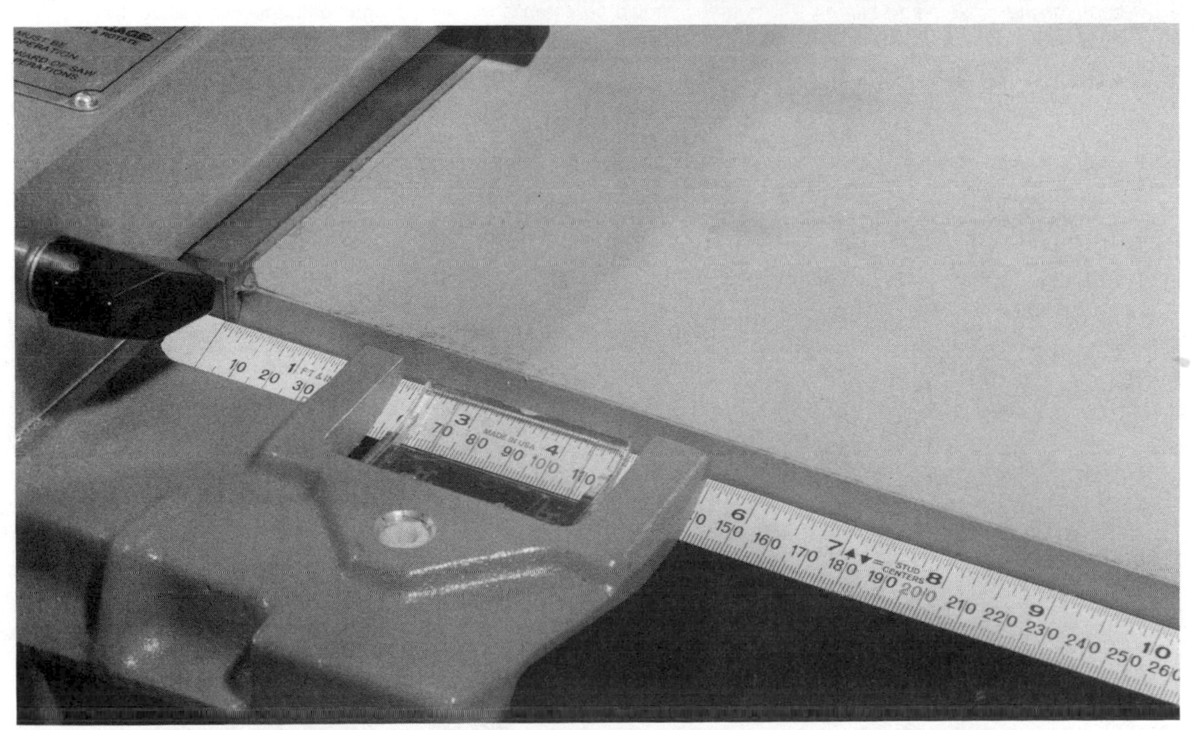

Fig. 1-22. Rip-fence supports, whether they are bars or tubes, often will have a scale that is used to determine settings from blade to rip fence. If testing indicates that your scale is not accurate, use it for approximate settings only. Make final adjustments by actually measuring between the blade and fence.

Fig. 1-23. The head of the miter gauge can be pivoted for cuts other than 90 degrees. A good miter gauge will have adjustable stops for the most-used settings. The settings should be established with gauges and checked frequently.

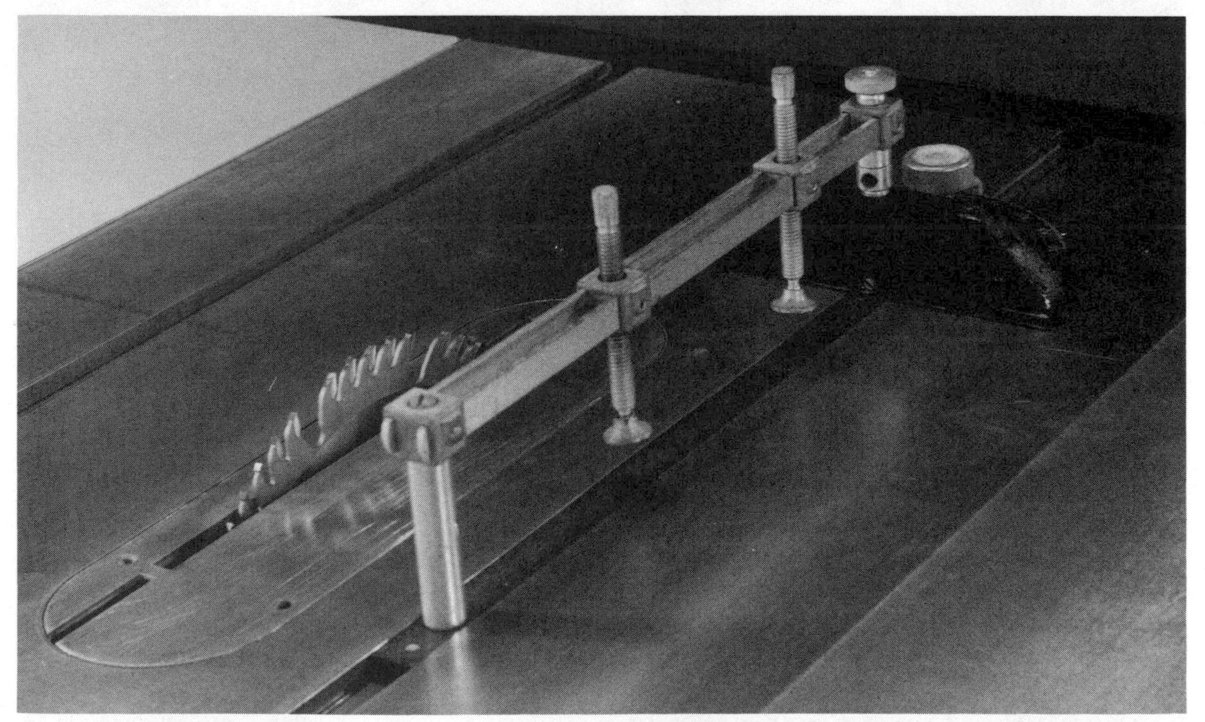

Fig. 1-24. A miter-gauge hold-down provides a valuable assist for crosscutting and mitering. It keeps the work securely against the head of the miter gauge and flat on the table as you make the cut.

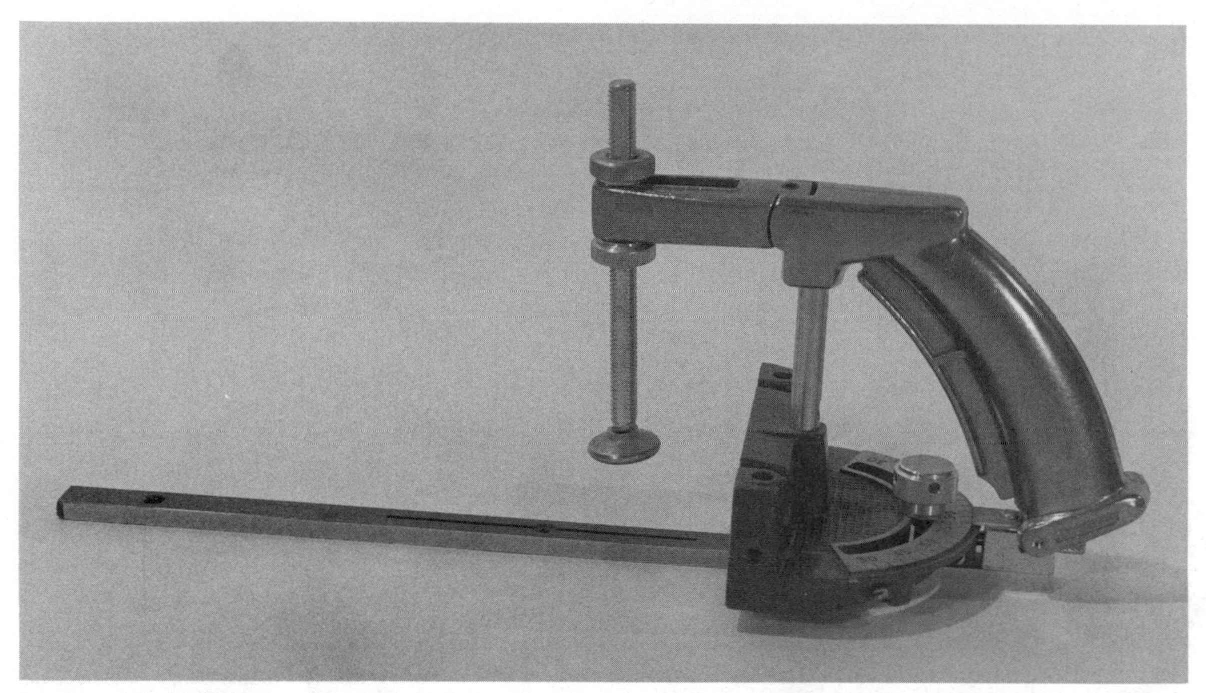

Fig. 1-25. This hold-down, a Shopsmith product, bears down on the work as the handle is squeezed. The vertical rod is adjustable for various stock thicknesses.

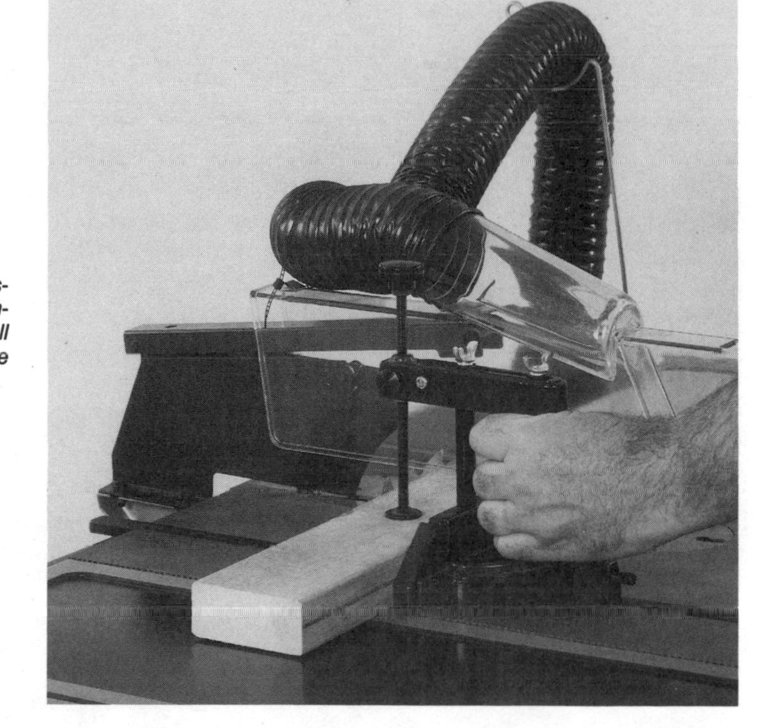

Fig. 1-26. This hold-down, a Craftsman product, works in a similar fashion to the Shopsmith model. All hold-downs contribute to accurate sawing and safety.

Fig. 1-27. On most large table saws, the blade elevation control is a large hand wheel situated on the front of the machine. An integral device, in this case a triangular knob, is used to lock the setting. Built-in stops keep the blade from being lowered or raised beyond extreme points.

Fig. 1-28. A separate control on a side of the machine is used to adjust the tilt of the blade. It, too, has a locking device. Adjustable auto stops, which should be checked at the beginning of a project and frequently thereafter are provided for 90- and 45-degree blade settings.

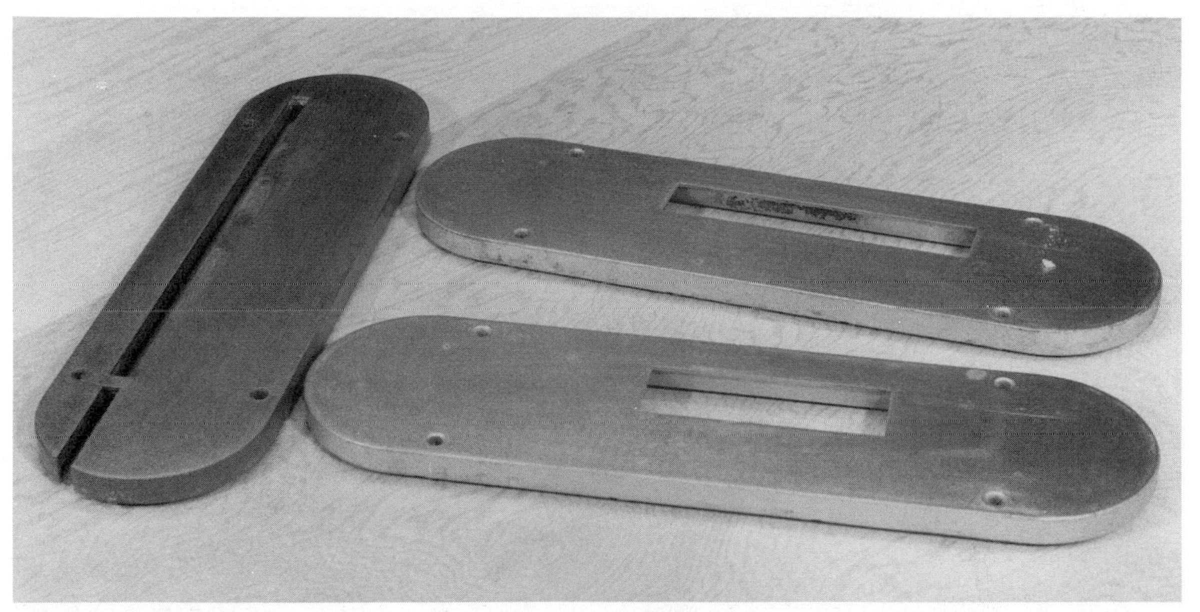

Fig. 1-29. Table inserts are removable so you can get into the machine to mount saw blades, and they are designed so the cutting tool can project above the table's surface. The one used for a saw blade (at the left) comes with the tool. The others, for a dadoing tool and a molding head, are available as accessories.

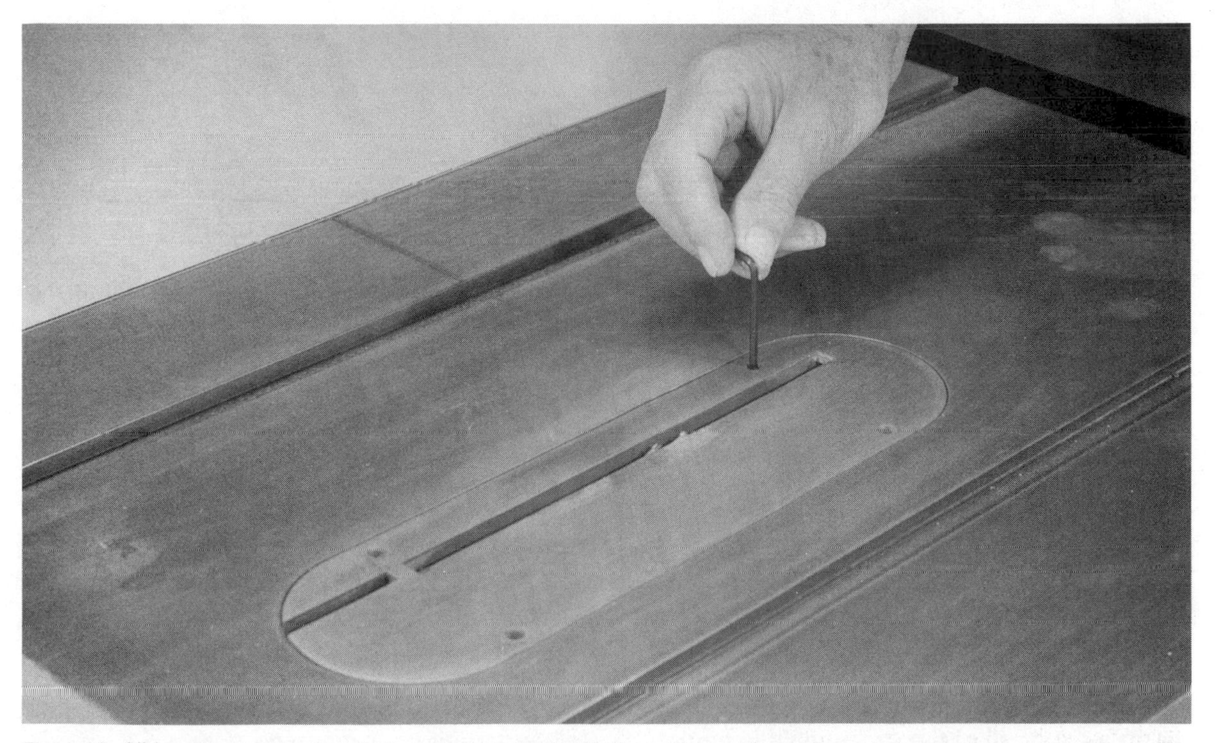

Fig. 1-30. All inserts must be installed so they are level with the table's surface. Some just snap into place; others, like this one, are adjustable.

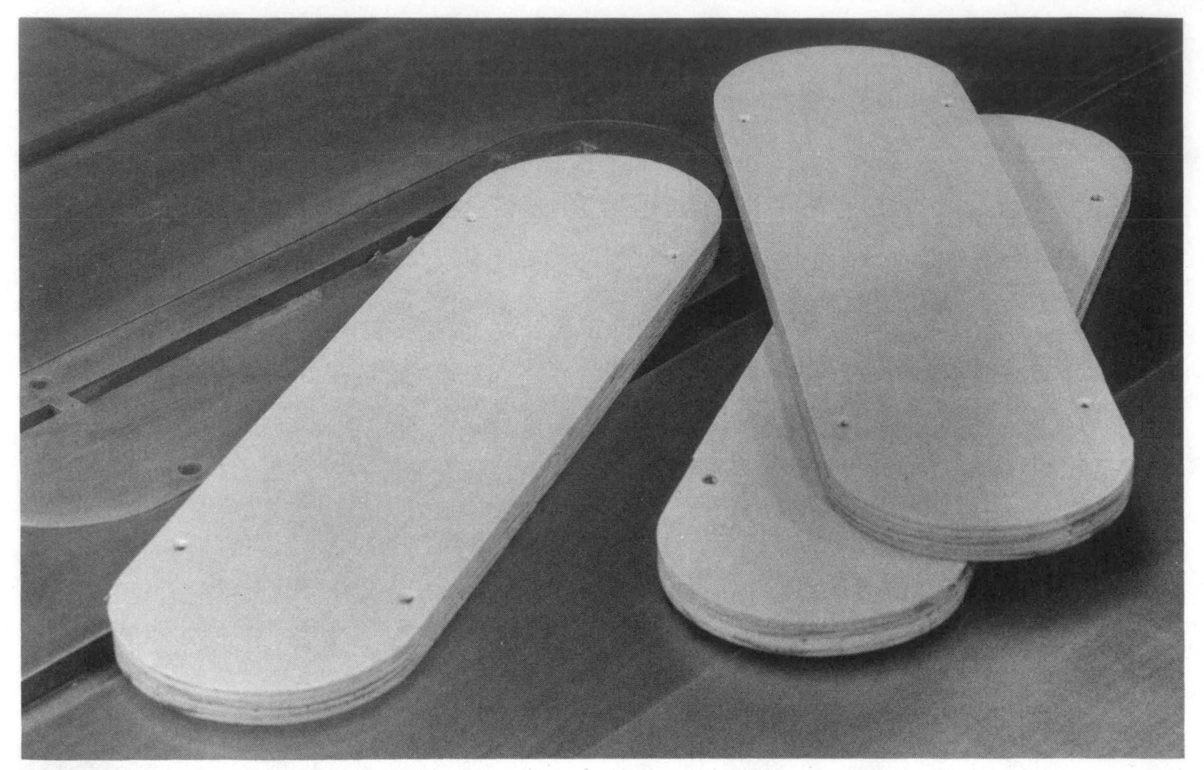

Fig. 1-31. It's often necessary, or wise, to make special inserts so the opening around the cutting tool you are using is nil. Use the standard insert as a pattern for your homemade one. Making several of them right off is a good idea.

or saber saw, or even by hand with a coping saw.

Figure 1-32 shows how the opening for the cutting tool is formed, in this case for a saw blade. First, lower the blade so it does not project above the table. Then, put the homemade insert in place and secure its position. You can accomplish this step with the rip fence, using slim wedges between it and the insert if necessary. Next, turn on the machine and very, very slowly raise the blade so it will cut its own slot.

When custom-making inserts, form them a fraction oversize so they will be a tight fit in the table's opening. Oversizing isn't necessary if the saw is designed to secure inserts with screws.

Sliding Tables

On some saws (mostly industrial equipment), part or all of the tool's table is mounted on a roller design of some type so that the table and the work are moved together past the saw blade. This method makes cutting much easier, especially when workpieces are wide or heavy. Such designs incorporate stops that limit the movement of the table, which prevents the sliding part of the table from moving completely off the tracks.

Some companies, like Delta for its Unisaw, offer a sliding table that is actually an extra-cost accessory (Figs. 1-33 and 1-34). Since this is an add-on unit, it does add to the floor space required for the machine, but it allows a lone operator to efficiently handle oversize workpieces without special outboard supports or extra hands.

Chapter 7, which deals with angular sawing, will include some ideas for making sliding tables to suit your machine.

Fig. 1-32. To make the opening in a special insert, have the cutting tool, in this case a saw blade, below the table and then be sure the insert is secure in the table opening. Then, turn on the machine and very slowly raise the cutter until it pokes through.

Fig. 1-33. A sliding table can make many operations easy to do without outside help. This mitering operation on a piece of heavy stock is an example. The clamping attachment helps to keep the stock firmly in place during the cut.

Fig. 1-34. Here, the same device is used to cut a compound miter, which normally requires both a miter-gauge setting and a table tilt. The ability to move the work easily contributes to accuracy and to safety.

EXTRA ACCESSORIES

There are two accessories everyone should add to their table saw. One is a dadoing tool, an example of which (actually a *dado assembly*) is shown in Fig. 1-35. The other, a typical one (Fig. 1-36), is a molding head, sometimes called a *shaping head*. Both tools contribute considerably to the applications of table saws. Dadoing will be dealt with in depth in Chapter 8, and the applications of the molding head will be explained in Chapter 9.

Sanding Disk

Sanding disks for table saws are available in simple, circular, steel form or as cast aluminum plates. When the disk is mounted on the saw's arbor in place of the saw blade, the machine can be used for many of the sanding operations normally done on an independent disk sander (Fig. 1-37).

Use care when selecting a sanding disk. Important factors are the diameter of the disk, the size of its arbor hole, and its allowable maximum rpm. If a disk is rated at, say 3,000 rpm, then it must not be used on a saw that drives a blade at a greater speed. Always follow the manufacturer's instructions very carefully.

Abrasive papers made specifically for the disk are available, or you can make your own from conventional square-cut sheets. The easi-

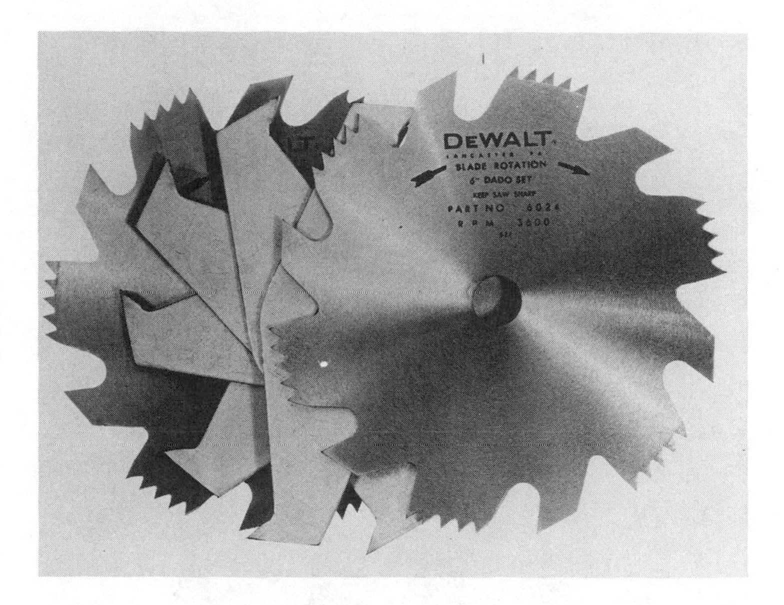

Fig. 1-35. Accessories like this dado assembly are used to make cuts in a single pass that are wider than can be accomplished with a single saw blade. They are fine aids for joinery work.

est way to mount the abrasives is to buy those that are self-adhesive. They simply press into place and are reasonably easy to remove when they must be replaced. Other mounting methods for nonadhesive papers include contact cement and special disk cements.

Fig. 1-36. A molding head works with matched sets of knives to do the type of edge and surface forming normally done on a special shaping machine.

Because of the high rpm of a disk when used on a table saw, you must be very careful when you are working with a fine abrasive. Sawdust will not clear the abrasive particles fast enough, and the result can be a glaze on the disk that makes the paper useless and causes the wood to burn. It is always best to start a job with a coarse abrasive and then finish with a finer one.

Some disk designs allow you to mount abrasive sheets on both sides. Thus, you can progress from a coarse abrasive to a finer one in minimum time. Some operators have two or even three disks on hand so they will have a variety of grits ready to use.

Good practice, no matter what grit paper is being used, calls for a light touch. Abrasive particles are actually tiny cutting tools and must be allowed to work at their own pace. Forcing them, to remove a lot of material quickly will merely cause burning and quick clogging of the paper.

Sawdust Collector

The accessory shown in Fig. 1-38 is not available for all saws, but something like it can be fitted to almost any machine. Two types are

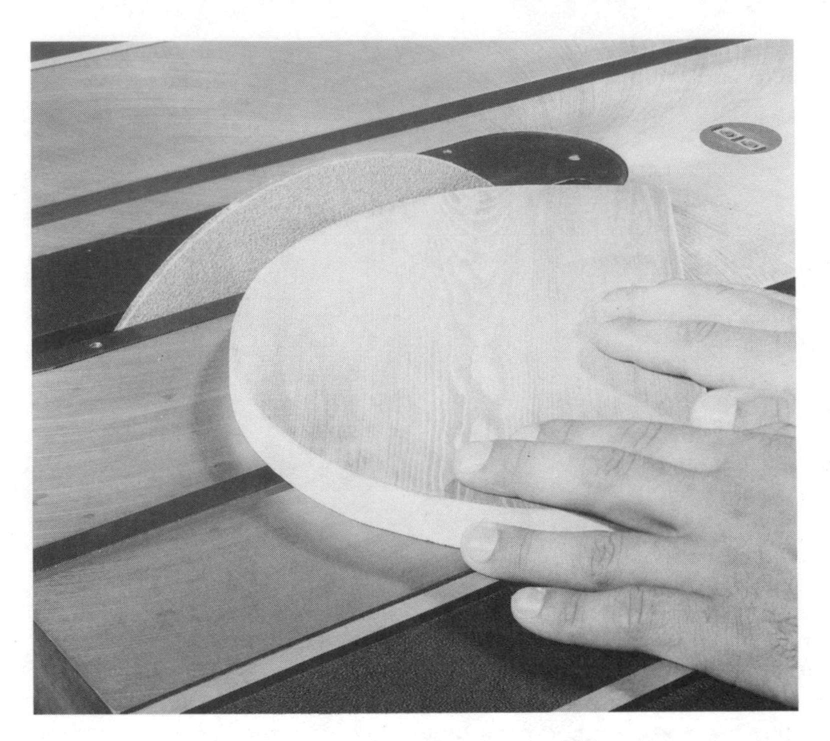

Fig. 1-37. A sanding disk enables you to use the table saw for many practical edge-sanding operations. You must be sure that any sanding disk you buy can be used safely on your table saw.

Fig. 1-38. This sawdust collection unit is made especially for some Craftsman table saws. It is possible that you can adapt it to a tool you own, but be sure before you buy.

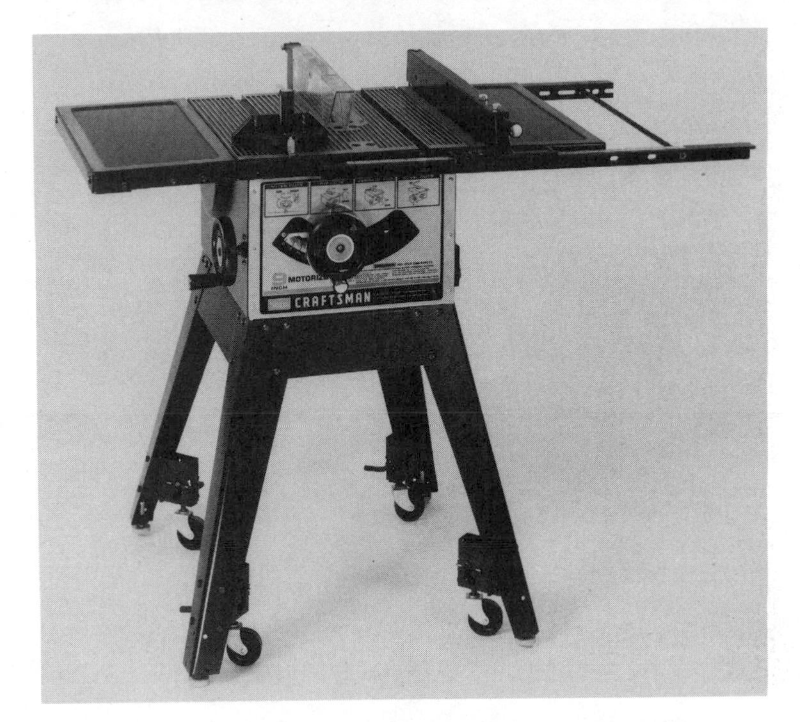

Fig. 1-39. It's okay to equip a table saw with casters so you can move it about, but not any set of casters will do. They must be able to be locked or raised above the floor so the tool won't move about when you are sawing.

available. One, for direct-drive saws, collects sawdust above and below the cutting level and deposits it into a cloth filter bag. The other, for belt-drive saws, works with an impeller and uses the saw motor to draw dust into the filter bag.

Casters

Casters (Fig. 1-39) enable you to move a table saw about easily, which can be an asset in a small shop. Many types are available, but be sure to select a design that allows the tool to sit solidly on the floor when the casters aren't used. Such designs might include a locking device to keep the wheels still, or a height mechanism so the casters can be elevated above the floor. You don't want the machine to wheel away when you are sawing!

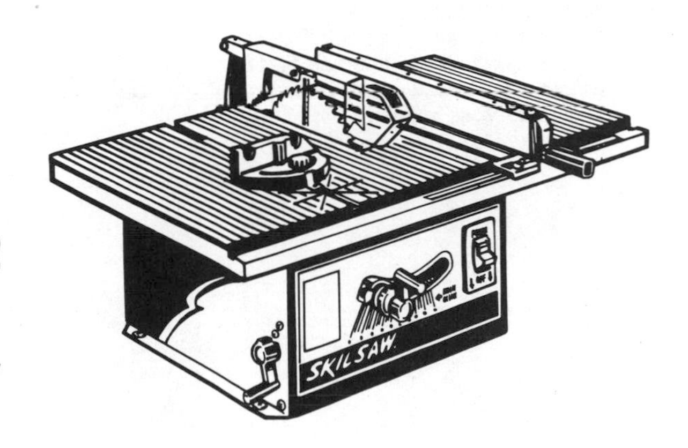

Chapter 2

Working Safely

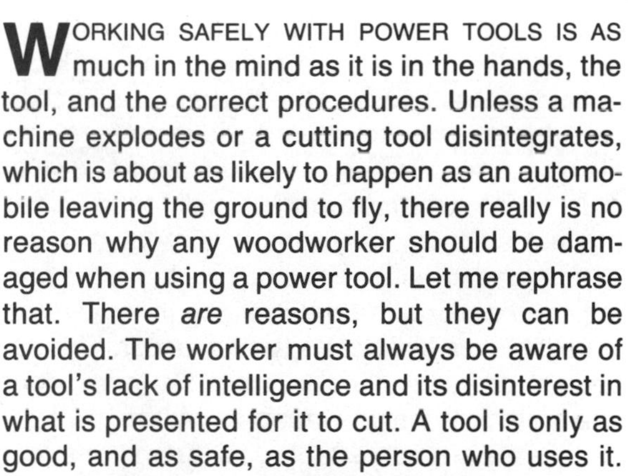

WORKING SAFELY WITH POWER TOOLS IS AS much in the mind as it is in the hands, the tool, and the correct procedures. Unless a machine explodes or a cutting tool disintegrates, which is about as likely to happen as an automobile leaving the ground to fly, there really is no reason why any woodworker should be damaged when using a power tool. Let me rephrase that. There *are* reasons, but they can be avoided. The worker must always be aware of a tool's lack of intelligence and its disinterest in what is presented for it to cut. A tool is only as good, and as safe, as the person who uses it.

Acquiring a detailed knowledge of the tool—how it functions, what it can do, how to use it—and knowing its limitations are important safety factors. Be aware, however, that expertise does not shield you from harm. Statistics prove that as many professionals as amateurs have accidents. Expertise can breed overconfidence, which creates a dangerous state of mind. An experienced operator has an accident usually because he has become careless or feels confident enough to ignore good, safe practice.

Hanging on to a degree of fear of the tool, no matter how well you learn to use it, is as logical as being sure that your car's steering mechanism is not faulty. Potential hazards when working with power tools are no different than those of everyday living. If you are not careful, you can fall off a ladder, walk into a tree limb, slip off a stairway. Accept that, when you use a tool, you alone control the risk factor. Accept that tools can't think, and don't be complacent because the tool has a guard.

Safety rules are everywhere—in books, owner's manuals, and magazine articles—but we're so anxious to get to the doing that we might be inclined to give them minimum attention. I hope you don't ignore them.

GUARDS

The guard in Fig. 2-1 is a fairly common design on table saws. It consists of a pivoting blade cover, usually a see-through type, that is

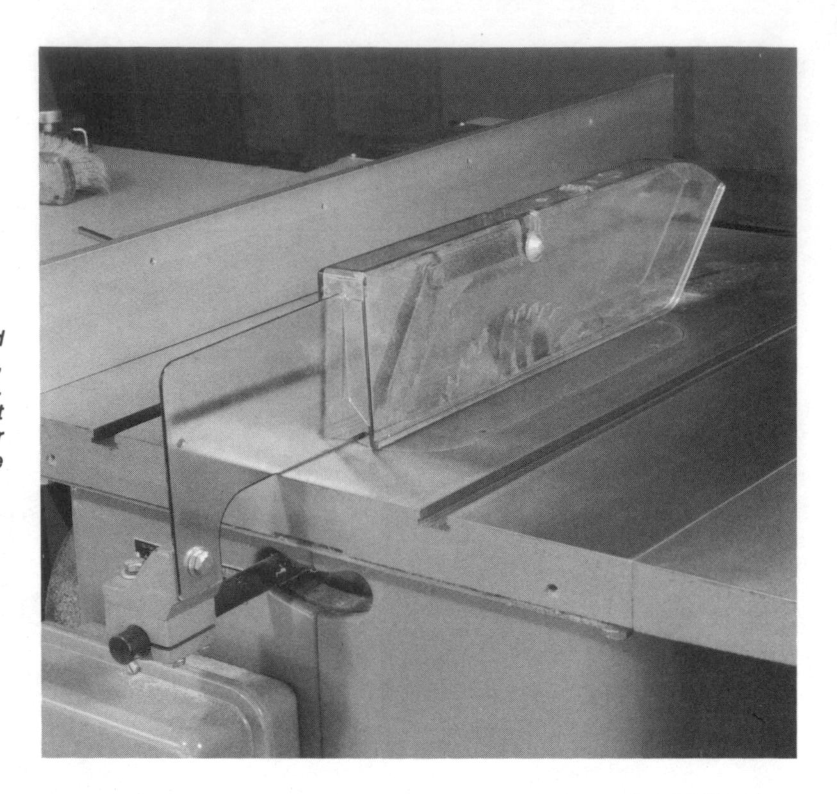

Fig. 2-1. A typical table-saw guard consists of a blade cover, or hood, a splitter, and anti-kickback fingers. The assembly is mounted so that it will tilt with the saw blade when, for example, cross miters or bevels are being cut.

mounted on a *splitter*, which in turn is attached in some particular fashion to the mechanism of the saw so the assembly is in line with the saw blade and will tilt as the blade is tilted. Guards should be mounted very carefully, following the manufacturer's instructions, and should be maintained in perfect working order.

The blade cover should lift easily when work is moved against it and should settle nicely on the surface of the work as the pass is made. These actions should occur regardless of the thickness of the stock and the cutting operation, whether you are crosscutting or ripping (Figs. 2-2 and 2-3). On some tools, notably multipurpose concepts, where the blade is exposed under the table, a secondary guard is provided and attached as shown in Fig. 2-4. Most times, the lower component cannot be ignored since the splitter and thus the blade cover rely on it for installation.

It is critical for the splitter to be in perfect alignment with the saw blade so it will ride in the groove (kerf) formed by the blade. Its purpose is to keep the kerf from closing and thus binding the blade, which can result in the work being moved back toward the operator—an action called *kickback* that must be avoided. An antikickback device, the serrated blades shown in Fig. 2-5, is mounted on the splitter to guard against such an occurrence. The blades are spring-loaded so they normally ride on the surface of the work, but they will dig in to hold the work should a kickback situation occur. This aspect of power sawing is more likely to occur when stock is being ripped to width.

There are some table-saw operations that do not allow the use of the standard guard; examples include dadoing and molding head operations. At such times you must be especially alert and take special precautions to ensure safety. Often, as you will see in many places in this book, a homemade device provides a necessary safety factor when work is done without a conventional guard.

Industry often takes workers a step closer to safe table-saw operation than might be

Fig. 2-2. When the guard is installed correctly, cutting can be done without interference from the hood. It will lift nicely as the wood makes contact and then settle down on the surface of the stock throughout the pass.

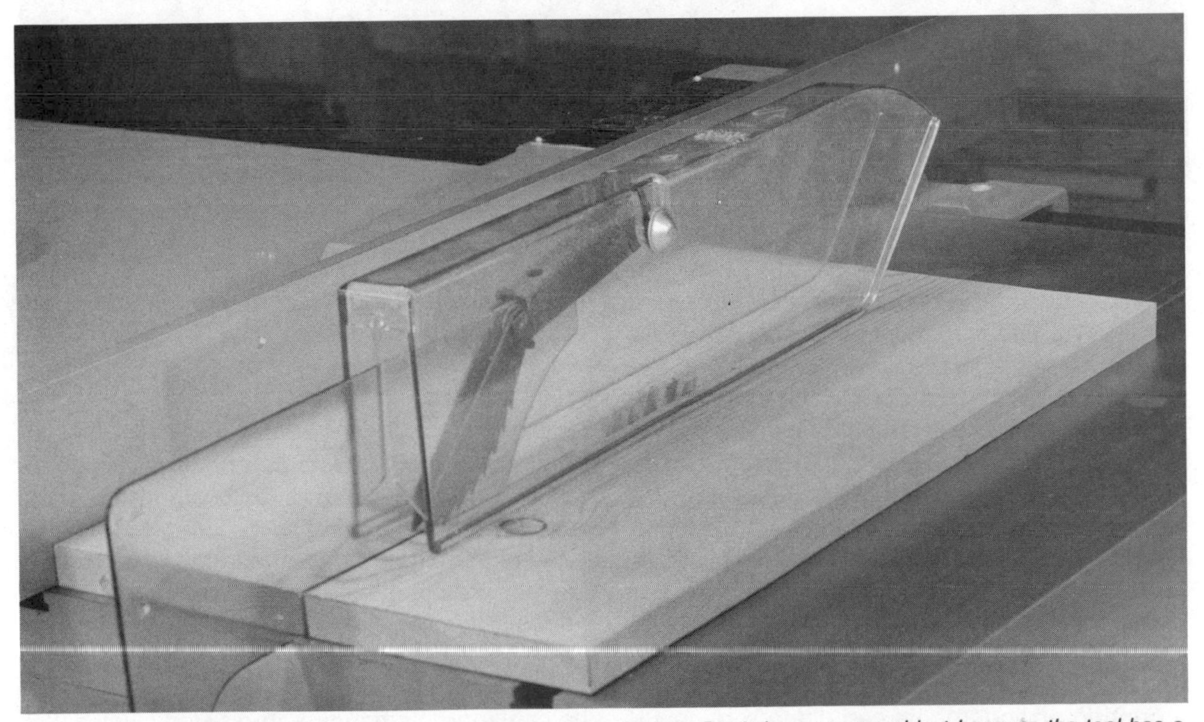

Fig. 2-3. The smooth action should also occur when you are ripping. Don't become casual just because the tool has a guard. Always place hands so they are well away from the cutting area.

Fig. 2-4. On some multipurpose tools, the blade is exposed under the table, so a secondary, stationary cover is provided. Note that the splitter, which supports the top, pivoting hood, is attached to the lower guard.

provided by a conventional guard by equipping the tool with special devices. A popular example, which is finding its way into many amateur shops, is the Brett-Guard, which is demonstrated in a few of its possible setups in Figs. 2-6 through 2-9. Unlike a standard guard, it is usable for special operations like coving and panel-raising, and for work involving dadoing tools and molding heads, even sanding disks. The special guard protects hands, guards against kickback, and because of its see-through plastic cover, keeps chips and dust from flying about. The components of the Brett-Guard are standard, but mounting arrangements can differ from saw to saw. If you order one, tell the manufacturer the dimensions and the thickness of the saw table on which you will use it.

HOLD-DOWNS

Miter gauge hold-downs contribute as much to safety as they do to accuracy. When the work is mechanically secured so it is snug against the head of the miter gauge and kept flat on the table throughout the pass, the only job for the hands is to move the work (Figs. 2-10 and 2-11). Since the operator doesn't need to strain, the possibility that his hands might slip is nil, and he can keep his hands and his body well away from the path of the saw blade.

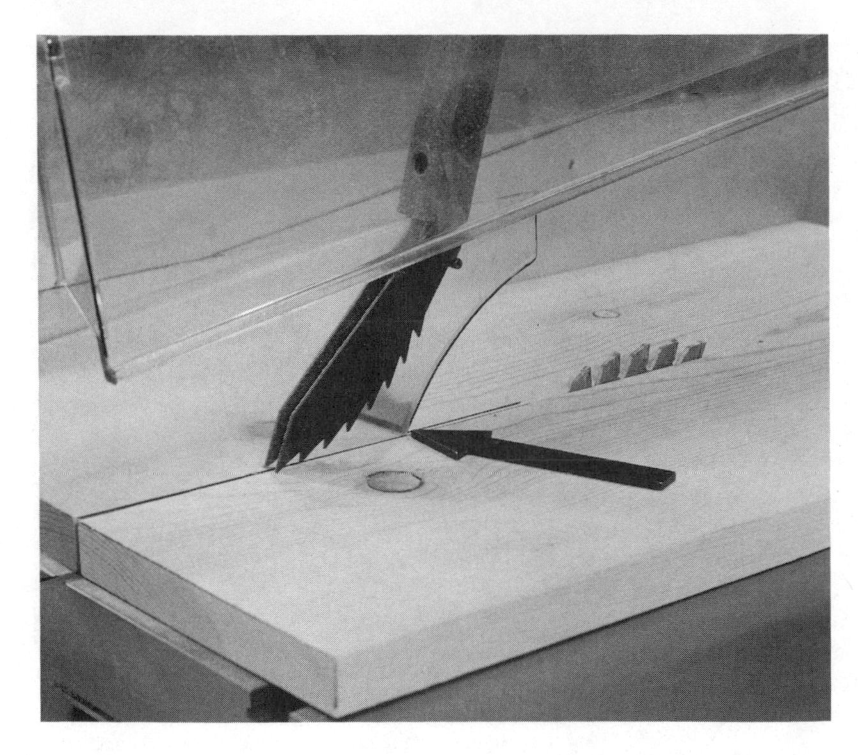

Fig. 2-5. The splitter (see arrow) will work correctly only if it is aligned precisely with the cut path of the blade. The serrated fingers act to prevent kickback by digging in should the action of the saw blade try to move the work back toward the operator.

There are different hold-down concepts. The one shown in Fig. 2-12 employs a rod that bears down on the work as the handle of the tool is pressed. The rod is vertically adjustable to accommodate various thicknesses of stock.

Not all miter gauges are designed to accept a hold-down, but since gauges are similar, it might be possible to modify an on-hand tool to accept a hold-down made by another company. Also, quite often the table slots for miter gauges on different saws have the same dimensions, so you might be able to add an "alien" miter gauge/hold-down combination to your equipment.

RIPPING AIDS

An essential rule of table-saw work is to employ a substitute whenever the operation requires that fingers be used close to the saw blade. This is especially critical on ripping operations. The narrower the piece you are sawing, the closer your fingers would come to the saw blade as you moved the work through. To state in inches the minimum distance between blade and rip fence before you should reach for a "pusher" would be arbitrary and does not allow for the difference in sizes of hands. Use a pusher instead of your fingers whenever possible.

Many operators reach for a piece of scrap to use as a pusher, but it is better practice to keep on hand a special device for the purpose, like the homemade one shown in Fig. 2-13. This fence-straddling pusher/hold-down keeps the workpiece flat and moves it past the saw blade while permitting hands to be kept safely out of the way. Chapter 6, which deals with ripping operations, contains construction details for several types of pushers.

A commercial product called a Ripstrate (Fig. 2-14) has found wide acceptance because it contributes considerably to safer and more accurate ripping. The tool employs two rubber wheels mounted on spring-loaded arms that are situated so the wheels are tilted slightly in the direction of the rip fence. The tilt of the wheels keeps the work against the fence as it is moved

Fig. 2-6. The Brett-Guard, shown with its inventor, Mr. Bretthauer, is a unique safety device that is much more flexible than a conventional guard. Here, it is in place for a routine crosscutting job.

Fig. 2-8. The Brett-Guard can be used on many operations that don't allow positioning of a conventional guard. Note the complete coverage of the cutting area. The unit can be used with saw blades, dadoing tools, molding heads, and even a sanding disk.

Fig. 2-7. The Brett-Guard is used this way for ripping operations. A crank, which is on top of the control component, is used to raise or lower the see-through shield. Since it can be lowered on the work, the shield also serves as a hold-down.

Fig. 2-9. It's also completely compatible with homemade jigs, like this sliding table for mitering. The tool is already established in many industrial facilities, and is finding its way into amateur workshops.

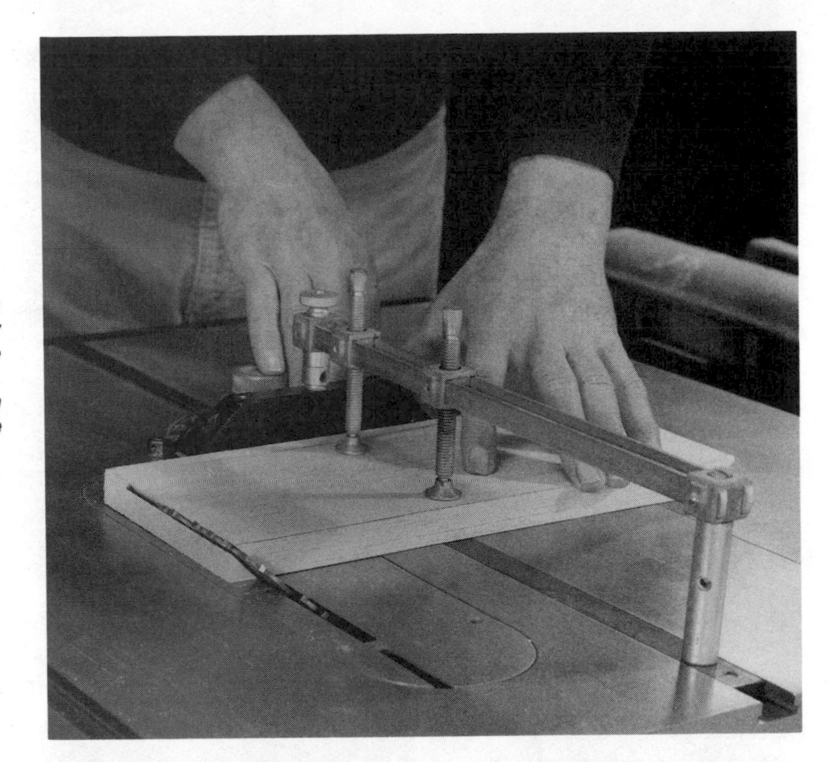

Fig. 2-10. A miter-gauge hold-down ensures that the workpiece will stay flat on the table and snug against the gauge's head throughout the pass. The operator doesn't need to strain with his hands, and he can place them away from the cutting area.

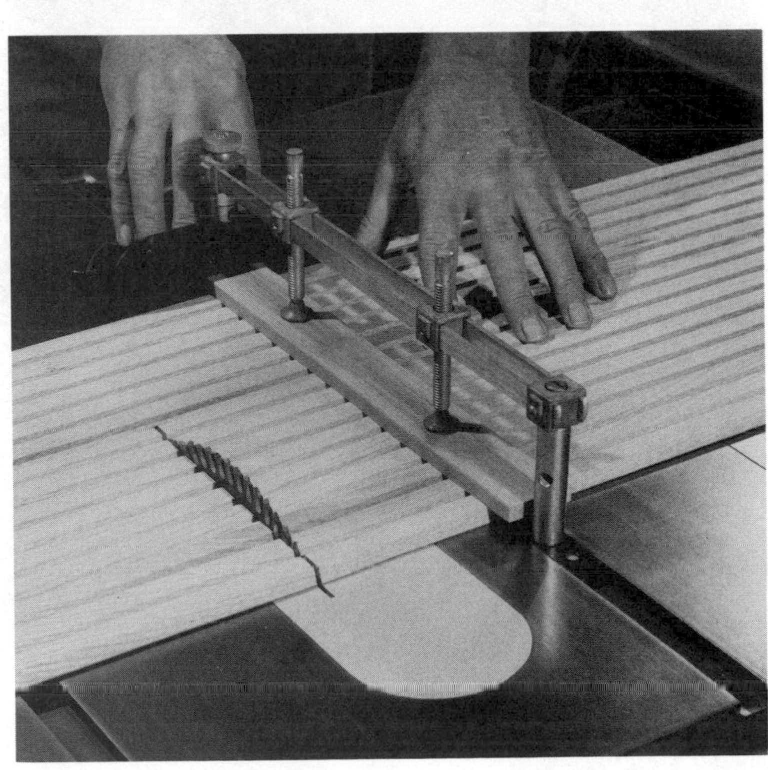

Fig. 2-11. This type of hold-down, a Delta concept, aids considerably when you are sawing very thin or flexible material, like this prefabricated sheet tambour. The strip of wood under the posts keeps the material from undulating while being cut.

Fig. 2-12. The Craftsman miter-gauge hold-down employs a vertically adjustable rod that bears down on the stock to keep it steady as it is sawed. Accessories like this contribute to accuracy as well as safety.

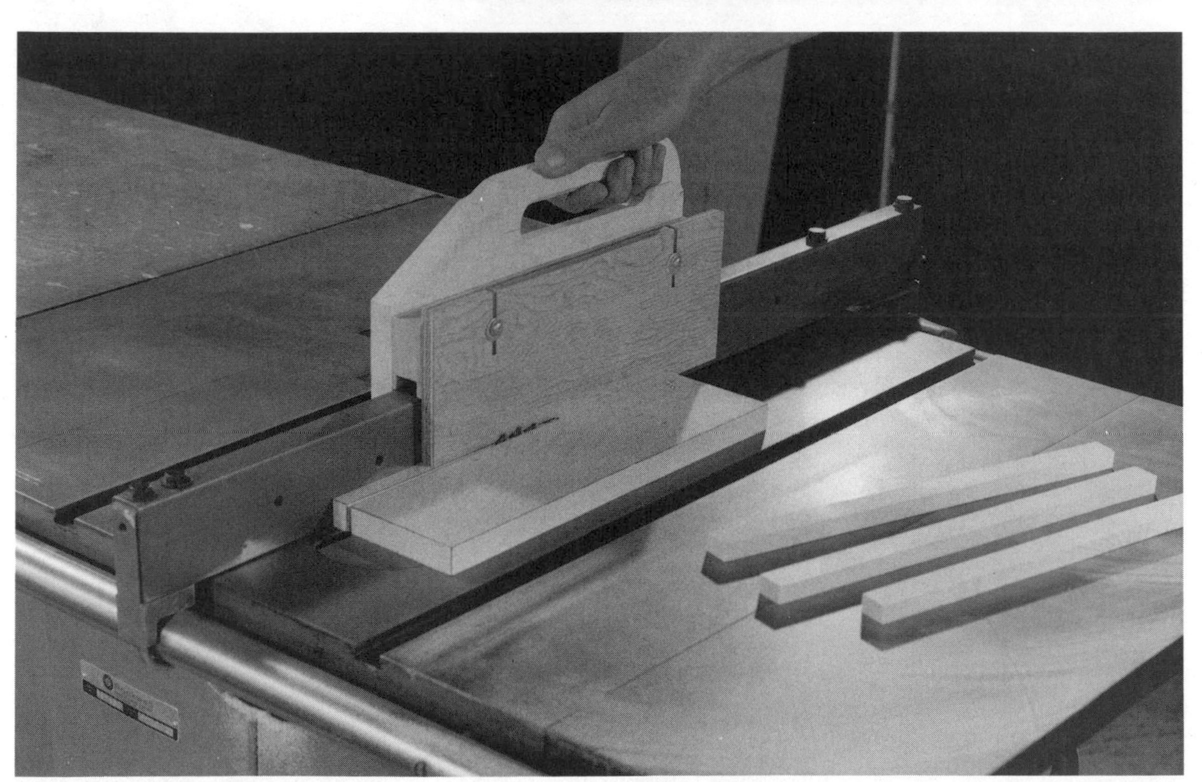

Fig. 2-13. This homemade, fence-straddling pusher/hold-down lets you safely rip narrow pieces. The face of this version is adjustable vertically so it can be situated to bear down regardless of the stock's thickness.

Fig. 2-14. The Ripstrate is a commercial product that can be used on any table saw. It aids when ripping because the wheels that bear down on the work are slightly tilted to keep the work against the rip fence.

Fig. 2-15. Spring steel arms and adjustable brackets allow this unusual tool to hold workpieces in the correct position. The arms must be situated so they won't snap into the cutting tool when the pass is complete. A push stick is used to get the work past the cutter.

past the saw blade. The wheels turn only in one direction. A pawl will lock the wheels to hold the work should a kickback situation occur. The antikickback action is aided by the arms, which tend to press down harder as they are forced back.

The arms adapt themselves to various stock thicknesses so the need for adjustments between jobs is minimized. The width of the cut you need is adjusted for in normal fashion, and it can be as wide or as narrow as you wish.

The hold-down in Fig. 2-15 consists of a set of spring steel arms and adjustable brackets that allow the arms to be situated so they keep the work snug against the fence and flat on the table. It can be used for molding head work, as well as for ripping.

INSERTS

Chapter 1 mentioned table inserts as basic components of table saws, but it is important to stress that using the right insert for a cutting tool contributes to safety as well as good practice. Basic inserts include one for a saw blade, which is supplied with the machine, and those designed for a dadoing tool and a molding head, both of which are extra-cost accessories. The purpose of any insert is to allow the cutting tool to project above the table while providing work support and minimizing the open space the cutter needs.

You will find that standard inserts are not always ideal for the chore at hand. Professionals make special inserts that are exactly right for specific jobs. The size and shape and thickness

Fig. 2-16. Special inserts are merely solid wood, plywood, or plastic duplicates of original equipment. It's a good idea to have several, blank inserts on hand since they will be required for safety and accuracy on many operations.

of the customized insert should duplicate a standard one, except for the size and shape of its opening.

The way you attach the insert depends on the machine you own. In some cases, the component is secured with screws (Fig. 2-16). Other times it might be held with spring clips or merely sit on ledges at the base of the table opening. In any case, design the homemade insert so the clearance space around the cutter being used is practically nil (Fig. 2-17). The business of special inserts and the way they are made and used will be discussed again in various chapters in this book.

SPRINGSTICKS

Springsticks, like those shown in Fig. 2-18, are sometimes called *featherboards* or *fingerboards*. No matter what the name, these homemade accessories can be used as horizontal or vertical hold-downs to increase safety and accuracy on many table-saw operations (Figs. 2-19 and 2-20). You can make these units in various width and lengths and with various shapes at the end of the fingers. The fingers are the result of parallel cuts made with a regular saw blade. In use, clamp the accessory to the table so the fingers bear against the work to keep it flat on the table or snug against the rip fence, as the case might be. Never place it so that it forces a workpiece against a saw blade or other cutter. Practical applications for these devices will appear in various parts of this book.

PROTECTION FOR EYES, LUNGS, AND EARS

Most woodworkers do not need much coaxing to wear safety goggles or a face mask to protect their vision, but it is often a chore to convince them that lungs and ears are also vulnerable to damage. It is accepted today that headphone-type hearing protectors are as important as any safety device. High frequencies

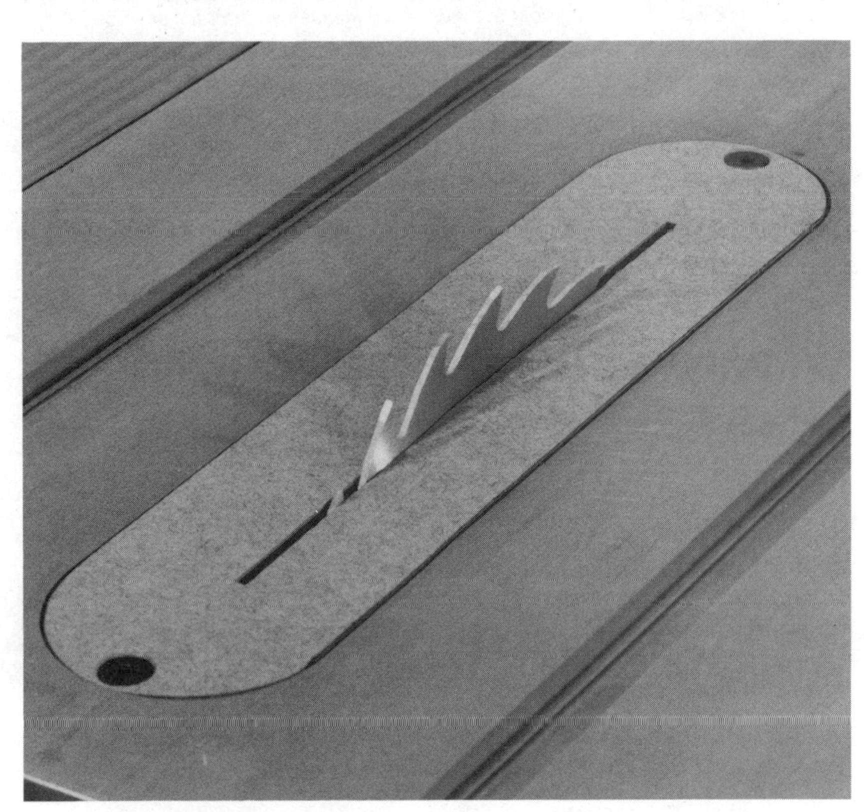

Fig. 2-17. Customizing an insert is a question of allowing the cutter to form its own opening.

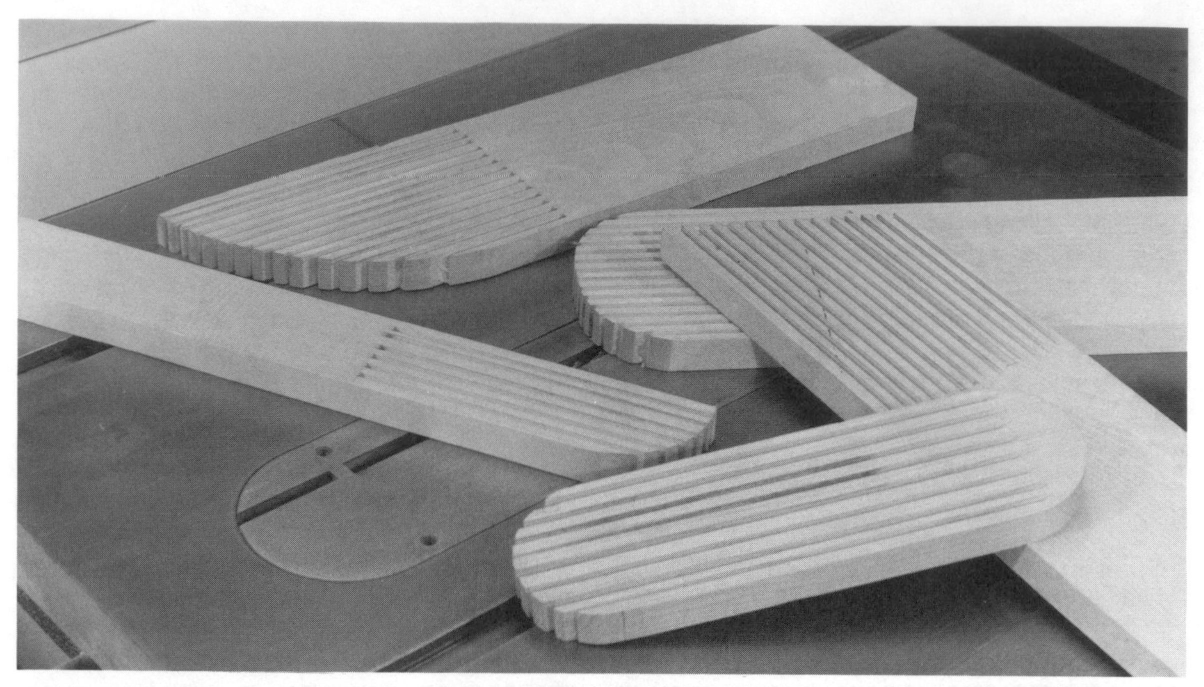

Fig. 2-18. Springsticks are homemade devices that are clamped to the saw table so the flexible fingers can bear down on work to keep it flat, or against the side of workpieces so that, for example, they will stay snug against the rip fence.

Fig. 2-19. When used this way, a springstick helps to keep the work against the fence while you make the cut. The device may be placed ahead of the cutter or at the back, but never so that it bears against the cutter itself.

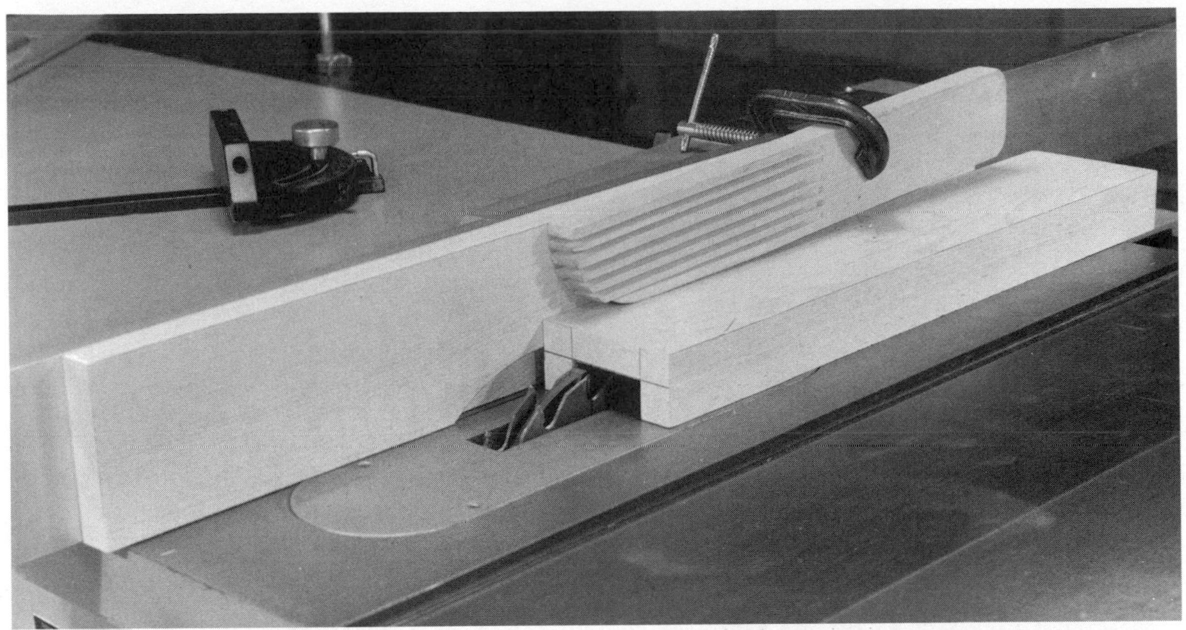

Fig. 2-20. Springsticks are often clamped to a fence to keep the work from trying to climb the cutter. Press the fingers down on the work as you clamp the accessory. Position it so it can't be damaged by the cutting tool.

can be generated by high-speed electric motors, by air movements generated by saw blades, and even by many standard woodworking operations. The effects are cumulative; each exposure contributes to possible hearing damage. Good ear protectors will screen out the damaging high frequencies, but will still allow normal conversation and the normal woodworking noises you should hear.

Don't assume that you need a dust mask only when you are doing sanding chores. Many sawing, dadoing, and molding-head operations can produce waste particles that are best kept from your lungs. A dust mask is only as good as its filter, so be sure to replace the filter as often as necessary.

A dust mask, safety goggles or a face mask, and hearing protectors are essential components of good, safe craftsmanship.

PROPER ATTIRE AND MIND-SET

It is not out of line to have a special uniform to wear for shop work. Heavy, nonslip shoes, preferably with steel toes, and tight-fitting shirts and trousers make sense. I don't believe in wearing gloves, and a necktie or any loose-fitting clothing that might snag on a tool whether it's idle or in use is a no-no. Jewelry—rings, wristwatches, bracelets, and similar items—are adornments for outside the shop. Inside the shop they are hazards. It is not a bad idea to cover your hair, whether it is long or short, for safety and protection from dust.

Treat your shop as if it were the kitchen. Tables, benches, tool surfaces, and the floor should be maintained in a pristine state. A shop-type vacuum cleaner is a good investment (Fig. 2-21). They are available in various sizes, and most will have an exhaust port so the unit also can be used as a blower.

The surfaces of table saws often should get attention since they can become dirty and gummy. Sometimes, a cleaning solvent, used carefully by following the directions on the container, is enough to return the table to proper condition. If not, you can go over the table with a pad sander fitted with a very fine emery abrasive (Fig. 2-22). The weight of the sander will ap-

Fig. 2-21. A heavy-duty vacuum cleaner is a practical addition for any shop. It's more likely that you will pay more attention to keeping the shop clean if the unit is handy. Having one equipped with casters is a good idea.

ply all the pressure you need. Don't force; keep the sander moving. Wipe the table with a lint-free cloth and then apply a generous coating of paste wax (Fig. 2-23). Rub the wax to a fine polish after it is dry. The waxing procedure should be repeated frequently so that workpieces will move smoothly and easily when you are sawing.

GOOD SHOP PRACTICE

It is never a good idea to overreach, no matter what the operation or the tool being used. Do not try to prove your self-reliance when it is necessary to saw extra-large workpieces like panel materials. When necessary, have someone provide extra support, but be sure to explain the procedure to the assistant. You will always have extra support available if you add special stands to your equipment. Some are available commercially (Figs. 2-24 and 2-25), but there are homemade versions you can duplicate. Construction details for customized ones are given in Chapter 6.

Never work with dull tools. Cut results will not be acceptable and, since dull tools require the operator to apply extra feed pressure, your hands might slip.

Do not have the tool plugged in when you are changing to another cutting tool, cleaning the machine, or doing alignment checks. Check to be sure the switch is in the "off" position before you plug in. Don't leave the tool running when you need to attend to another chore, regardless of how little time is involved. It's a good idea to wait for the cutting tool to stop before you move away from the machine.

Having to force a cut is a warning signal. It can mean that the cutter is dull or that you are trying to cut too deep in a single pass. It is usually best to accomplish extra-deep or oversize cuts by making repeat passes.

The shop is not for socializing. Don't visit and try to work at the same time. You should warn friends and neighbors not to barge into your shop when they hear a tool running. You don't want to be startled.

The primary rule is to always be alert. The job on hand should occupy your entire attention. If you are tired or upset, have taken medicine that makes you drowsy, or have had an alcoholic drink, substitute television or a good book for shopwork.

GOOD TOOL PRACTICE

Get to know all aspects of your table saw. Don't just scan the owner's manual; study it! Check the security of cutting tools, the miter gauge, and the rip fence before making cuts. Carefully check installation instructions for saw blades and other cutting tools. Don't use saw blades that are larger in diameter than the tool can take. Always be sure of the blade's rated maximum rpm. It should never be less than the tool's speed.

Always move work so the pass is made against the cutter's direction of rotation, and be

Fig. 2-22. It's good practice to occasionally use a pad sander fitted with a very fine emery paper to bring the table's surface to pristine condition. Keep the sander moving and don't apply a lot of pressure. Don't use a belt sander.

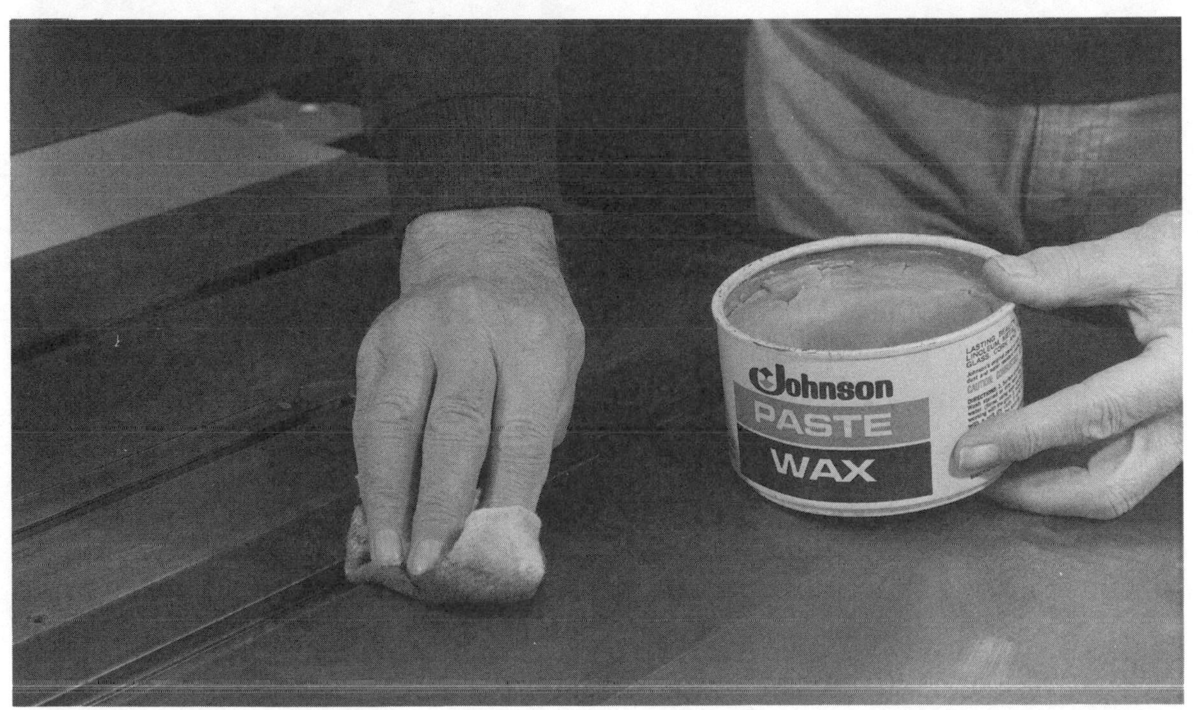

Fig. 2-23. An occasional application of paste wax helps to keep the table neat and smooth so workpieces can slide easily. The bearing surface of the rip fence and the miter gauge bar also should be waxed.

sure the cutter is mounted so its teeth point toward the front of the machine. It is important for accuracy and for safety to check frequently to be sure machine components are in correct alignment. The procedures to follow will be outlined in Chapter 3.

It is dangerous to work with wood that is too small for safe handling. For example, when a small piece is needed, do the work on a large piece of stock and then cut off the part you need.

Use the guards and be sure they are installed correctly. Many operations in this book are demonstrated without the guard in place. This is done for photo purposes only. If the guard had been used, you would not be able to see what was going on. There are particular procedures in table-saw woodworking where guards can't be used in normal position. When involved in one of these, you must be extra cautious about how you proceed. These are times when a preliminary dry run, going through the steps required but with the machine turned off, makes a lot of sense. Always preview a chore, even simple ones. Become "tool-wise."

Never forget that the machine and the cutting tools it drives cannot think for you. There is no way they can distinguish between wood and flesh.

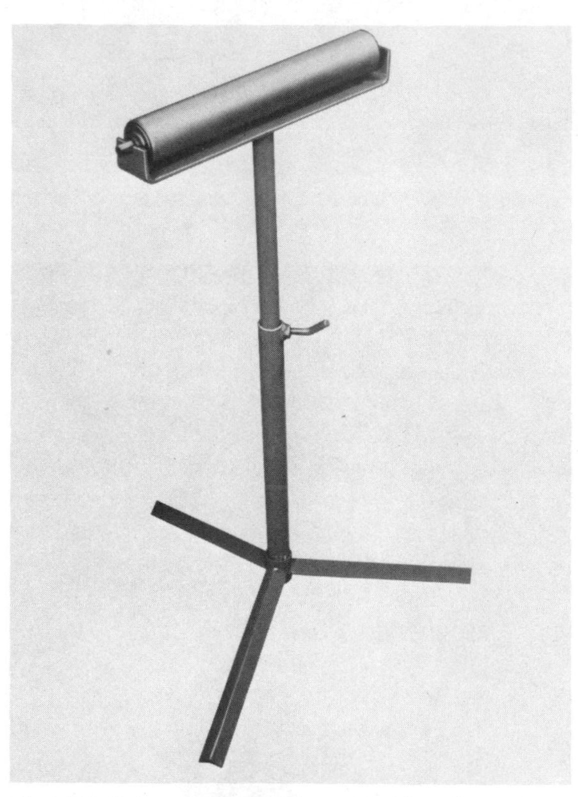

Fig. 2-24. Outboard supports that are height-adjustable and have roller tops are available commercially for use in ripping or crosscutting unwieldy workpieces.

Fig. 2-25. Units like this, called a Supportable, can be positioned so cumbersome materials can be handled conveniently and safely. The unit measures 18 × 60 inches and has 10 plastic rollers. Its height is adjustable, and it can be folded for storage so it takes up floor space only when needed.

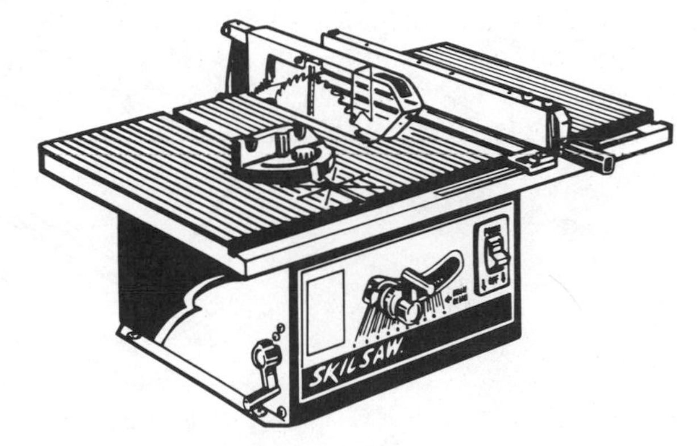

Chapter 3

Alignment

ALIGNMENT IS A GENERAL TERM THAT APPLIES TO the relationship between the adjustable components of a table saw. Since the components—the table, the saw blade, the miter gauge, and the rip fence—must work together in various ways, it is essential that they be organized for precise settings. For example, if you wish to cut across a board so that the cut edge will be square to adjacent *edges*, the angle between the miter gauge and the saw blade must be 90 degrees. It is also necessary for the angle between the saw blade and the table to be 90 degrees for the cut edge to be square to adjacent *surfaces.*

If the slots in the table in which the miter gauge slides are not parallel to the saw blade, you will have a heeling problem, which will result in excessive vibration, inaccurate cuts, and damage to the blade and the work. If the position of the rip fence forms a closed angle with the saw blade, it will be difficult to do ripping operations, the possibility of kickback will be increased, the blade will be stressed on one side, and cut edges probably will have burn marks.

There can be some slight variation in methods of adjustment among saws, but the correct relationship of components and how to check for accuracy is standard. Check the owner's manual for specifics that apply to your table saw. There is no reason why a novice can't set up his machine just as efficiently as anyone.

THREE IMPORTANT RULES

The important alignment factors shown in Fig. 3-1 apply to any table saw.

- ☐ The slots in the table must be parallel to the saw blade.
- ☐ The face of the rip fence, the head of the miter gauge, and the saw blade when in zero position must be perpendicular to the table surface.
- ☐ The miter gauge, when in normal crosscut position, must be at right angles to the saw blade and the rip fence.

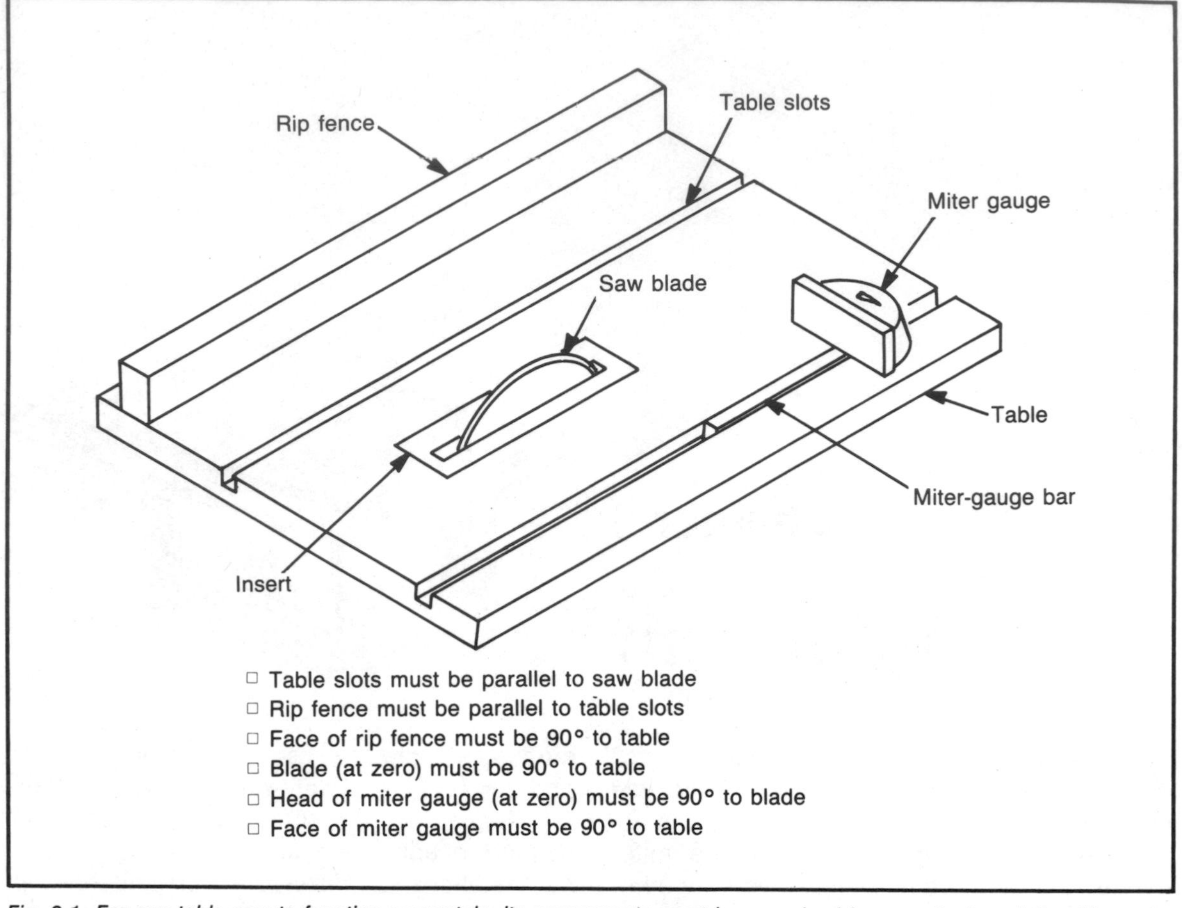

Fig. 3-1. For any table saw to function accurately, its components must be organized for a particular relationship.

Check these factors thoroughly as soon as you acquire a saw and regularly thereafter. Do not have the tool plugged in when you are checking alignment.

Because the saw blade, which locks on the arbor, is the one thing over which you have no control—you can't change how it is mounted—it is good practice to start all alignment checks by determining whether the table slots and the saw blade are parallel. There are various ways to check, but a system that is simple enough and reliable is demonstrated in Fig. 3-2 and 3-3.

Start by raising the blade to maximum elevation. Clamp a rod or some similar, firm item to the miter gauge so that it barely touches one tooth on the saw blade. If the blade has set teeth, work with one that points toward the miter gauge. Identify the tooth in some way; a felt pen will do. Next, rotate the blade by hand so that the same tooth is at the rear of the insert and then advance the miter gauge to see if the checking rod bears against the tooth as it did at first. If there is a gap, or if the rod is forced against the tooth, adjustment is required. The table must be rotated to the left or right to organize it correctly.

On some table saws, this adjustment is accomplished by loosening the four corner bolts that secure the saw to the substructure (Fig. 3-4). On other saws, it might be necessary to adjust

Fig. 3-2. To check for parallelism between the saw blade and the table slots, first raise the saw blade to its highest projection. Next clamp a rod to the miter gauge so it barely touches one tooth at the front of the blade.

Fig. 3-3. Rotate the blade so that same tooth is at the rear of the insert and then move the miter gauge forward so the rod bears against the tooth. Adjustment is required if there is a gap between the rod and tooth or if the rod must be forced against the tooth.

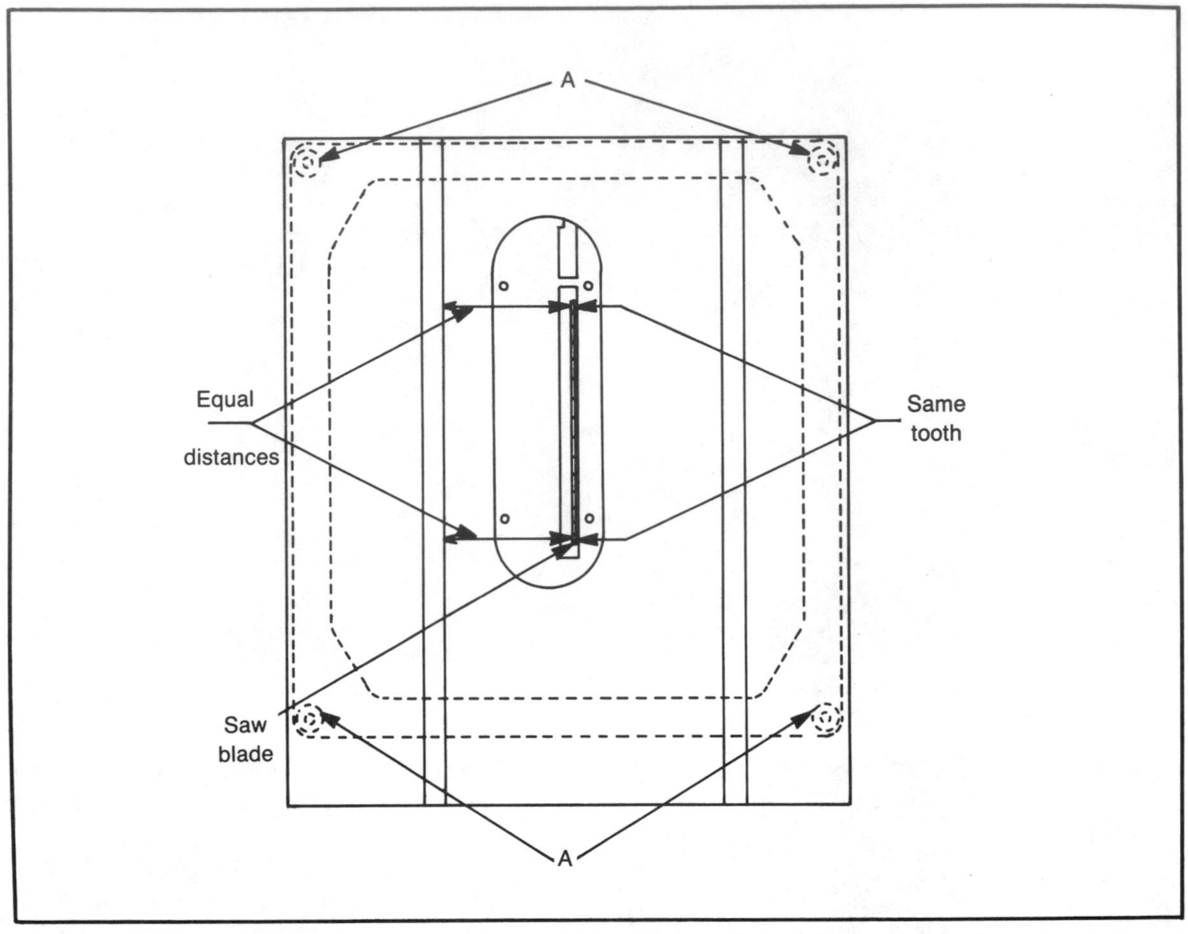

Fig. 3.4. To adjust for parallelism between saw blade and table slots, usually you will need to loosen four corner bolts and rotate the table to the left or right. The owner's manual for your saw will explain if there is a different method to follow. (Original illustration courtesy Delta International).

the internal mechanism that is bolted to the underside of the table. To discover which method to use, consult your owner's manual.

THE SAW BLADE

When the saw blade is in normal position (zero setting), the angle between it and the table must be 90 degrees. You can check this angle with a reliable square after you have raised the blade to its highest point (Fig. 3-5). Be sure to hold the blade of the square firmly against the side of the blade and between teeth. If the angle is other than 90 degrees, use the blade-tilt control to bring the blade to true verticalness and then check the owner's manual to determine how to adjust the automatic stop for this position.

Follow the same procedure, using a combination square or a draftsman's 45-degree triangle, to check the blade at its 45-degree tilt position. If this angle isn't correct, adjust by using the tilt-control wheel and then reset the 45-degree auto stop. If adjustments were necessary, then the bevel scale or its pointer on the front of the machine will also need to be reset. You can use the scale to read tilt settings that fall between 0 and 45 degrees, but be a bit leary of its accuracy. Use the scale for approximate settings and then check with a sliding T-bevel

Fig. 3-5. When the saw blade is in normal position, the angle between it and the table must be 90 degrees. Check with a square after you have raised the saw blade to its highest point. Be sure the blade of the square rests between teeth.

before you begin sawing (Fig. 3-6). A common practice is to check a cut made on a piece of scrap material before sawing good stock.

THE RIP FENCE

The major function of the rip fence is to control the passage of stock that is being cut with the grain. Primary uses are to saw material to a particular width or to true an edge. There are two schools of thought here. One is that the fence should be parallel to the saw blade (Fig. 3-7); the other is that the fence should be offset just a fraction at the rear. The justification expressed by those who prefer the second method is that it avoids having the "rear" teeth of the blade continue to rub on the wood after the "front" teeth have done the cutting. If you use this method, remember that it poses a problem when you must use the fence on the other side of the saw blade. Because there will then be a closed angle between the fence and blade, ripping operations will not go as smoothly as they should and the possibility of kickback increases. Of course, you can eliminate the problem by readjusting the fence.

To check for fence parallelism, bring it in line with one of the table slots. Then make a visual check or run your fingers along the slot and the bottom of the fence. Another method, which minimizes the possibility of human error, is shown in Fig. 3-8. Place a straight strip of wood in the table slot and hold the wood and fence together while you make any necessary adjustment.

It is assumed that parallelism between saw blade and table slots has been established. Therefore, if the fence is parallel to the slot, it will also be parallel to the saw blade.

THE MITER GAUGE

The miter gauge is used for normal crosscutting, for cross-miter cuts that require the blade to be tilted, and for simple miters that use the blade

Fig. 3-6. Also check the blade to be sure that the auto stop controlling its 45-degree position is correctly set. You can read angles between the extremes from the scale on the front of the machine, but you should always check them with a T-bevel or by making test cuts.

in zero position but have the head of the miter gauge pivoted to the angle needed.

Accurate, basic crosscutting requires that the angle between the head of the miter gauge and the side of the saw blade be 90 degrees. You can establish the right setting by checking with a reliable square or with a draftsman's triangle (Figs. 3-9 and 3-10). Be sure that the blade of the square or the edge of the triangle is flush against the side of the blade; situated so it rests between teeth. The best way to work is to place the checking device flush against the blade and then adjust the head of the miter gauge so it is snug against the adjacent edge of the device.

Another common adjustment for a miter gauge is for a 45-degree miter cut. This adjustment can be checked with a draftsman's 45-degree triangle (Fig. 3-11), or with a combination square. A good miter gauge will have adjustable stops so the head can be set to the most-used positions. Adjust these stops as you go through the alignment procedures. The purpose of the stops is to allow the head of the miter gauge to be returned to a standard, preset position after it has been used for an angular cut that falls between 45 and 90 degrees.

For odd-angle cuts, you can set the head of the miter gauge according to the scale on its base. Again, though, be leary of the scale's accuracy. It is always best to check the setting with a sliding T-bevel (Fig. 3-12) or to make a test cut before you saw good material.

Many operators will custom-make special checking gauges like the one shown in Figs. 3-13 and 3-14. Maybe they are being overcautious, maybe not. Workers operate in ways that are especially suitable for them. You must not, of course, rely on the table saw when you are customizing checking gauges. The idea is to lay them out accurately with instruments like

Fig. 3-7. The distance between the rip fence and blade at points A and B should be equal unless you offset the fence just a fraction at the rear to provide some clearance at that point. If you use a block to set fence parallelism, be sure the block has a uniform width.

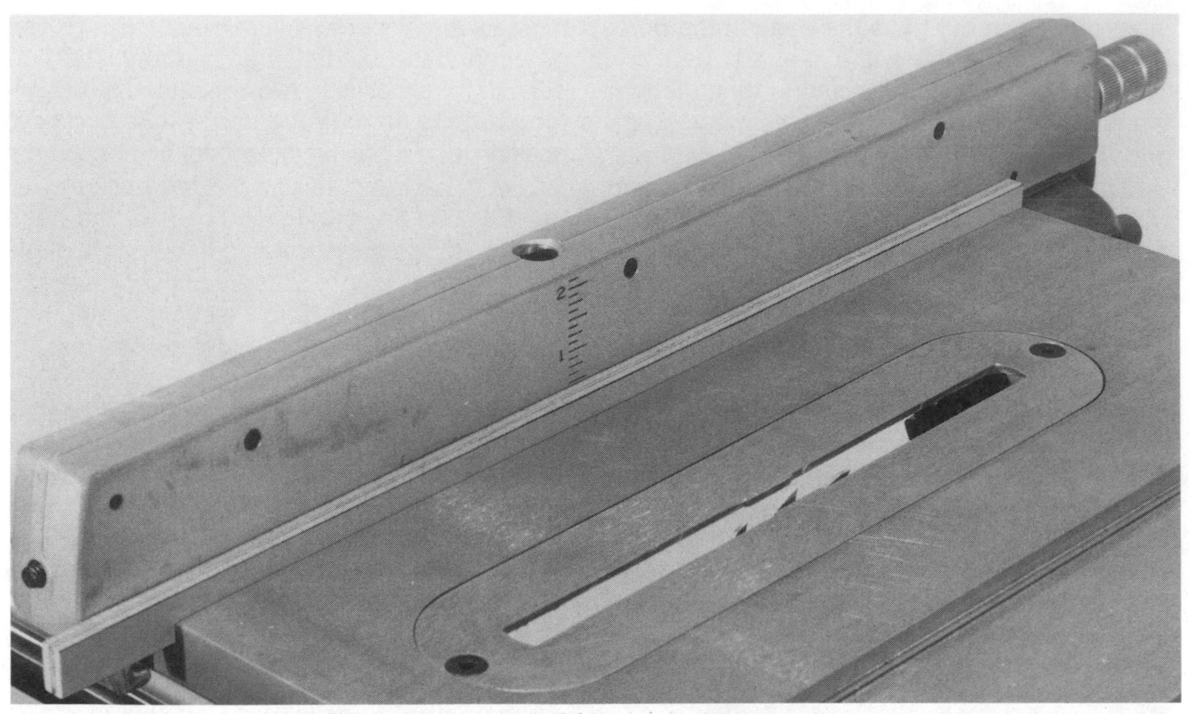

Fig. 3-8. This method of adjusting fence parallelism will eliminate the possibility of human error. Make any necessary adjustment while holding the fence against the strip of wood that sits in the table slot.

Fig. 3-9. When the miter gauge is locked for routine crosscutting, the angle between it and the saw blade must be 90 degrees. Set the blade at maximum elevation and be sure the blade of the square rests between teeth.

Fig. 3-10. You also can check the necessary 90-degree angle between the square and the saw blade, as well as other settings, using a draftsman's triangle. A 30/60-degree instrument is useful.

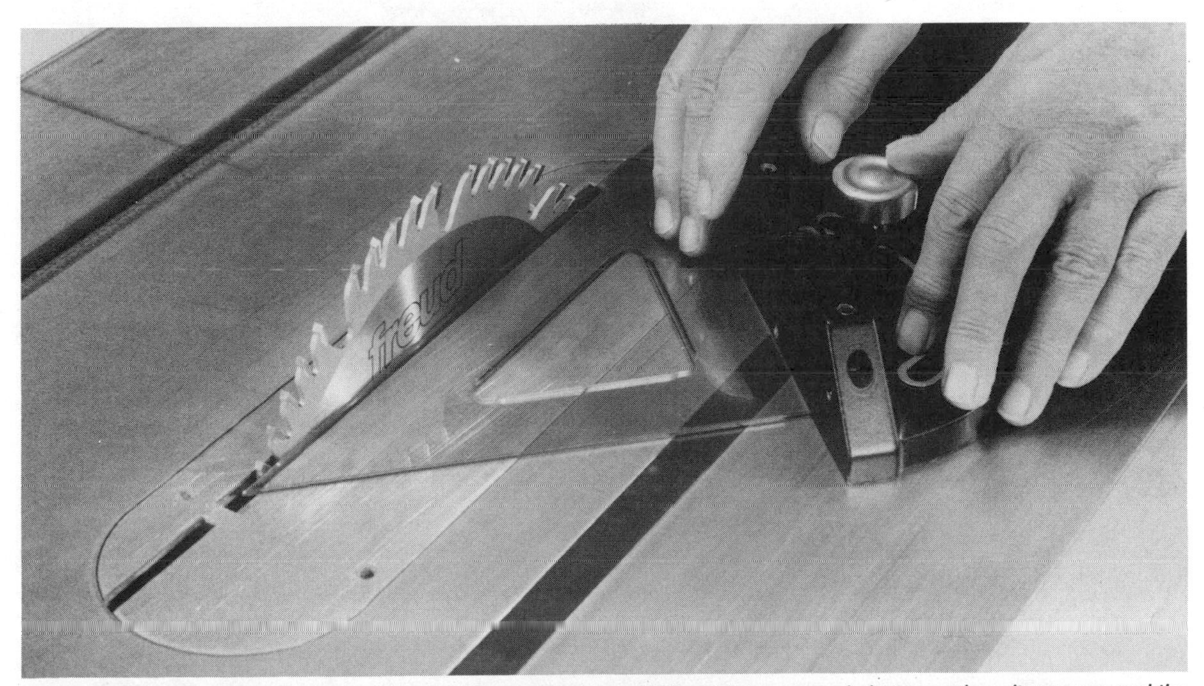

Fig. 3-11. You can use a draftsman's triangle to accurately establish a 45-degree angle between the miter gauge and the saw blade. A good miter gauge will have auto stops at the 90-degree and left and right 45-degree settings. Adjust these stops as you follow checking procedures.

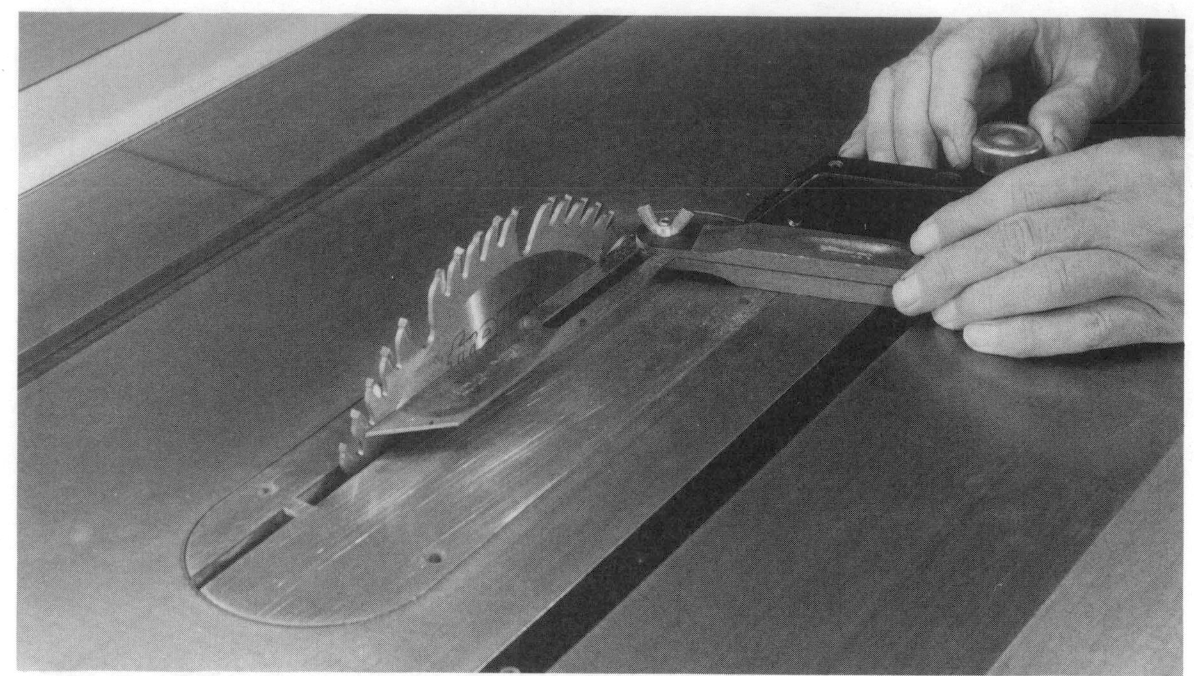

Fig. 3-12. You can read odd-angle settings—those between 90 and 45 degrees—from the scale on the miter gauge. Be sure though; check with a T-bevel before sawing or make trial cuts on scrap before sawing good stock.

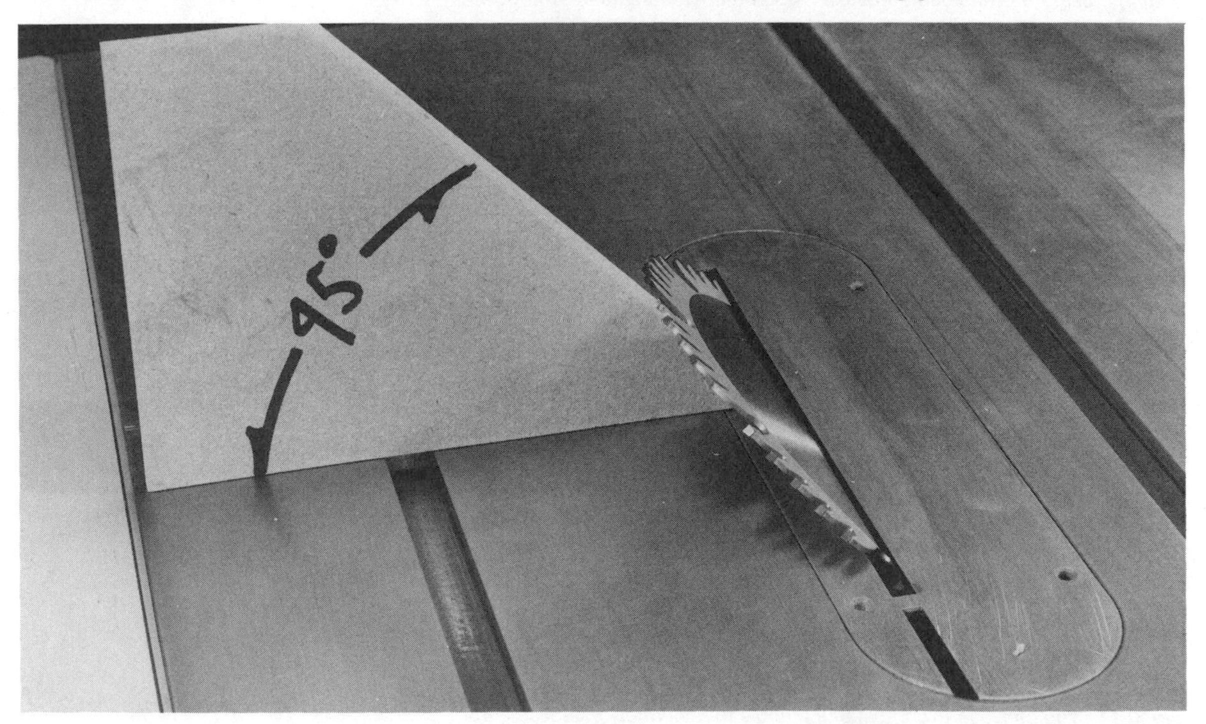

Fig. 3-13. Some operators like to make their own checking gauges. They can be designed for most-used angular settings or for odd angles that are peculiar to a particular project.

Fig. 3-14. The same gauge shown in Fig. 3-13 is used here, flopped over, to establish a 45-degree angle between the saw blade and miter gauge. Homemade gauges must, of course, be just as accurate as any high-precision checking device.

squares and triangles, and to sand them to precise size and shape after making approximate saw cuts. There is little point in making accessories like this unless you are determined that they will be precise instruments.

CHECK AS YOU GO

You can employ checks as you work, and they will feed you a constant stream of information on how component adjustments are working out. For example, if you mark a crosscut line with a square before sawing, you will know that something is wrong if the cut does not follow the length of the line. This is also a good method for checking angular cuts.

One of the most basic, ongoing methods of checking alignment accuracy is to frequently apply a square to the edge of stock you have just ripped or crosscut. If the square indicates that the cut isn't right—the edge, for example, might have a slight bevel—then you have to be a table-saw detective and find out why.

Being sure of correct alignment of table-saw components is an easy step to take toward respectable craftsmanship, and is one that can save time and effort, and avoid wasting material.

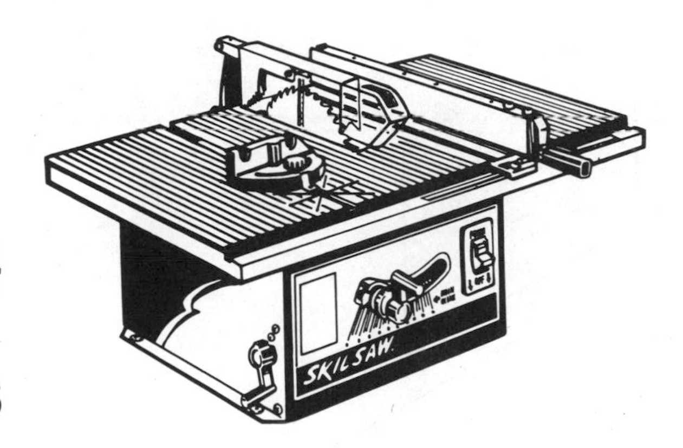

Chapter 4

Saw Blades

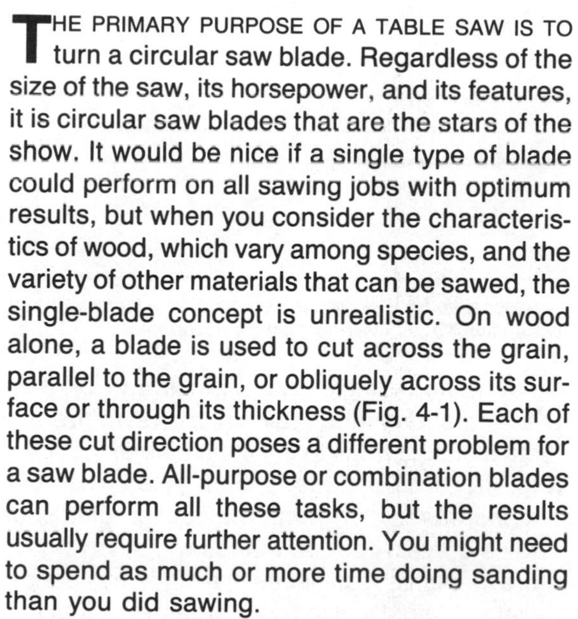

THE PRIMARY PURPOSE OF A TABLE SAW IS TO turn a circular saw blade. Regardless of the size of the saw, its horsepower, and its features, it is circular saw blades that are the stars of the show. It would be nice if a single type of blade could perform on all sawing jobs with optimum results, but when you consider the characteristics of wood, which vary among species, and the variety of other materials that can be sawed, the single-blade concept is unrealistic. On wood alone, a blade is used to cut across the grain, parallel to the grain, or obliquely across its surface or through its thickness (Fig. 4-1). Each of these cut direction poses a different problem for a saw blade. All-purpose or combination blades can perform all these tasks, but the results usually require further attention. You might need to spend as much or more time doing sanding than you did sawing.

Chances are that the blade supplied with the machine will resemble the combination blade being used for crosscutting in Fig. 4-2. It will be a good blade, but not intended as the only blade you should own. Some blades, notably the modern carbide-tipped varieties, come close to ideal, but the well-equipped shop will have an assortment of blades so the best results can be achieved on every job.

TERMINOLOGY

The slot or groove that is formed by the saw blade is called the *kerf* (Fig. 4-3). The width of the kerf will differ depending on the style of the blade. It must always be more than the thickness of the body of the blade, however, so the blade won't drag in the cut and cause friction, which can harm the blade and the wood. Most all-steel blades will have set teeth to provide for this clearance (Fig. 4-4). Alternate teeth are bent slightly in opposite directions. There are exceptions, notably the *hollow-ground* concept, which will be discussed later in this chapter.

The width of the kerf can be a telltale sign of the blade's condition. It should not be wider than the dimension across the blade's teeth (Fig. 4-5). If it is, chances are that the blade's teeth

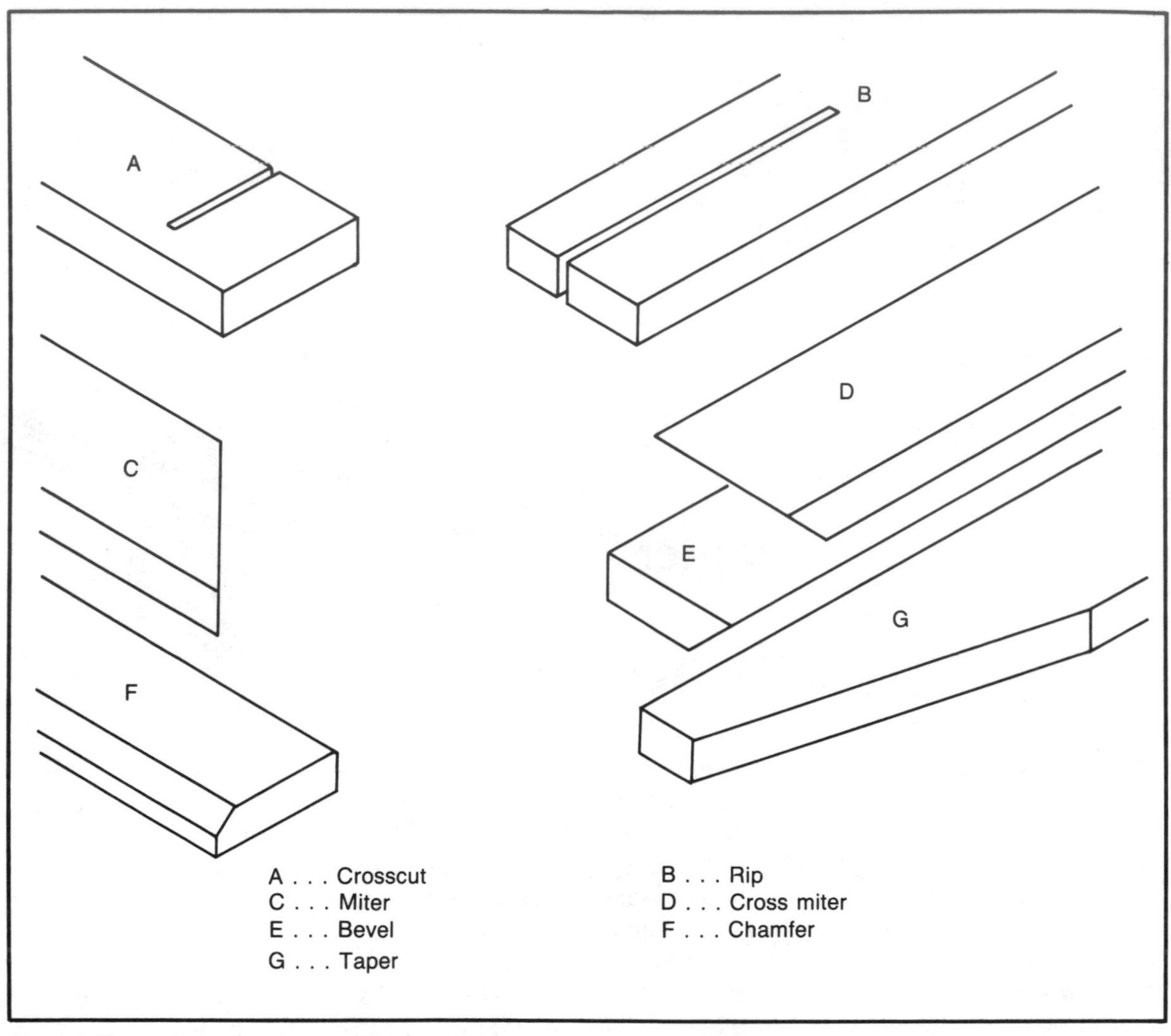

Fig. 4-1. The various ways that wood is sawed makes it virtually impossible for a single saw blade to do all jobs with optimum results. Some of the modern combination blades come close, but it's still wise to own an assortment.

are damaged, the blade is distorted, or it hasn't been correctly mounted on the arbor.

The *gullets* between teeth on a saw blade (Fig. 4-6) are designed to spew out the waste that is created while the blade is cutting. The gullet has a nicely rounded configuration since sharp corners would be more likely to crack, and provides bulk at the top area to back up the cutting edge of the tooth. Gullets on a rip blade are deeper than those on a crosscut blade because ripping creates larger waste chips than crosscutting.

The *hook angle* on a circular saw blade is the angle described by lines drawn from a tooth's cutting edge to the centerline of the blade (Fig. 4-7). The bite that a tooth takes when cutting increases along with the hook angle. Hook angles do vary, but rip blades will always have more than crosscut blades. If a blade has deeper gullets, it usually also has a greater hook angle. A negative hook angle can be found on some special blades, such as those designed for sawing used lumber where foreign materials might be encountered.

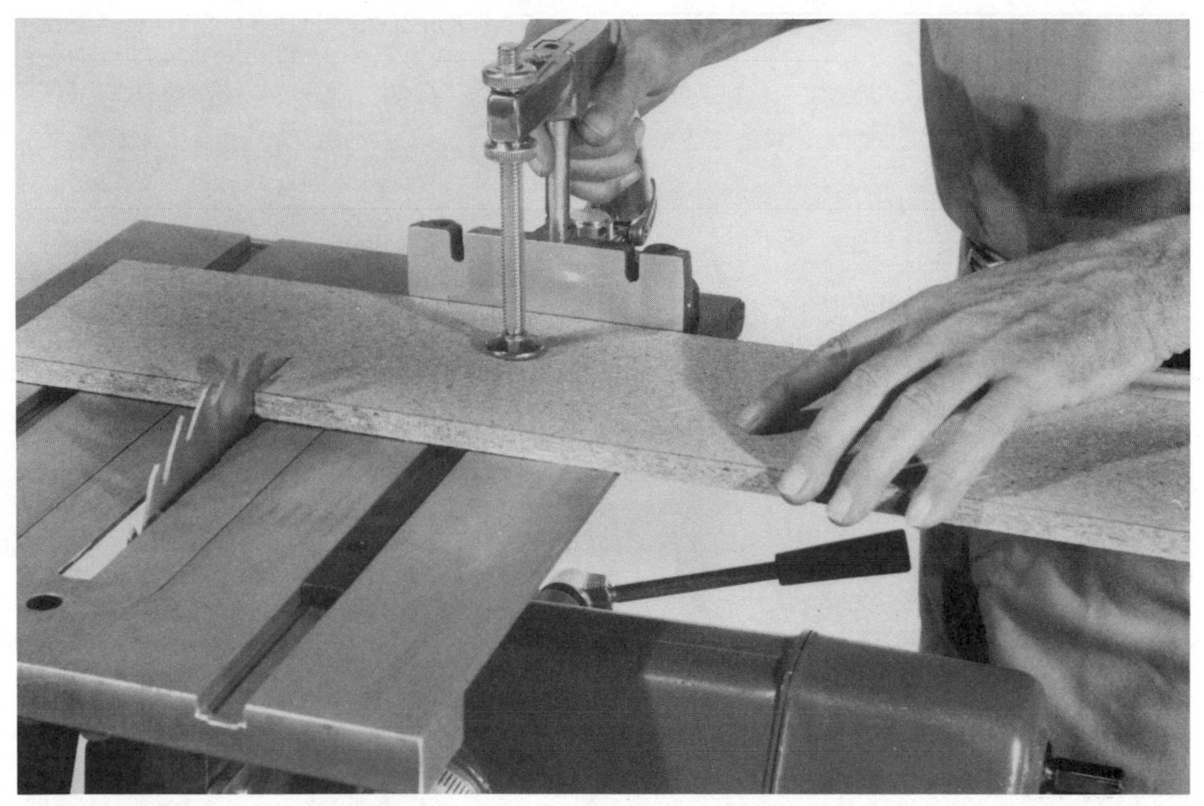

Fig. 4-2. Standard equipment with a table saw includes a combination blade, which probably will be all-steel with set teeth and deep gullets. It can be used for basic sawing chores, but cut quality won't be ideal.

Fig. 4-3. The width of the kerf is affected by the design of the saw blade.

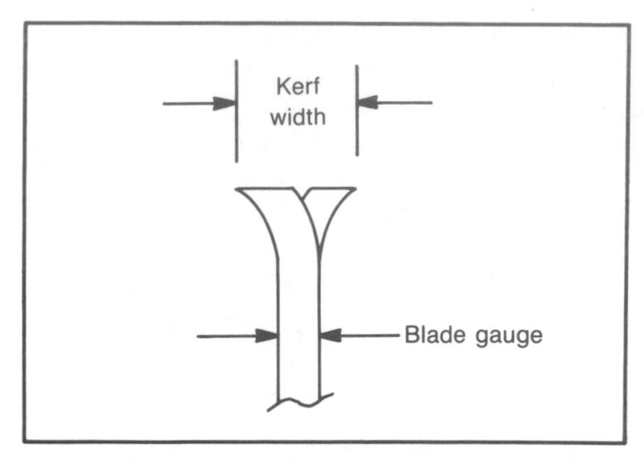

Fig. 4-4. All-steel blades will usually have alternate teeth bent slightly in opposite directions. This set provides a kerf that is wider than the blade's gauge so the blade can move freely though the wood.

The *pitch* of a blade (Fig. 4-8) is the distance between teeth. The greater the pitch, the fewer the teeth. Rip blades always have more pitch than crosscut blades.

Top clearance (Fig. 4-9) is the slope of the tooth from its cutting edge to its back edge. It ensures that the tooth won't keep rubbing on the work after the cutting edge has done its job.

A saw blade must oppose some pretty extreme forces. Just think about this slim piece of steel powered by a 1- to 3-horsepower motor spinning it at better than 3,000 rpm. The power is at the center of the blade, while resistance, which might be a really tough wood, is at its periphery. You don't need much imagination to see the stresses, the vibration, and especially

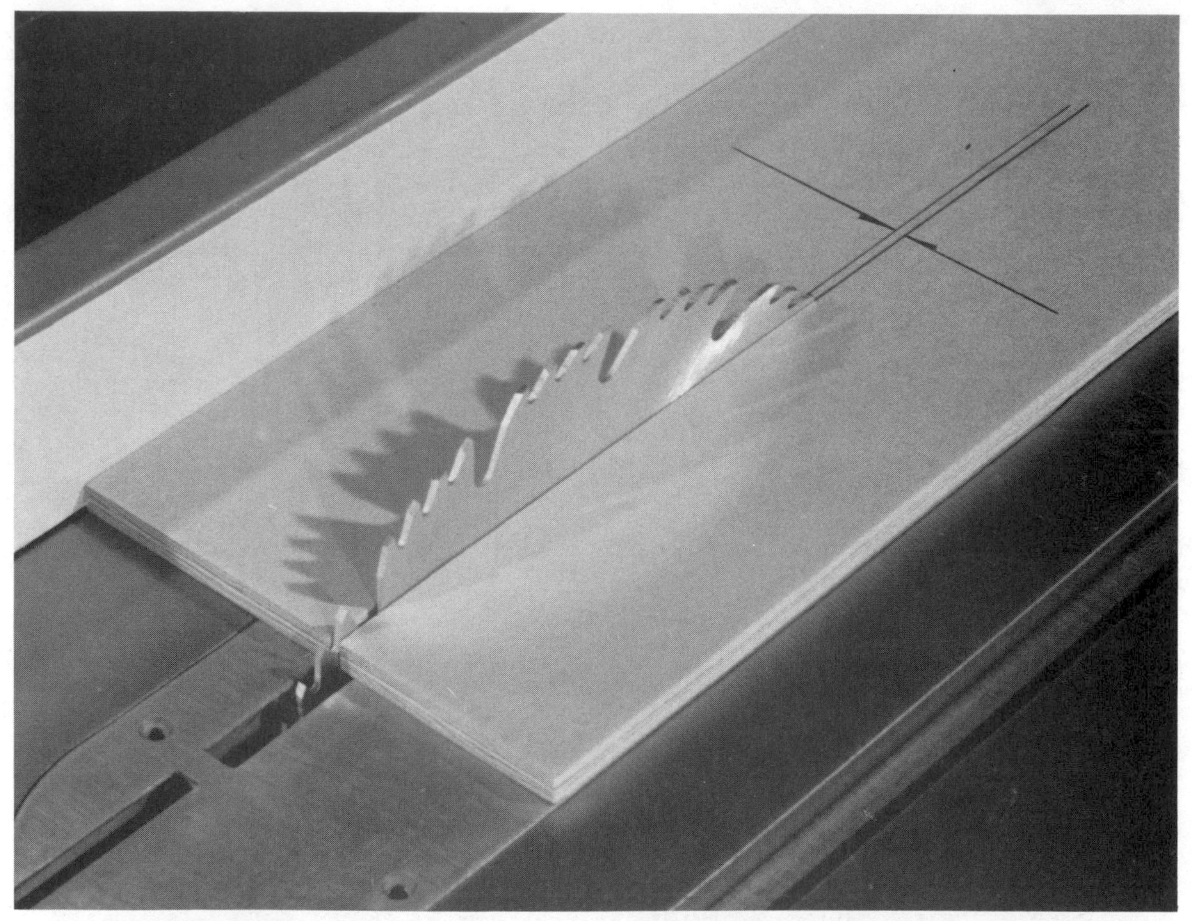

Fig. 4-5. The kerf should not be wider than the distance across the blade's teeth. If it is, check the blade or the way it is mounted on the arbor. This blade is a hollow-ground design.

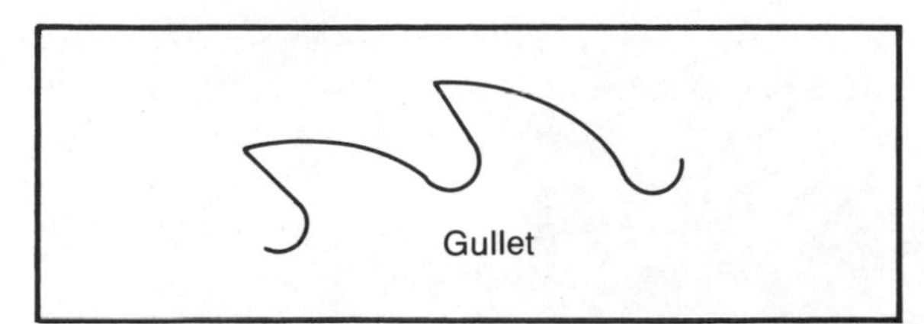

Fig. 4-6. Gullets are needed between the teeth on a saw blade to spew out waste chips or sawdust. Deep gullets are not needed on a crosscut blade, since the blade makes much finer waste than a rip blade.

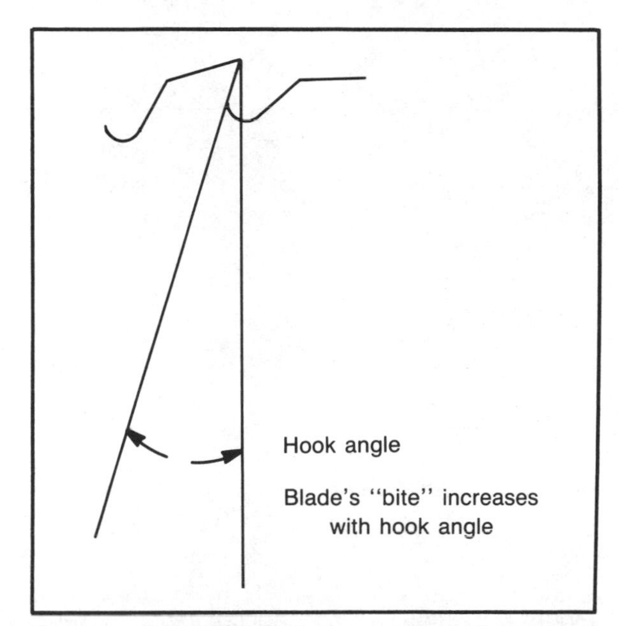

Fig. 4-7. A rip blade might have as much as 20 degrees of hook angle. Crosscut blades and combination blades do not require as much. A negative hook angle can be found on blades designed specifically for sawing metals.

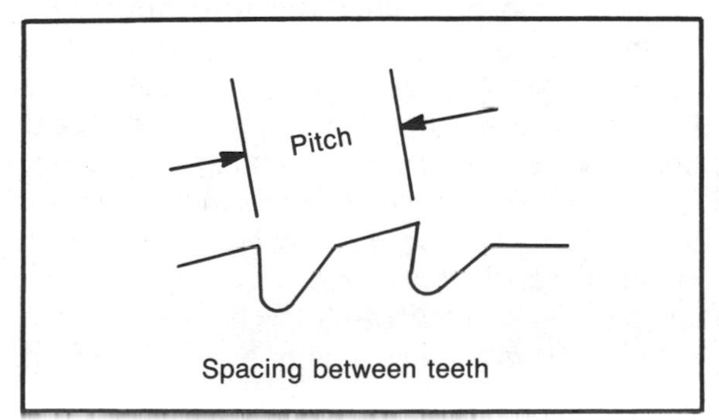

Fig. 4-8. The pitch is the distance between cutting edges; less pitch means more teeth. Except on some blades designed for a specific purpose, the more teeth a saw has, the smoother it will cut.

the heat that can be generated. That's why many blades have *expansion slots* like those shown in Figs. 4-10 and 4-11.

The purpose of the expansion slots is to prevent buckling under heat conditions generated during normal use. The hole at the end of the slot is included to guard against cracks. On some blades, usually all-purpose ones, deep gullets between banks of teeth serve a similar purpose.

TOOTH DESIGNS

The most popular tooth designs for circular saw blades are shown in Fig. 4-12. The *alternate top-bevel and raker* (ATB&R) has banks of teeth separated by deep gullets. The banks consist of five teeth each: two sets of teeth that alternate left and right and provide a shearing action, and a lone one, the raker, which is flat across the top and cleans out what is left in the kerf. This design is a good choice for respectable cuts both across and with the grain.

The *alternate top-bevel grind* (ATB) is a good

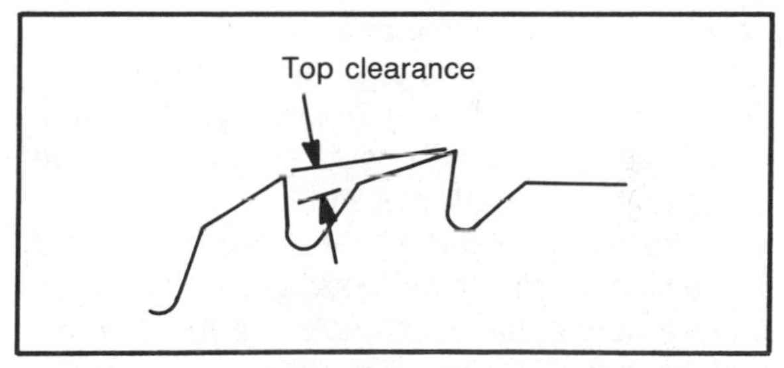

Fig. 4-9. Top clearance prevents the back of the tooth from rubbing on the wood. Without it, the tooth would not work efficiently.

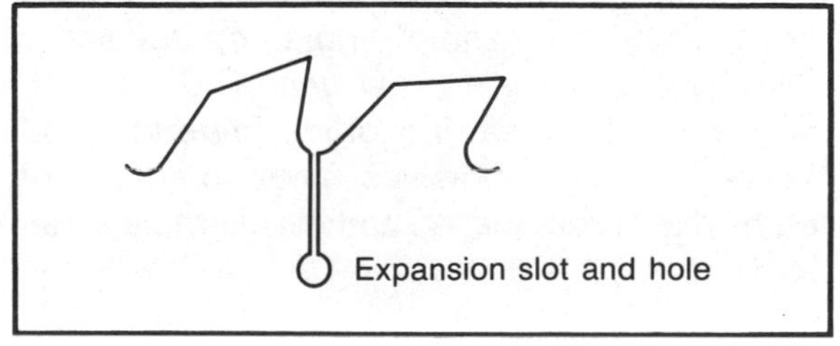

Fig. 4-10. Expansion slots prevent the blade from buckling under the heat generated by use. On some combination blades designed with banks of teeth, a deep gullet between the banks serves a similar purpose.

Fig. 4-11. Expansion slots are found on most, good carbide-tipped blades. There should be at least one slot for every 10 to 15 teeth. The hole prevents cracking and is often filled with a soft metal to reduce the noise factor.

choice for crosscutting and for cut-off and trimming operations. The top bevel on each tooth severs fibers with a clean, shearing action when crosscutting. The more teeth on an ATB blade, the higher the quality of the cut.

The *flat-top grind* (FT) is especially suitable for sawing parallel to the wood grain. Each tooth, square across the edge, acts like a tiny chisel, cutting away its own chip of wood. The teeth, like rakers, also do a good job of cleaning out the kerf. Blades having this tooth design also will have deep gullets to spew out the larger chips that are characteristic of ripping operations.

The *triple-chip and flat grind* (TC&F) is often selected for sawing brittle materials and those containing abrasives. The two shapes of teeth—flat across the top and the alternate triple-edge—provide a dual cutting action. The triple-edge teeth remove the bulk of the waste while the flat-top, raker teeth follow to clean out material from both sides. TC&F blades are often used to cut nonferrous metals. Usually, the teeth will have a negative hook angle to help prevent work from trying to climb the blade.

TYPES OF BLADES

Figure 4-13 shows an all-steel *combination blade* with set teeth. The blade is used for ripping, crosscutting, and even for mitering. This blade does not produce the smoothest cuts possible on all sawing operations, but it is not fair to downgrade it since it is not intended as the ideal all-purpose cutter. Generally, because of the tooth configuration and deep gullets, it does a better job ripping than crosscutting or mitering. It's a fair choice for general carpentry, construction work, and general sizing cuts. It's not a good choice for sawing plywood.

The *crosscut blade* (Fig. 4-14) has many small teeth with sharp points that sever wood fibers cleanly. It gets by with shallow gullets because its sawing action produces fine sawdust, rather than the chips a rip blade cuts away. Unless you have a special blade, the crosscut

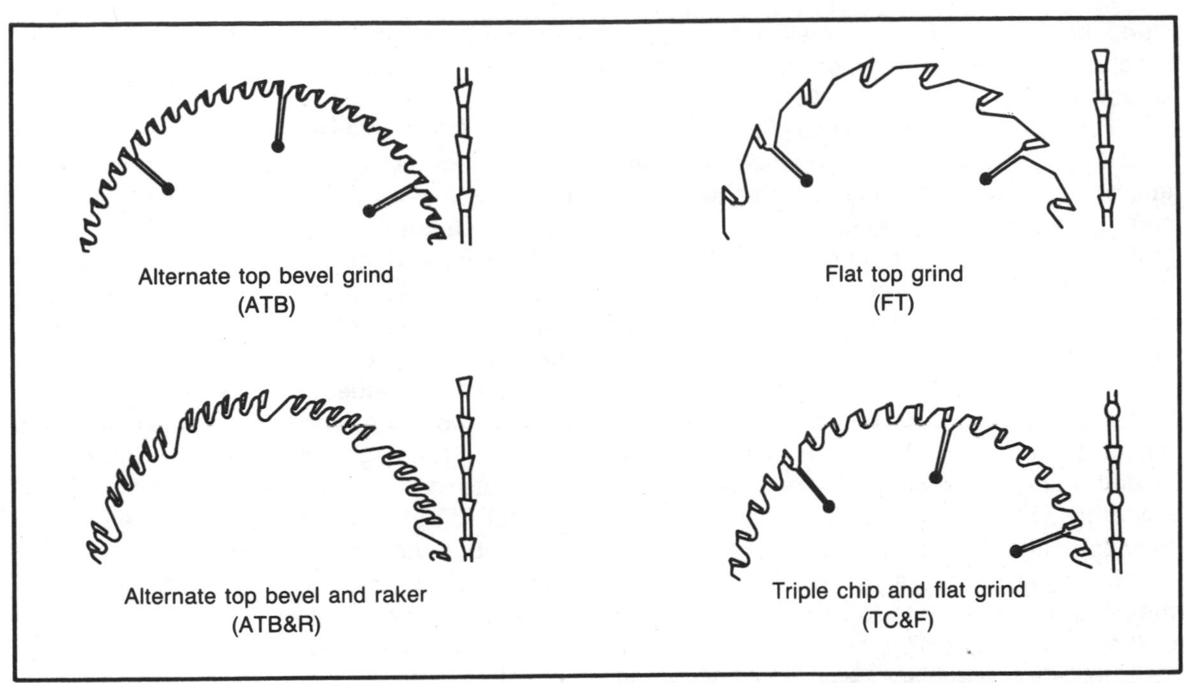

Fig. 4-12. These are the most popular tooth designs used on circular saw blades. Each one works best for a particular application.

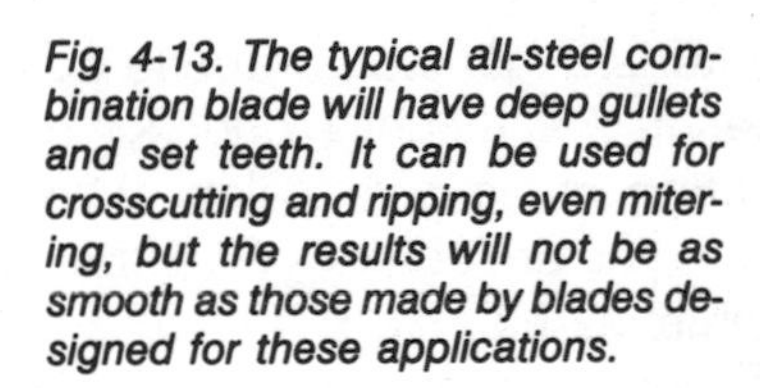

Fig. 4-13. The typical all-steel combination blade will have deep gullets and set teeth. It can be used for crosscutting and ripping, even mitering, but the results will not be as smooth as those made by blades designed for these applications.

blade, because of its many small teeth, is a better choice for sawing plywood than an all-steel combination blade or rip blade.

The teeth on a *rip blade* (Fig. 4-15) are usually quite large and are separated by deep gullets. Quite a bit of bulk supports the tooth's cutting edge, which acts like a miniature chisel to pare away its own chip of wood. A rip blade is a rip blade, and you shouldn't use it for crosscutting, mitering, sawing plywood and similar panel materials.

Hollow-ground blades are usually combination types which are designed to produce smoother edges whether crosscutting, ripping, or doing angular sawing. The reason they cut smoother is that, unlike some other concepts, they do not have set teeth. Clearance in the cut is obtained because the gauge is recessed from the points of the teeth to somewhere in the body of the blade (Fig. 4-16).

Hollow-ground blades work efficiently when their projection above the work is greater than is required for conventional blades (Fig. 4-17). Without this consideration, the blade will create excessive friction, which can cause both the blade and the work to burn.

Many special *plywood blades* are a combination of a crosscut and hollow-ground blade. They have many, very small teeth that have minimum set or none at all, and a limited recessed area (Fig. 4-18). The limited recess provides more bulk in the body of the blade, which keeps it more rigid while it is cutting, but it also dictates the maximum thickness of stock it can cut, which, logically, is usually ¾ inch. Blades of this type are usually specially tempered to stand up to the abrasive action of glue lines in the material.

Thin-rim blades are reduced in thickness at the periphery so kerf waste is minimized but blade rigidity is not sacrificed. The blades resemble the special plywood cutting blade but might differ in the number and configuration of teeth (Figs. 4-19 and 4-20). The concept is popular with amateurs and professionals since it produces optimum-quality cuts while reducing waste when cutting expensive, cabinet-grade, hardwood plywood.

Some Special Blades

One of my favorite blades is a fairly new Freud design that has about as many teeth as you get on the rim of a steel disk; for example, 80 teeth on a 10-inch blade (Fig. 4-21). The plate has a baked-on, self-lubricating, antigrip material that promotes smooth sawing while allowing the blade to be designed with minimum tooth clearance. A unique, concave configuration behind the teeth serves to remove waste and keep the blade running cool. Crosscuts and miters, even cuts on plywood, have edges that look and feel burnished. The blade also can be used for ripping but should not be used extensively for that operation. It does such a super job when crosscutting that it has become *the* blade to use in powered miter boxes.

The blade requires special sharpening considerations. For example, in order for it to keep working efficiently until its carbide teeth are completely consumed, the front of the tooth should be barely touched with a diamond wheel just once for every four sharpenings on the back of the tooth.

The *grit-edge blade* shown in Fig. 4-22 cuts with hundreds of particles of tungsten carbide that are bonded to its perimeter. It is nice to have for sawing products like fiberglass and synthetic marbles—materials that would abuse conventional all-steel blades. It does a good job on plywood, but cutting speed is very slow. It helps to have one when you are confronted with a piece of gritty, contaminated wood. It should not be used on soft wood and similar materials since they will soon gum up and clog the cutting particles.

It doesn't matter which way you mount it on the arbor since the blade will cut in any direction. In fact, you can extend its cutting life if you occasionally reverse it on the arbor.

Blades designed for cutting plastics (Fig. 4-23) will be of a special alloy that will withstand abrasion. The design of the blade is intended to

Fig. 4-14. The crosscut blade has many small teeth with sharp points that sever wood fibers cleanly. It makes much finer sawdust than, for example, a rip blade, so it doesn't require deep gullets. If you don't own a special blade for cutting plywood, use your crosscut blade.

Fig. 4-15. The rip blade works almost like a host of tiny chisels. Deep gullets are needed to throw off relatively large waste pieces. It should not be used for crosscutting or for sawing plywood.

Fig. 4-16. The hollow-ground blade, often called a planer blade, cuts smoothly because its teeth are not set. The gauge of the blade is reduced from the cutting edges of the teeth to somewhere near its center.

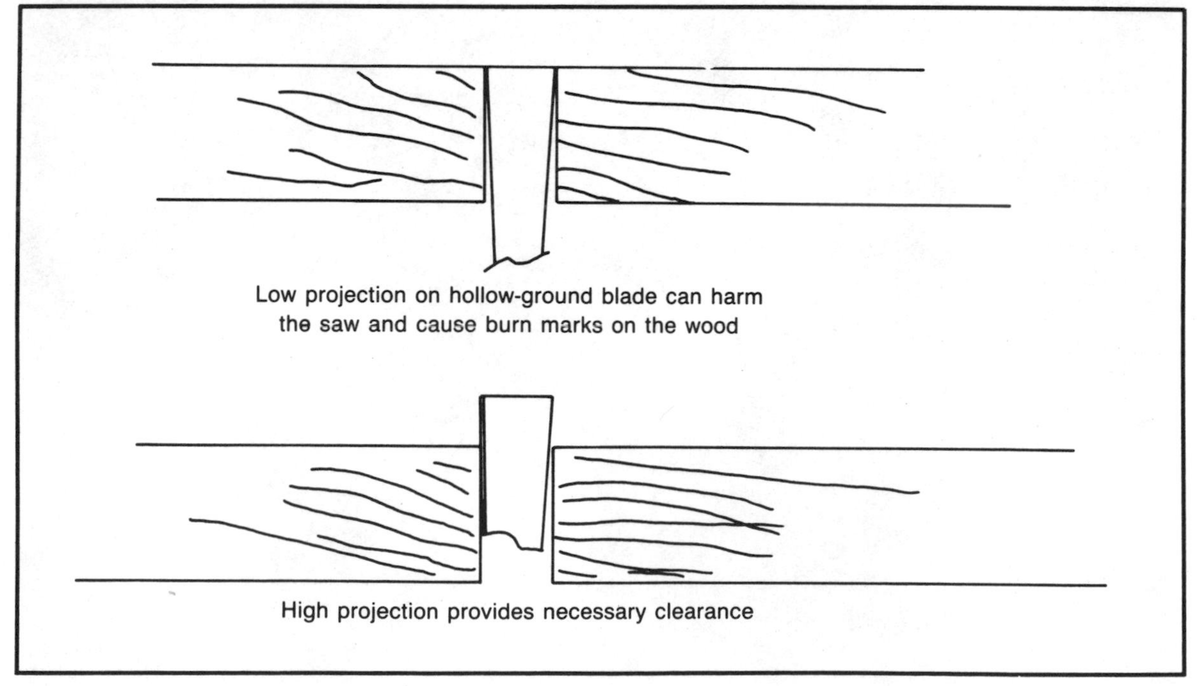

Fig. 4-17. The hollow-ground blade gets clearance in the cut because of the way it is ground. Therefore it requires more projection above the work than a conventional blade.

Fig. 4-18. Some special plywood blades have crosscut-type teeth and a limited area of hollow grinding. The design dictates the maximum thickness of stock it should be used on, usually ¾ inch.

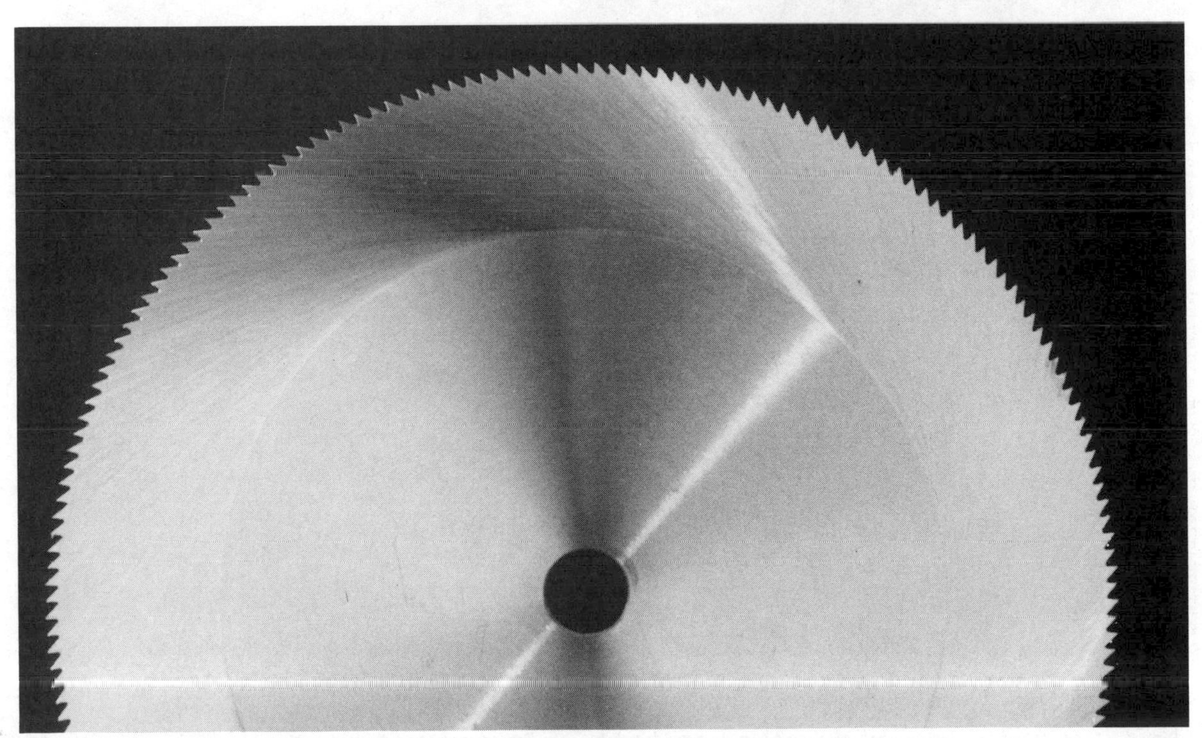

Fig. 4-19. The gauge of thin-rim blades is reduced in a limited area at the periphery, so kerf waste is reduced without affecting the rigidity of the blade.

Fig. 4-20. Thin-rim blades are also available with carbide-tipped teeth. Blades of this design are intended for finish work on expensive woods and plywood to minimize stock loss. Depth of cut is usually limited to ¾ inch. The kerf width on a 10-inch blade is about .097 inch.

Fig. 4-21. Most woodworkers have favorite table-saw blades. This Freud concept is mine. I use it almost exclusively for crosscutting and mitering since it makes featherless cuts that are correct enough for joining without further attention.

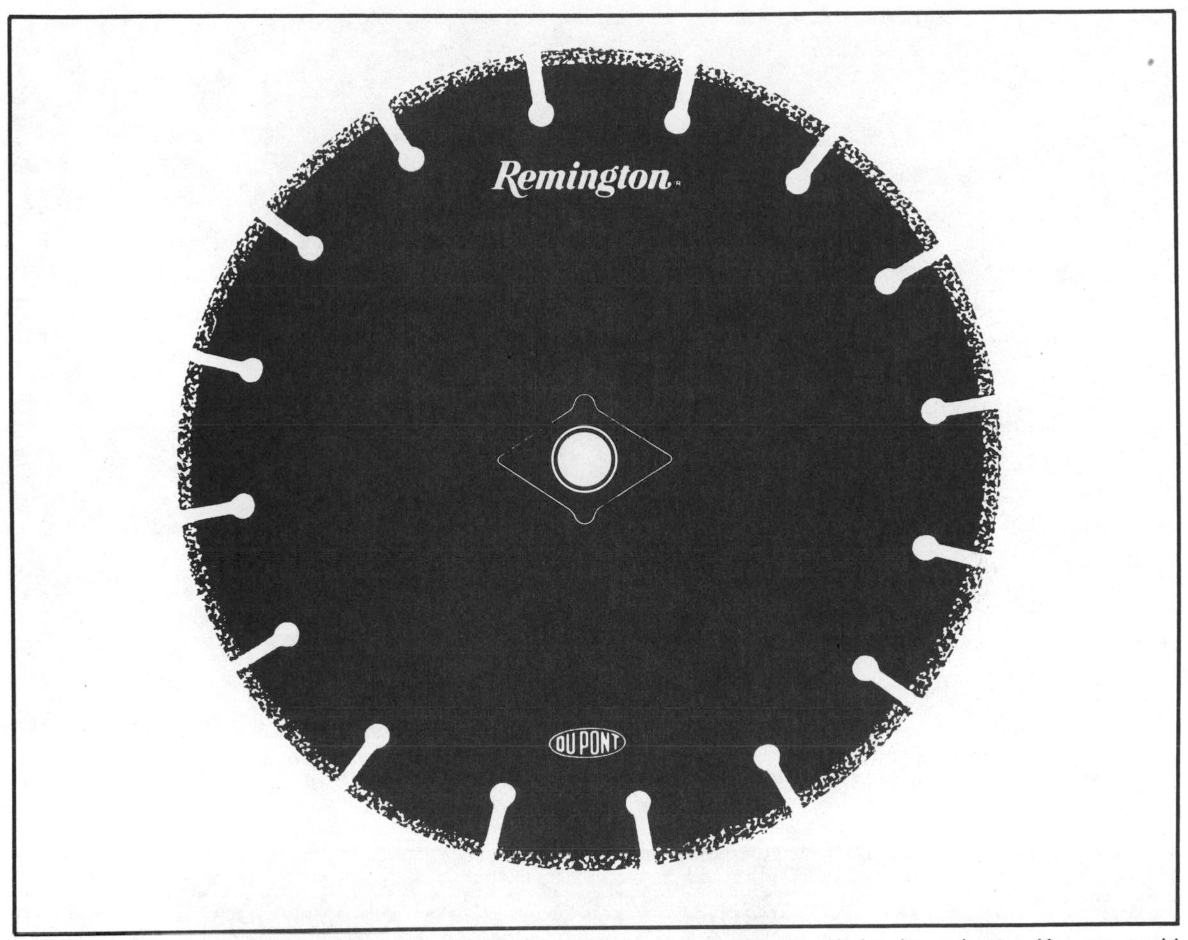

Fig. 4-22. The Grit-Edge blade cuts with hundreds of tungsten-carbide particles bonded to its perimeter. You can avoid abusing conventional blades when you need to cut abrasive materials by using this tool. It cuts in any direction, so it doesn't matter how you mount it on the arbor.

produce an edge that will require a minimum of polishing. It will run cool to keep the plastic from swelling and binding, and it will minimize chipping, cracking, and crazing. This is the kind of blade to buy only when you are extensively involved with plastics. For an occasional job, a multitoothed tungsten-carbide or all-steel crosscut blade will do.

Blades designed for metal cutting (Fig. 4-24) will be made of a special material, like a mix of high-carbon, high-chrome steel, so they will have great strength and durability and high resistance to abrasion. Such a blade will remain sharp for a long time during production sawing of nonferrous metals. As for special plastic-cutting blades, buying this type of blade is justified only if you are involved extensively with metal cutting.

A recently introduced, novel saw-blade concept, called the *Sanblade*, can be viewed as a combination saw blade and disk sander. The blade itself is a respectable 40-tooth, tungsten-carbide design, but it has 80-grit aluminum oxide abrasive particles bonded to both sides of the body. The abrasive is set back from the arbor hole so the blade is mounted in routine fashion. The intent is for the cutter to sand the edges that it saws, and it works. Those that were originally introduced had the abrasive set back too

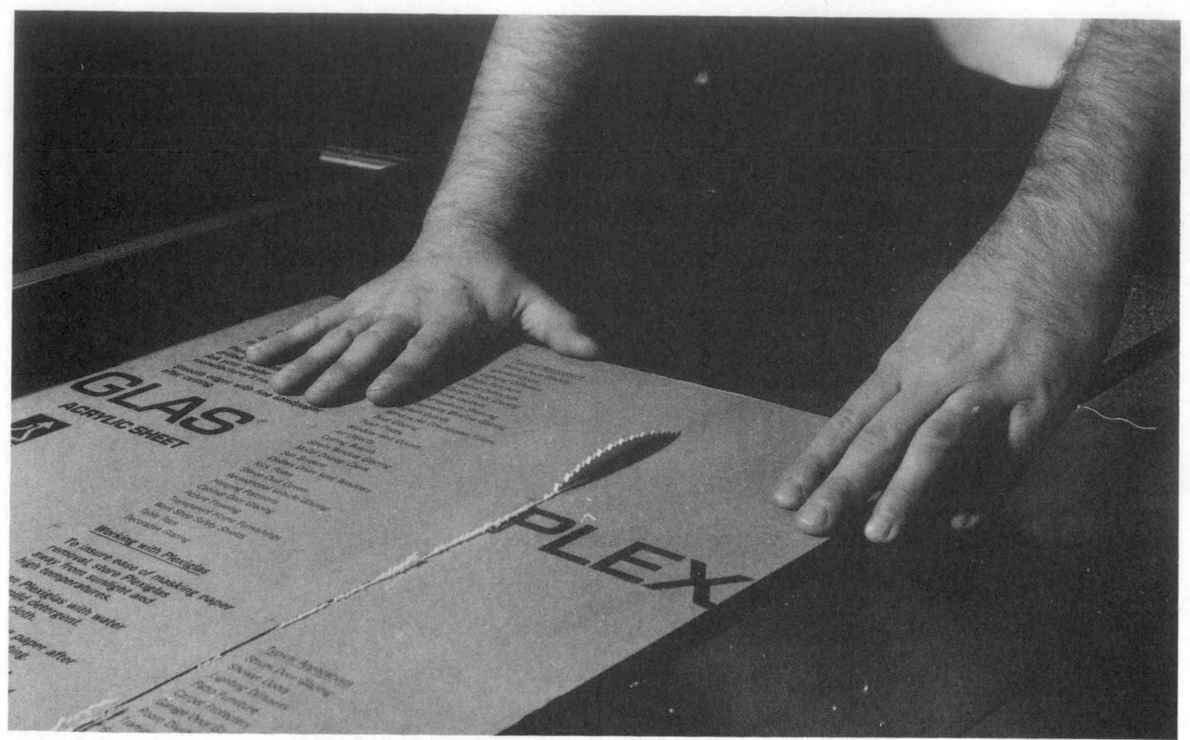

Fig. 4-23. A blade designed specifically for sawing plastics will withstand abrasion and leave edges that require a minimum of polishing. For occasional sawing of plastics, you can get by with a good crosscut blade or a multi toothed tungsten-carbide unit.

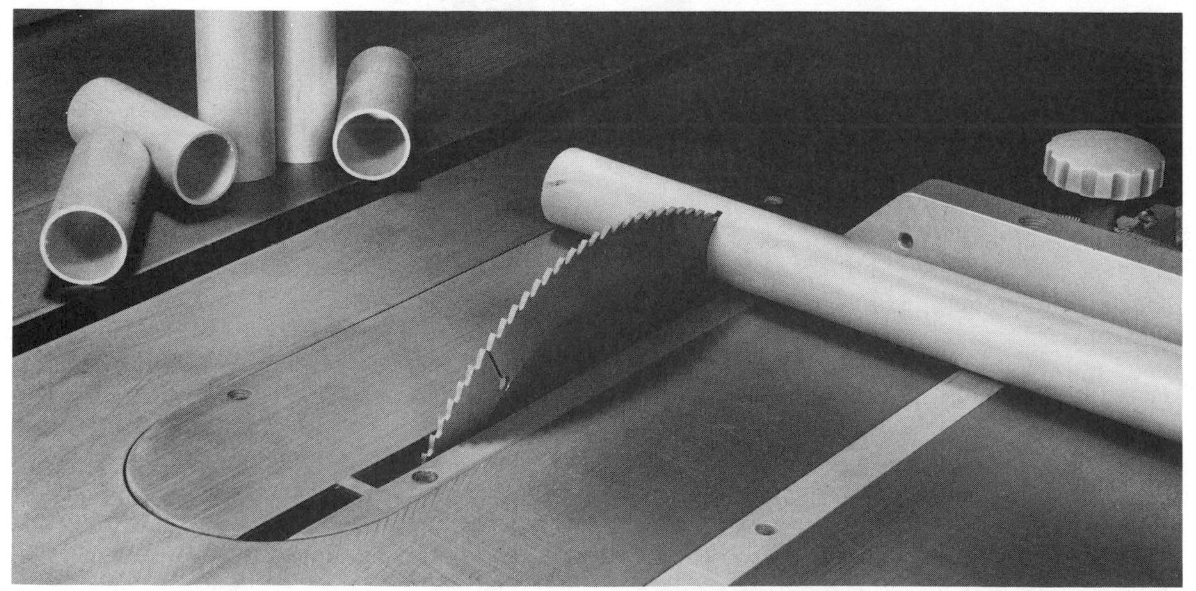

Fig. 4-24. Blades designed especially for sawing metals must have great strength, durability, and high resistance to abrasion, which is why the plate is a special alloy. For occasional nonferrous metal sawing, especially on a material like Do-It-Yourself Aluminum, you can get by nicely with a conventional crosscut blade.

Fig. 4-25. Blades with tungsten-carbide teeth are available in as many designs as all-steel blades. There are types designed specifically for ripping, crosscutting, and for all-purpose use. Tungsten-carbide blades cost more, but stay sharper through much longer periods of use than all-steel blades.

far from the gullets, so a high blade projection was needed for the abrasive to be effective. Current products have the abrasive extending right to the gullets, so blade projection can be minimized.

The blades are available in sizes ranging from 8 to 14 inches with arbor holes from ⅝ to 1 inch. Kerf width ranges from .150 to .170 inch depending on the blade's diameter. What should you do when the blade requires reconditioning? The company, United Saw Technologies Int'l, has solved this problem by offering resharpening, retipping, and reabrasing services.

Saw Blades with Tungsten-Carbide Teeth

It wasn't too long ago that *carbide-tipped saw blades* became generally available. Now they are available in as many concepts as you will find in all-steel blades. The basic ones are designed as combination units, or specifically for crosscutting or ripping (Fig. 4-25). One of the major advantages of a carbide-tipped blade is that, if it is correctly used and maintained, it will stay sharp longer than a steel blade. Another asset is that the teeth on a carbide blade cut a wider kerf than the blade's gauge. So, since the teeth are not set, they generally will produce smoother cuts than comparable all-steel blades that must have set teeth in order to function efficiently.

What Is Tungsten Carbide?

If you use the words *cemented carbide* to describe the tooth material on a carbide-tipped saw blade, you will be technically correct. Tungsten carbide is man-made, an alloy of powdered tungsten and carbon permanently bonded by *vacuum sintering*, a combination of high temperature and extreme pressure. The final product can contain as much as 94 percent tungsten carbide, with the balance composed of a binder such as cobalt powder.

In order to understand what a carbide-tipped saw blade is and what it can do, you must know that there are many grades of cemented carbides used in the saw-blade industry. The most common of these are designated as C1,

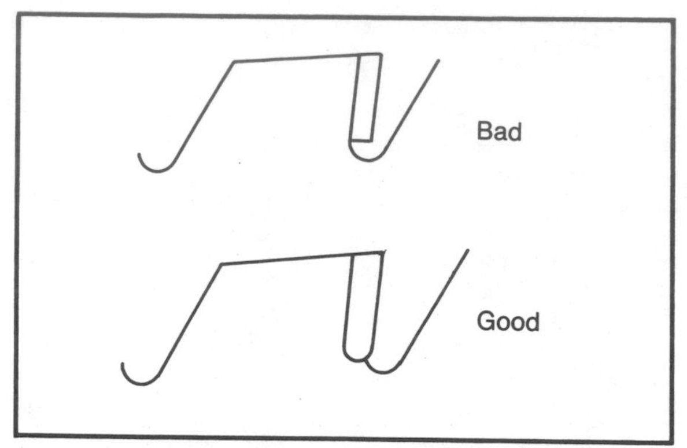

Fig. 4-26. The teeth on a good tungsten-carbide blade will be set in a special niche, not just brazed to an edge. Check the thickness of the carbide. The thicker it is, the more times the blade can be sharpened before teeth must be replaced.

C2, C3, and C4. The difference between the grades has to do with resistance to shock and wear. C4 has the lowest shock resistance, but the highest wear resistance. C1 is low on wear resistance, but high on shock resistance. A C2 grade, which is about medium in both areas, is often used on special ripping saws that have a flat-top grind and extreme hook, and on blades with a triple-chip grind and minimum hook that are designed for sawing nonferrous metals. C4 seems to be the proper choice for general-purpose and crosscut blades.

Not all carbide-tipped saw blades are manufactured to optimum specifications. Some important aspects to examine before you buy follow:

☐ The size of the carbide tips, since the larger the tips, the more times it can be sharpened before the tips must be replaced.

☐ The braze connection between the carbide and the blade. If *pit marks*, tiny holes, are evident, then the blade has not been manufactured to high standards.

☐ The way the tooth is mounted. As shown in Fig. 4-26, it should be seated in its own niche, rather than simply abutted against an edge.

You can expect much from a single carbide-tipped saw blade, but not everything. The

Fig. 4-27. This general-purpose tungsten-carbide blade has an alternate top-bevel grind and a 13-degree hook angle. The 10-inch blade has 40 teeth and can be used for ripping, crosscutting, and mitering in hard or soft woods. It is also a good blade to use on plywood and other man-made panel materials.

characteristics of wood and other materials and the various methods of sawing affect the design of the super blades, just as they do all-steel blades. There is a clear difference, for example, between teeth that are shaped specifically for ripping and those that do an optimum job when crosscutting. You will find differences even among the blades designated as "all-purpose" and those listed as "combination" types. Examples of carbide-tipped saw blades are displayed in Figs. 4-27 through 4-31.

An important point to remember is that, while tungsten carbide is a very tough material —second only to diamonds in hardness—it is also very brittle. Any carbide-tipped saw blade should be handled, and used, with tender, loving care.

CARING FOR BLADES

Blades perform best when they are maintained in like-new condition. This has to do with cleanliness as well as the sharpness of the teeth. Because blades do get hot in use, it is not unusual for pitch to accumulate at the rim and on the top of the and behind cutting edges (Fig. 4-32). If this pitch is not removed, but allowed to build up, the blade will dull faster, and sawing will not go as smoothly as it should. I feel that blades should be cleaned frequently, even if a visual check does not indicate they need it. Figures 4-33 and 4-34 show blades that have been spot-cleaned, just to demonstrate in a before-and-after way the results of blade housekeeping.

Commercial pitch removers are available. Before you use them, carefully read the instructions on the container. Some workers use solvents like kerosene, paint thinners, even gasoline, but I feel that such methods should be avoided. Often, a blade that has been immersed in warm water made soapy with a detergent and rubbed with a soft cloth will emerge sparkling clean. Don't work with an abrasive material or with steel wool. After the treatment, dry the blade thoroughly with a soft towel and, if storing it, apply a coating of wax or a film of light oil.

For stubborn cleaning chores, I have found that a common oven cleaner does a good job, but I wear a mask when following the procedure. I place the blade on a thick layer of newspaper and spray one side and then the other. After a few minutes, I hose off the blade and dry it. Sometimes, because of a really tough deposit, I work on the teeth with an old toothbrush. It's very important when using products of this type to obey the instructions on the container.

A dull blade is a "sore" blade. Dull blades cause problems, but no sawing problem is as easy to solve. Just be sure that cutting teeth are kept in keen condition. A dull blade makes it necessary to apply more feed pressure, which creates a situation where your hands might slip. A dull blade won't cut as it should, so you will be unnecessarily reducing cut quality.

Sharpen your own blades? I don't think so. It's much better to find a good, professional sharpener who will have the proper equipment and the expertise to bring the blade to prime condition, and costs are not out of line. Remember that, although tungsten-carbide teeth remain sharper for longer periods of time than all-steel ones, they will eventually get dull.

Warning signs of a dull blade include the following:

☐ The work has a tendency to climb over the blade and cut edges will have burn marks. In extreme cases, you might even see smoke or smell the odor given off by burning.

☐ The blade might tend to move off the cut line.

☐ You will feel the increased effort that is required to move the work during the pass.

STORING BLADES

Placing blades in a drawer so they rest on or bang against each other is obviously a bad idea. If you must use a drawer, wrap the blades in heavy layers of newspaper that have been sprinkled with drops of light oil. A better storage method is to hang the blades individually on hooks in a cabinet, but this calls for more space than you might care to provide.

Fig. 4-28. This type of general-purpose blade has a triple-chip grind and 7-degree hook angle. The 10-inch blade has 48 teeth and can be used for general sawing. It does an especially good job on plywood and composition materials, whether making sizing cuts or trimming.

Fig. 4-29. This blade design has banks of teeth, each consisting of twin sets of alternate top-bevel teeth and a single raker. The raker works like a planer tooth to clean out the kerf. The 10-inch blade has 50 teeth and a hook angle of 7 degrees.

Fig. 4-30. Carbide-tipped saw blades are also available for special purposes. This 10-inch blade with 80 teeth and an alternate top-bevel grind is suitable for sawing plastics and other materials, such as double-faced laminates.

Fig. 4-31. This carbide-tipped blade is sometimes called a safety blade or a control cut blade. It can have 8, 10, or 12 teeth, which are installed so cutting edges are slightly above the plate's periphery. It's offered as an economy blade for use where a smoothly finished cut is not necessary.

Fig. 4-32. Sawing creates heat, so a buildup of pitch near the perimeter and on areas of the teeth is not unusual. You can avoid this buildup by cleaning blades even more frequently than a visual check indicates they need it.

Fig. 4-33. Allowing a blade to get to this condition indicates poor sawing practice and bad storage. A reliable sharpening shop brought it back to usable condition. Note the spot-cleaned area on the right. (No, this was not my blade).

Fig. 4-34. To show how just a bath in warm, soapy water can clean a blade that hasn't been abused, I used the system on a small area to have a before-and-after picture. If deposits on the teeth are stubborn, work on them with an old toothbrush.

The best system is to make a special case like the one in Fig. 4-35. The unit is designed for 10-inch blades, but will also be usable for smaller ones. Each blade has its own felt-lined shelf that slides in grooves in the sides of the case (Fig. 4-36). A materials list for a case of this type that will safely store five blades is offered in Table 4-1.

Another idea for a blade-storage case is shown in Fig. 4-37. Here, the blades are held in grooves cut into the bottom and the ends of the case. Spacing of the grooves ensures that the blades can't contact each other. A thick, foam rubber pad cemented to the underside of the cover keeps the blades firmly in place when the lid is closed.

MOUNTING A SAW BLADE

First steps in mounting a saw blade are to unplug the saw, remove the table insert, and move the guard so it can't interfere. The operation is usually easier if the arbor, on which the blade is mounted, is raised to its maximum height. To remove a blade, use a piece of soft wood, as shown in Fig. 4-38, to keep the blade from turning as you use the wrench provided with the machine to turn the arbor nut clockwise.

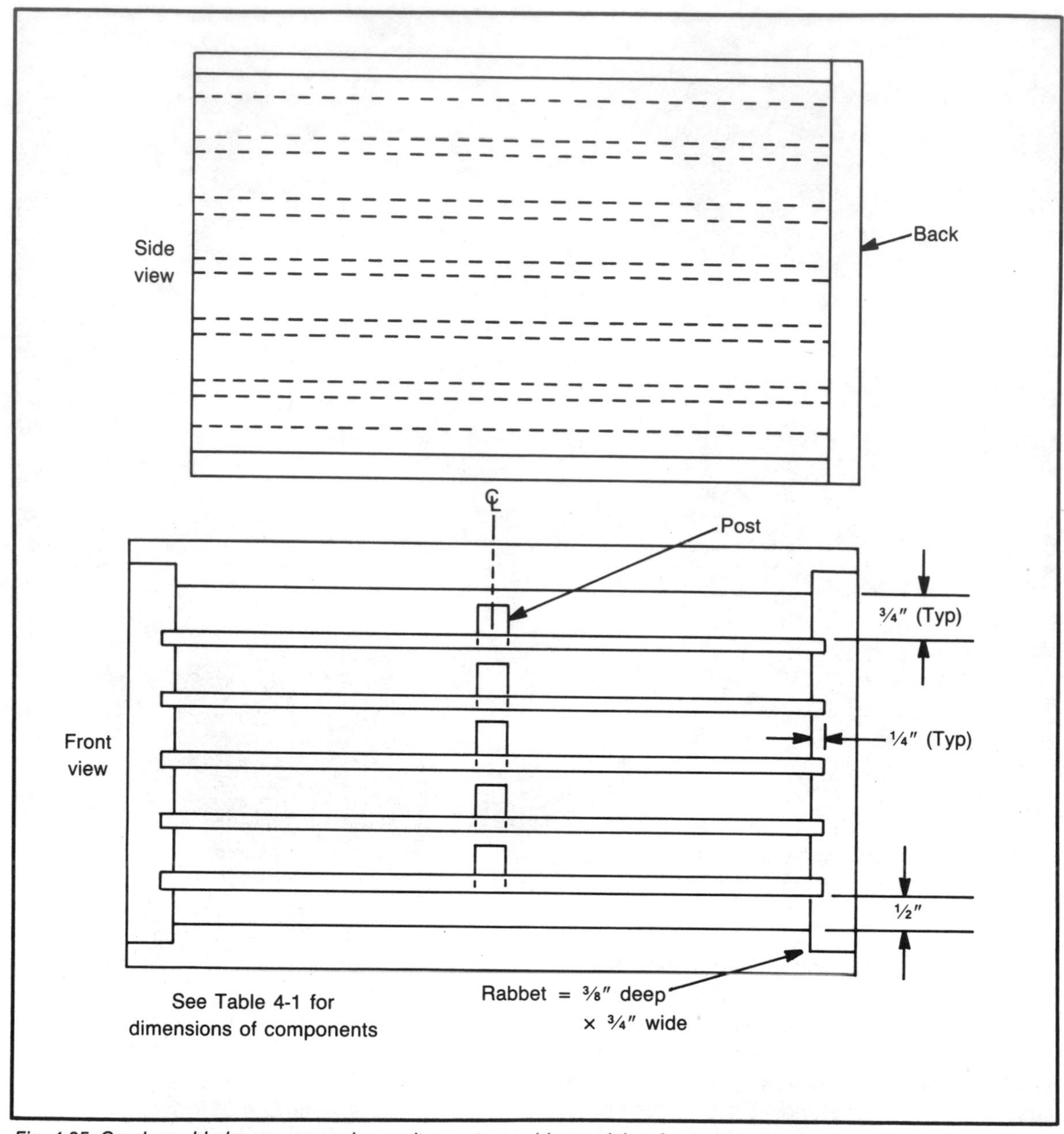

Fig. 4-35. Good saw blades are expensive, so it pays to provide special, safe storage units for them. This project will hold five blades, but you can modify it to contain more.

To secure a blade, do the opposite. Hold the blade still by using the strip of wood at the back of the insert and turning the wrench counterclockwise. On some saws, the arbor is designed with a flat so that a second wrench, also provided with the machine, can be used to hold the arbor still while either tightening or loosening the arbor nut (Fig. 4-39).

Before you mount blades, inspect the arbor threads and the arbor washers to be sure they

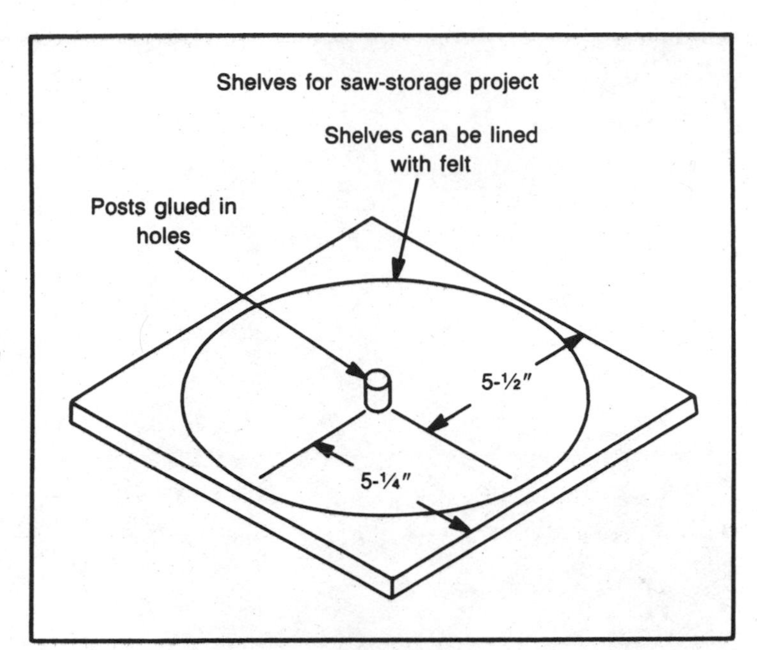

Fig. 4-36. Shelves slide in grooves cut into the sides of the case. Each shelf holds a single blade. Providing a felt base for the blades is a good idea.

are clean. Just a small chip of wood between the washers and blade can cause inaccurate sawing. Tighten the arbor nut just enough to secure the blade. Overtightening can cause problems when you need to change a blade.

Many operators use blade collars, or *stabilizers*, on one or both sides of the blade. These accessories, which resemble oversize metal disks, are placed against the blade to provide additional saw-blade rigidity. Generally, the assembly consists of a regular arbor washer, then a stabilizer, followed by the saw blade, a second stabilizer and washer, and finally the locknut.

It is possible that stabilizers can change the normal position of the blade on the arbor. Be

Table 4-1. Materials for the Five-Blade Storage Case.

PART	#PCS.	SIZE	MATERIAL
Top and bottom	2	¾″ × 10½″ × 12″	Pine
Sides	2	¾″ × 6¼″ × 12″	Pine
Shelves	5	¼″ × 10½″ × 11″	Hardboard
Posts	5	⅝″ D. × ¾″	Dowel
Back	1	½″ × 7″ × 12″	Plywood

Fig. 4-37. This storage case keeps blades safe because individual ones rest in grooves cut into the case ends. The blades rest on foam rubber.

Fig. 4-38. To remove or install a blade, you must keep the blade from turning as you use the wrench on the arbor nut. To do so, use a strip of soft wood as a lever. Here, the nut is being loosened by being turned clockwise.

Fig. 4-39. Some table saws, such as the Delta Unisaw, have flats on the arbor so one wrench can be used to keep the arbor still while a second wrench is used to tighten or loosen the locknut. Always work carefully when you are mounting or removing a blade.

Fig. 4-40. Many times, the projection of the blade or other tools like dado assemblies and molding heads must be specific. Special commercial gauges are available so the projection can be accurate, but you can make your own, like this one.

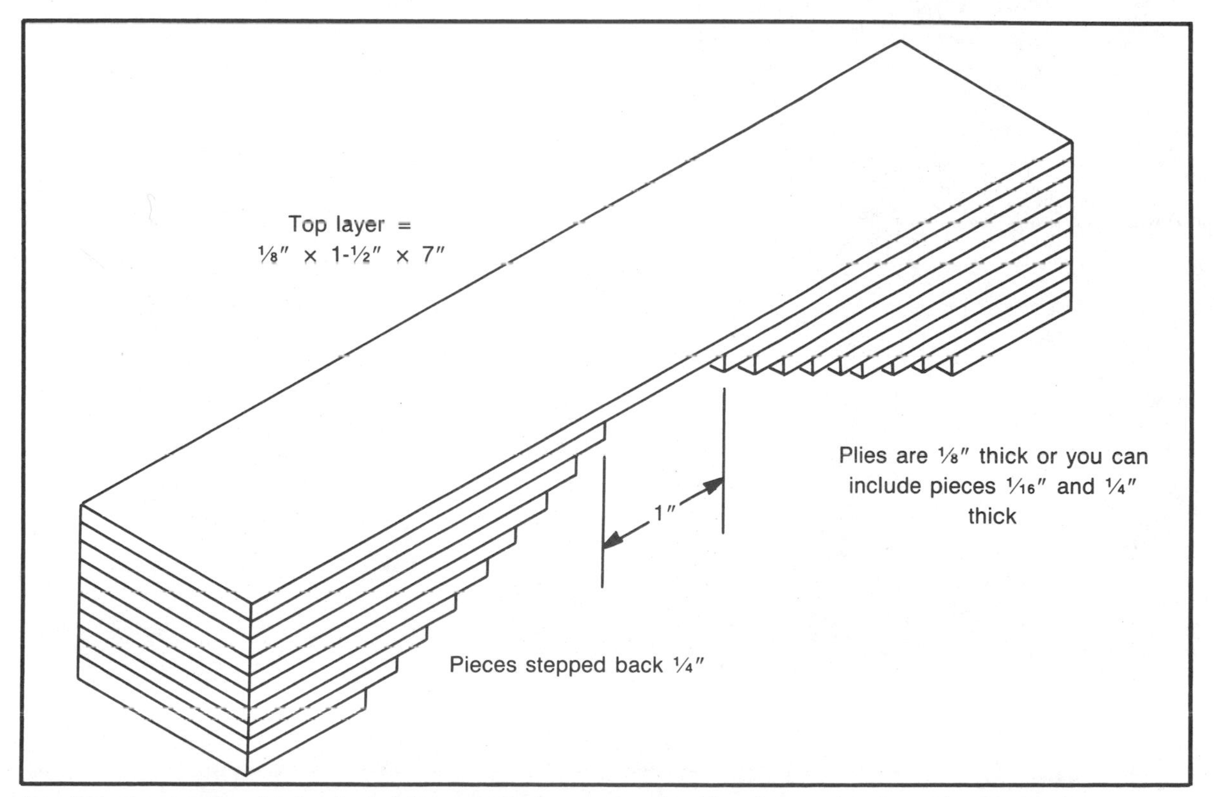

Fig. 4-41. Construction details of a height gauge that you can make. The plies can be plywood or hardboard.

sure that the blade will still poke through the table insert as it should. Also, you will need to change the location of the splitter so it will line up with the saw blade.

BLADE ELEVATION

Blade elevation deals with the projection of the blade above the surface of the table. The projection for cutting through material must, of course, always be more than the thickness of the stock. The amount of projection above the stock is based on efficiency and safety. A high projection shortens the blade's cutting arc in the wood so it meets less resistance, but more of the blade is exposed above the workpiece. Minimum projection creates more drag because of the longer cutting arc, but it reduces blade exposure, which can be a safety factor. Many operators work so the blade barely pokes through the work. A generally accepted rule that takes both safety and efficiency into account is to set projection to not more than the deepest gullet on the blade.

There are many times (as you will see when we get into various areas of sawing techniques) when projection is set to less than the stock's thickness. This setting can be needed with a saw blade and with accessories like dadoing tools and molding heads. This limited projection can be measured with a scale or square but it is best to have a special gauge that will help you work more accurately. Tools for this purpose are available commercially, but you can make your own by duplicating the project shown in Figs. 4-40 and 4-41.

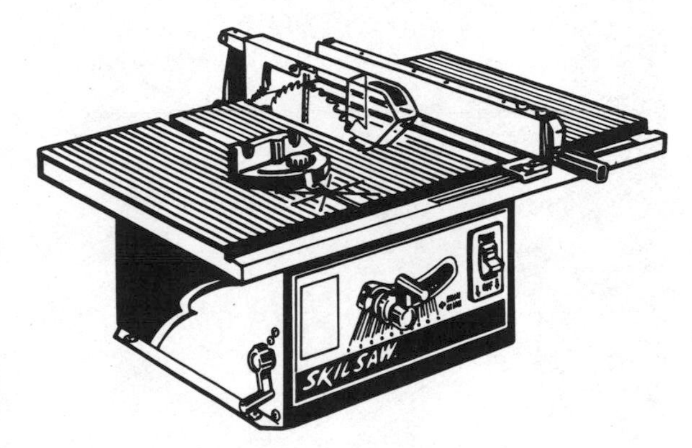

Chapter 5

Crosscutting

TWO OF THE BASIC USES FOR CROSSCUTTING ARE sawing stock to a specific length and squaring the end of workpieces. To make a simple crosscut, place a good edge of the stock against the head of the miter gauge and then move gauge and work, as a unit, past the saw blade (Fig. 5-1). Most workers prefer to use the miter gauge in the left-hand slot, but the table does have two slots, so a choice can be influenced by operator preference, by the size of the workpiece, and by the nature of the cut.

Generally, use one hand to move the miter gauge, and the other to keep the work secure. A good stance is directly behind the gauge so as to avoid being in line with the saw blade. Position your hands so they are well away from the cut area. Move the work at a reasonable speed and without pausing until the cut is complete. How fast you move the work will depend a lot on the design of the saw blade and the hardness of the material. Feed speed should not be too fast or too slow, but just fast enough to allow the blade's teeth to cut as they should. It is a good idea to slow up a bit at the end of the cut to minimize the feathering that can occur as the blade breaks through the workpiece.

When the cut is complete, return the work and the gauge to the starting position. Do not remove the part that has been cut off until you have turned off the machine and the blade has stopped turning.

It is bad practice to use a free hand to push against the cutoff end of the workpiece. This method can pinch the blade in the kerf and create kickback. On some workpieces, particularly long ones, you can use your free hand to provide extra guidance, possibly some support when the work overhangs the table, but you should never use it to force the work forward.

When crosscutting, remove the rip fence or lock it far away from the cutting area. You don't want to trap cutoffs between the fence and blade so they become projectiles thrown up or back. Also, don't allow scrap to accumulate on the table.

It is good practice to use a miter gauge hold-

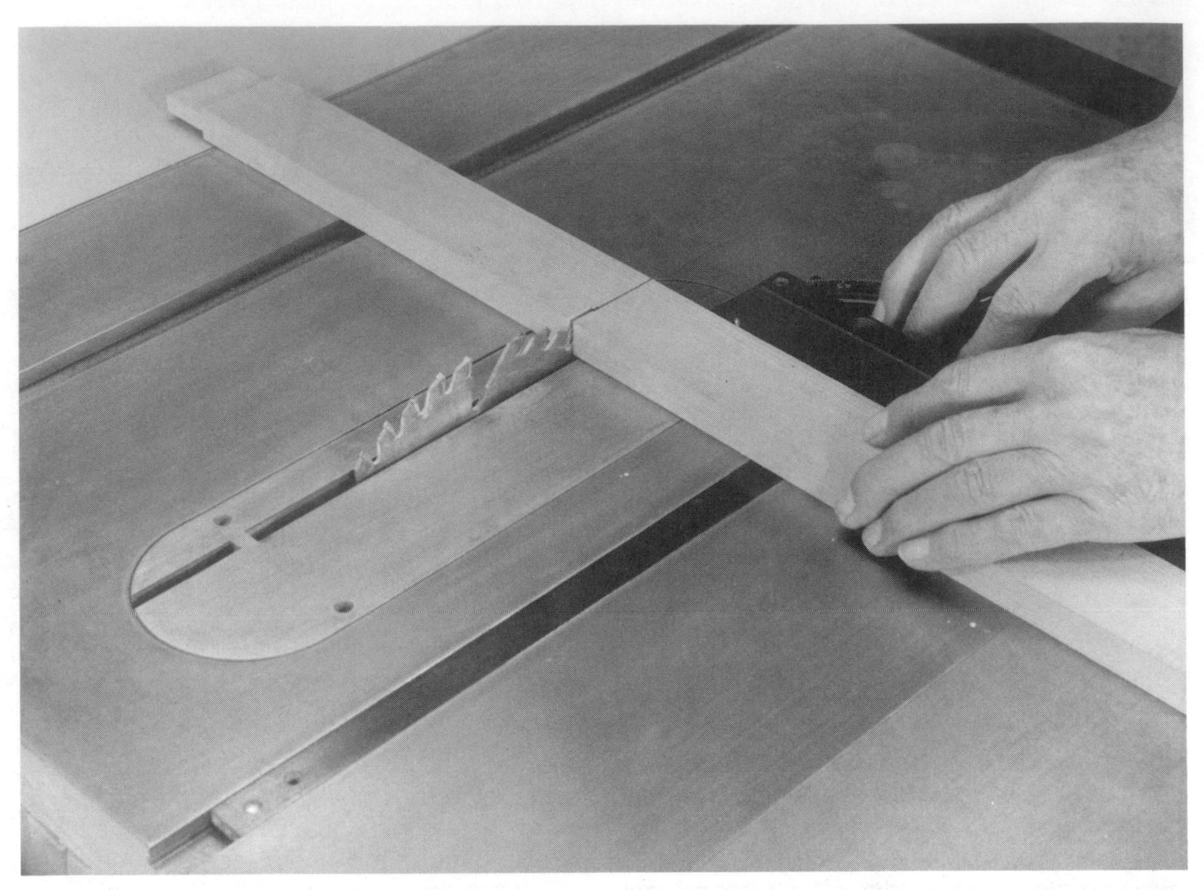

Fig. 5-1. To do simple crosscutting, hold the work against the miter gauge and move both, as a unit, past the saw blade.

down whenever possible. This accessory keeps the work firm and correctly positioned throughout the pass so hands are used merely to move the work.

It is not unusual to see a professional use a miter gauge backward when the width of the work is greater than the distance from the blade to the front of the table. This procedure requires that one hand be on the gauge while the other hand pushes against the opposite edge of the workpiece. This method places the operator in a strange position, so extra care is required to do the job safely.

Alternate methods are available. If the workpiece is overly wide, it might be best to think about ripping instead of crosscutting. A second solution is to clamp or tack-nail a strip of wood to the underside of the work so that it can ride against an outboard edge of the table, thus serving as a control for the cut.

SIMPLE MITER-GAUGE EXTENSION

The purpose of a miter-gauge extension is to provide extra support for workpieces and to make accurate sawing easier. The most elementary concept is simply a straight piece of wood that is much longer and a bit higher than the head of the miter gauge. It is rare to find a miter gauge that isn't designed to accept extensions. The attachment method might be nuts and bolts that pass through holes or slots in the head of the gauge, or just round-head wood screws (Fig. 5-2).

After you have secured the extension, move it forward so the blade will cut a kerf, which you can use to position the work for accurate crosscutting. Use a square to mark the cut line on the work and align it with the kerf in the extension so there can be no doubt as to where the cut will occur (Fig. 5-3). You can use a pencil to mark the cut line, but many workers prefer to use a sharp knife, feeling that this will sever surface fibers and minimize feathering. It's not a bad way to go. If the knife line is difficult to see, you can go over it with a sharp pencil.

More sophisticated versions of miter-gauge extentions that you can make will be demonstrated late in this chapter.

EXTRA-THICK STOCK

There is a maximum height to which a saw blade can be elevated. This distance limits the thickness of stock that can be crosscut in one pass. Figure 5-4 suggests a solution. Use a square to mark the cut line on all four surfaces of the workpiece. Make the first cut with the saw blade elevated to a bit more than half the stock's thickness. Then flip the stock and make the second cut. Results will be good if you place the stock carefully for each of the cuts and if you have checked to be sure the angle between the saw blade and the table is 90 degrees. You can obtain more support by using a miter-gauge extension.

CROSSCUTTING TO LENGTH

Many woodworking projects require that several components be exactly the same length. You can work by marking stock and then visually placing the piece for the cuts, but there is much possibility here for human error. It is much better to create a mechanical setup so the length of the pieces will be gauged automatically. You can accomplish this operation using a commercial miter-gauge stop like the one in Fig. 5-5.

Most stops consist of assemblies of two rods and a movable block through which the rods pass. Either of the rods can be secured in the head of the miter gauge while the other, against which one end of the workpiece is abutted, is adjusted for correct cut length (Fig. 5-6). If the saw blade has set teeth, be sure to measure from a tooth that points toward the rod.

Thereafter it is just a matter of placing the work as demonstrated in Fig. 5-7, and cutting as many pieces as you need. Every piece you cut will be exactly the same length.

If you lack a commercial stop rod, you can work just as efficiently and accurately by clamping a stop block to a miter-gauge extension (Fig. 5-8). On production-type situations like this, workers are inclined to move too fast. Make each cut as if it is the only one you need. Don't allow the cutoff pieces to accumulate around the saw blade.

THE RIP FENCE AS A STOP

Understand this critical point right off. You must never use the rip fence alone to gauge the length of cutoffs! If you do, the work will be captured between the blade and the fence and twisted n such a way that it will be tossed up or back at you. The only role the rip fence plays in this application is as a support for a stop block that is used to gauge the length of the cutoff (Fig. 5-9). The stop block, which can be just a block of wood secured with a clamp, is positioned well to the front of the table so workpieces will not bear against it as the cut is made (Figs. 5-10 and 5-11). As you can see, there is ample room between the blade and the rip fence to prevent jamming when the cut is complete.

The distance from the rip fence to the blade must equal the length of the cutoff plus the thickness of the stop block. Any piece of wood can be used as a stop block, but it is best to have special ones, which you can make, as ever-ready, accurate tools. Knowing the exact thickness of the stop block, without needing to measure, makes it easier to set up the tool.

Figures 5-12 through 5-16 demonstrate and give construction details for rip-fence stop blocks that I have made for use in my own shop. You might not want to make them all right away, but having the ideas on hand might prove useful in the future.

Fig. 5-2. Most miter gauges are designed to accept an extension. This one has holes through it so wood screws can be used. The length of the screw should equal the thickness of the head plus about three-fourths the thickness of the extension.

Fig. 5-3. An extension provides additional support for the workpiece. The kerf through the extension makes it easy to position work for accurate sawing. Be sure to position the work so the cut will be on the waste side of the line.

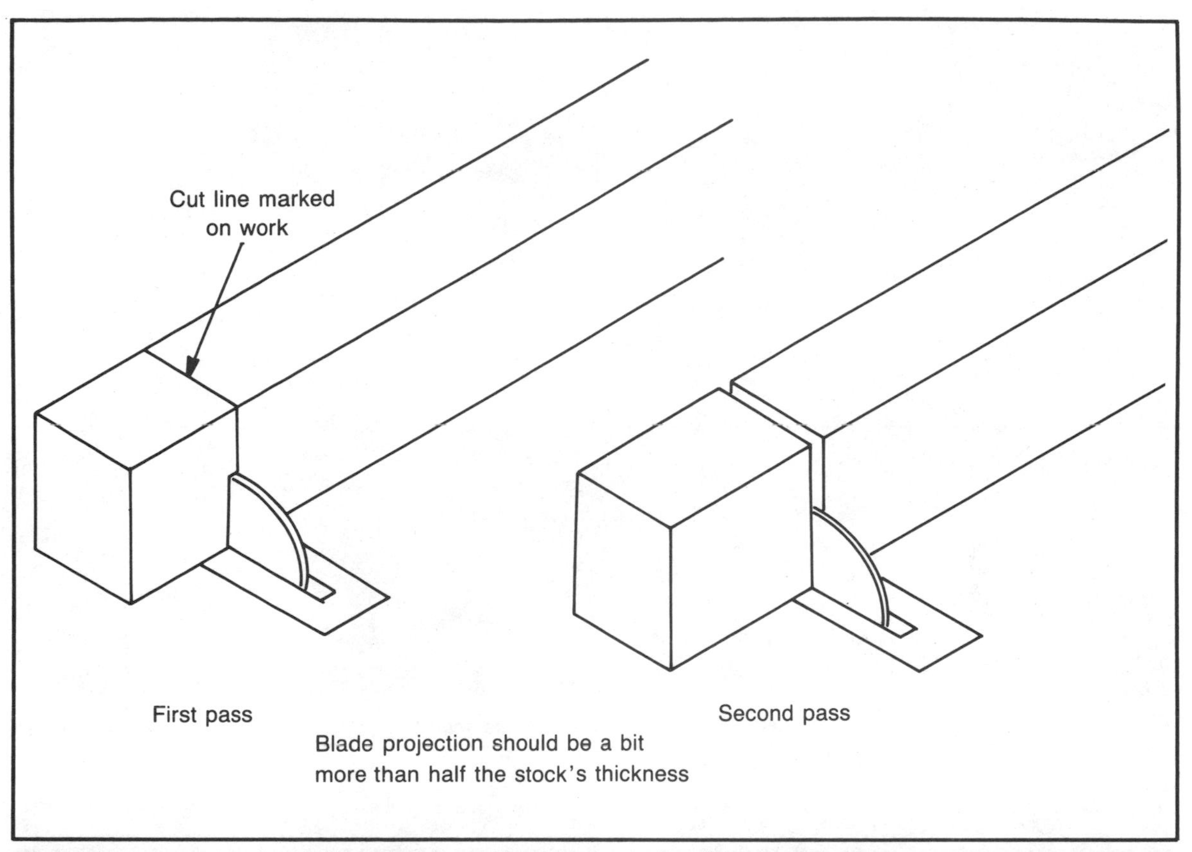

Fig. 5-4. You can crosscut extra-thick stock by making two passes. Results depend on how accurately you place the work for the second cut, and whether the angle between the side of the blade and the table is 90 degrees.

Fig. 5-5. A typical miter-gauge stop rod includes a long and a short rod, plus a connecting block. Either of the rods can be secured in the miter gauge, making it possible to set the accessory for long or short cuts.

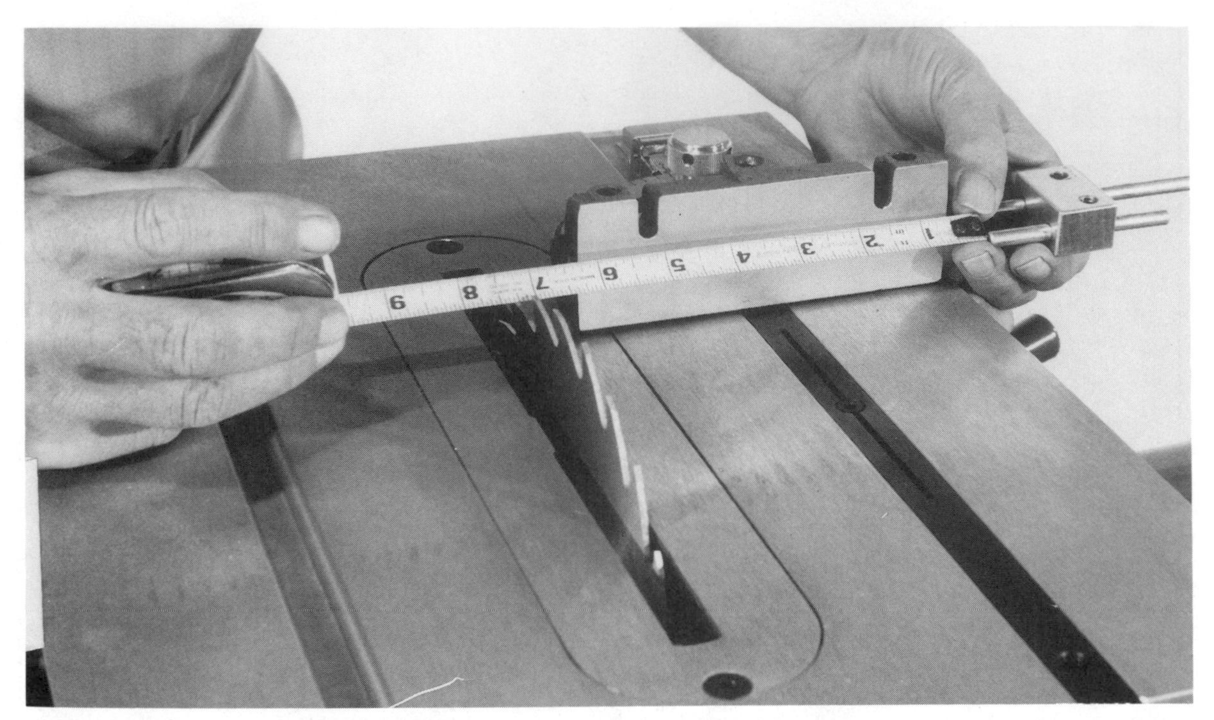

Fig. 5-6. To determine the length of the part you need, measure from the end of the rod to the saw blade. When the blade has set teeth, be sure to measure from one that points toward the rod.

Fig. 5-7. After the setup is established, you can crosscut as many pieces as you wish, knowing that each will be exactly the same length. Be sure to return the miter gauge and work to the starting position after each cut.

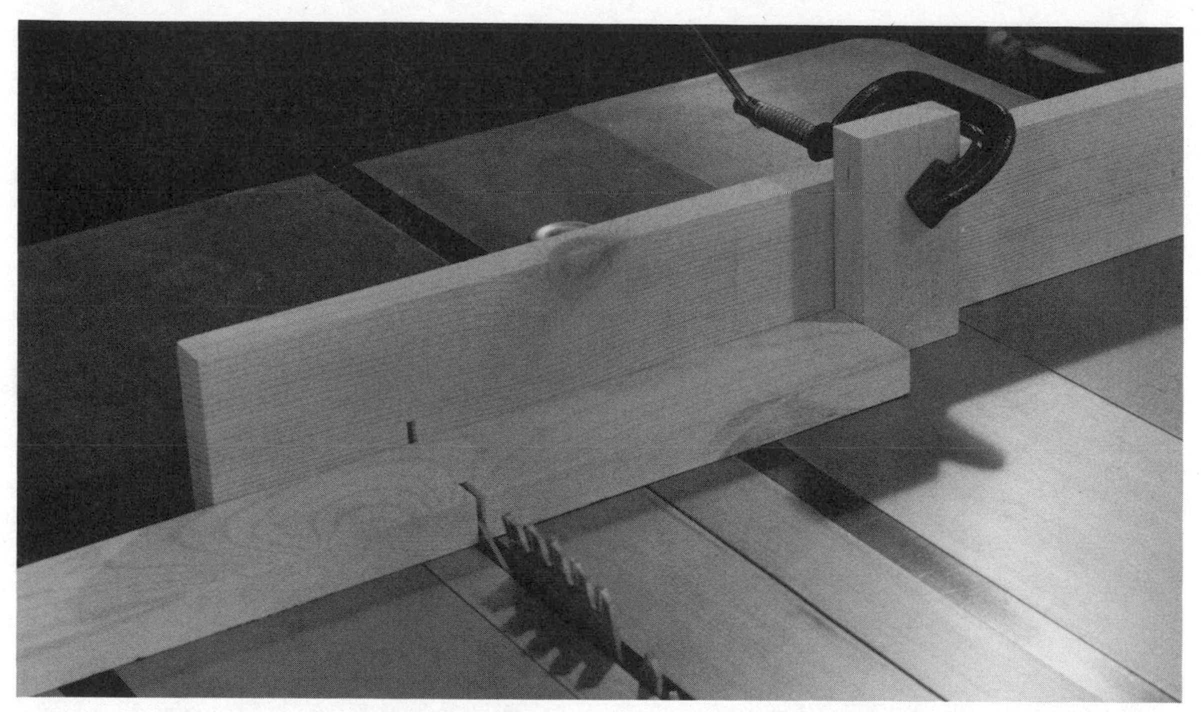

Fig. 5-8. You can also use a simple miter-gauge extension to gauge the length of workpieces. A block of wood clamped to the extension serves as a stop.

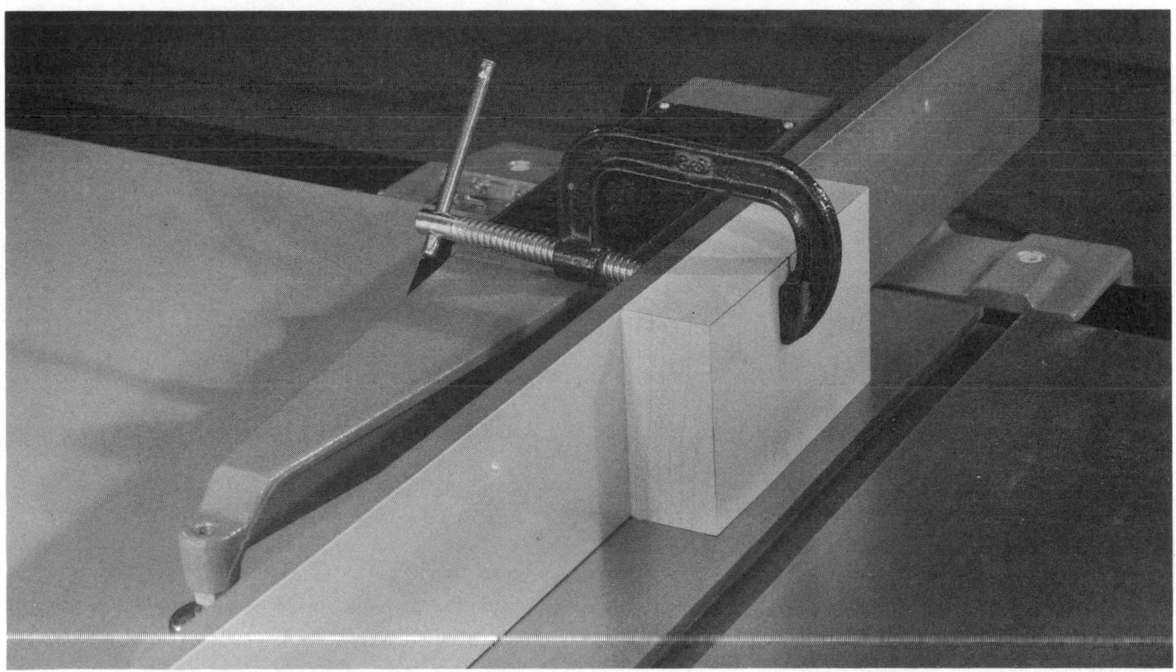

Fig. 5-9. The easiest stop block to use on the rip fence is simply a block of wood held in place with a clamp. The distance from the face of the block to the saw blade determines the length of the pieces being cut.

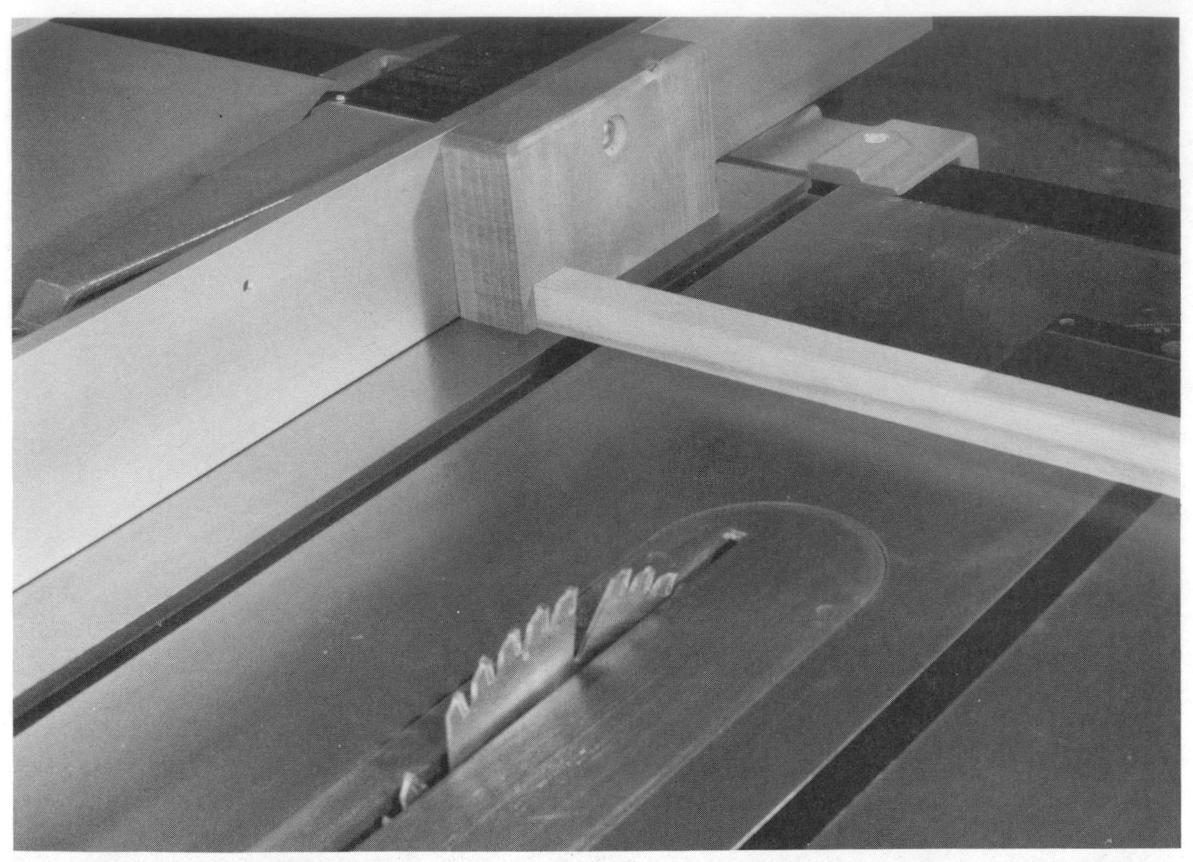

Fig. 5-10. Always position the stop block so it is close to the front of the table. Thus the work will be free of the stop block before making contact with the saw blade.

Fig. 5-11. The purpose of the stop block is to supply room for the cutoffs. Working this way ensures that workpieces can't jam between fence and blade, but don't allow the cutoffs to pile up.

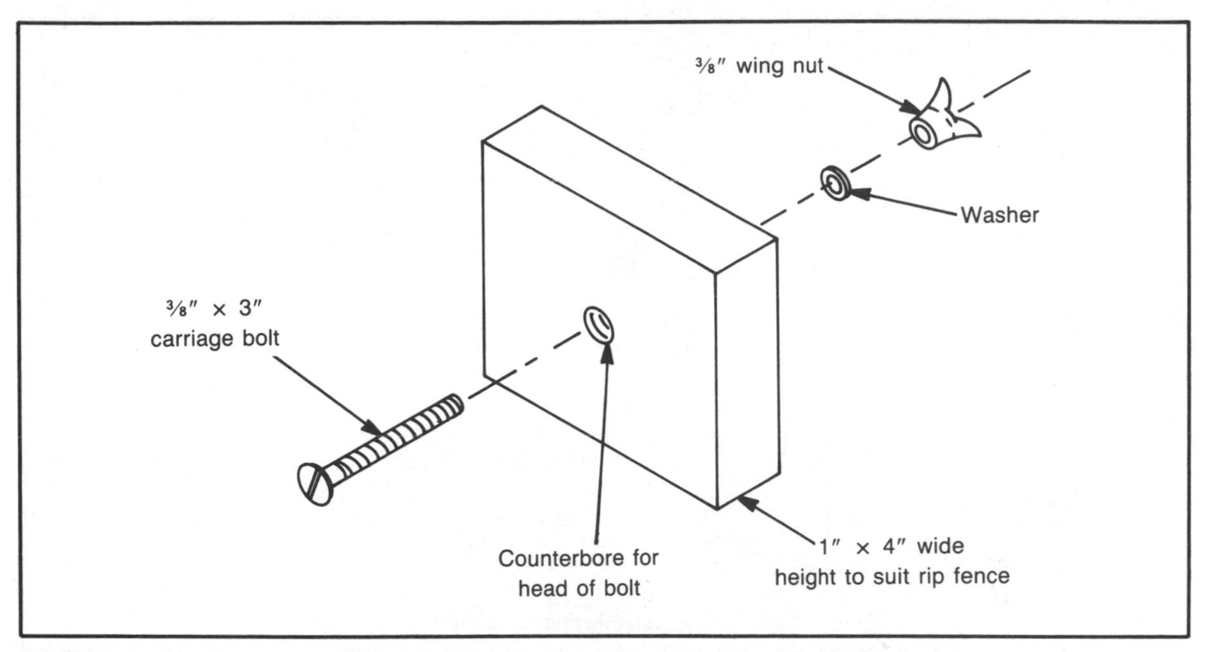

Fig. 5-12. You can make a stop block this way when the rip fence has holes through it.

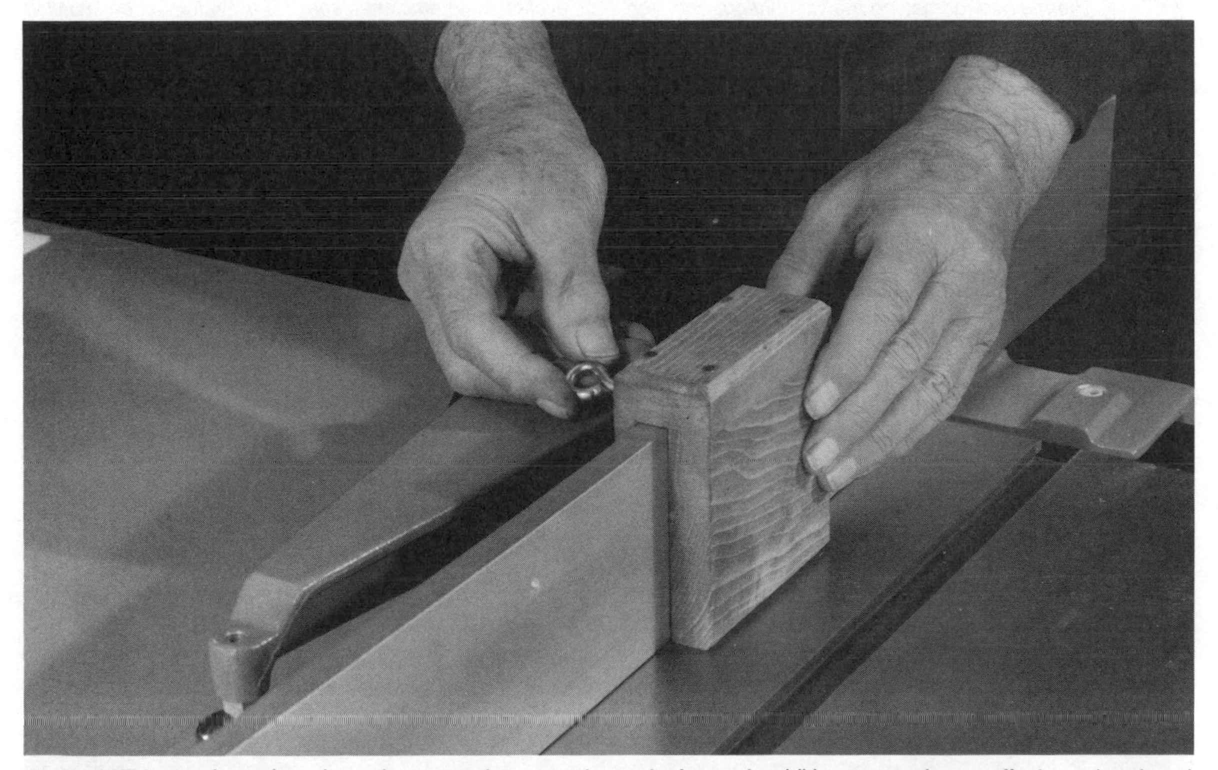

Fig. 5-13. This stop is made so it can be secured at any point on the fence. It addition to gauging cutoffs, it can be placed behind the blade to limit the length of rip cuts, a procedure that is used for stopped kerfs or grooves.

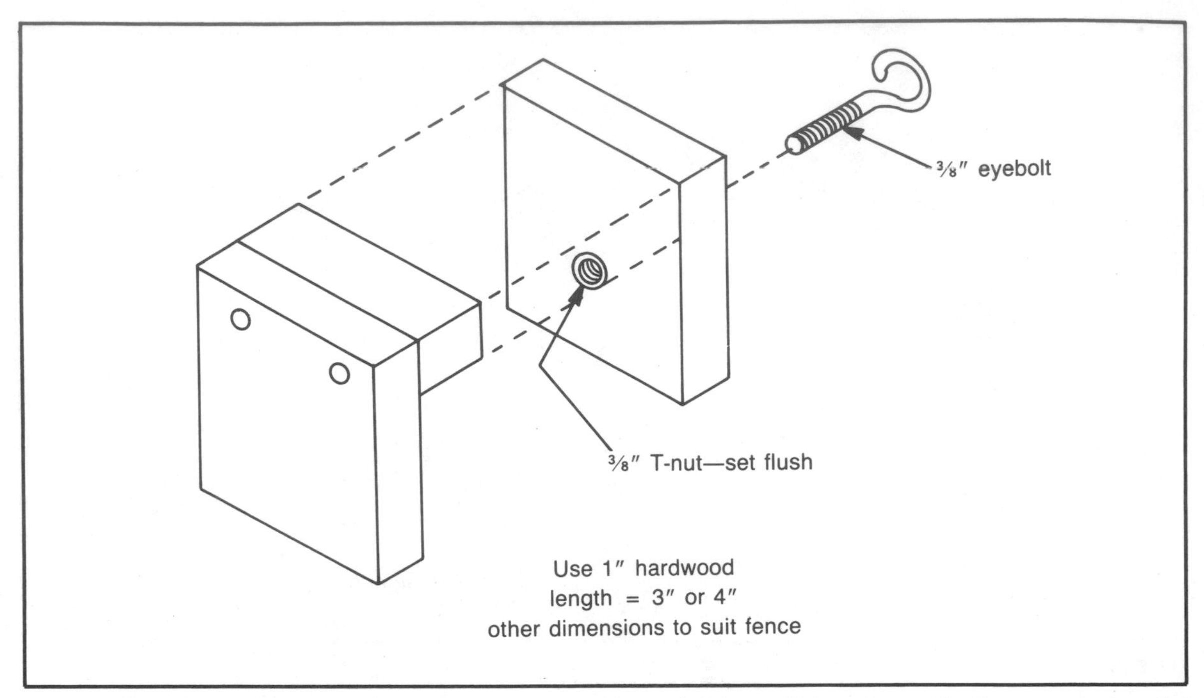

Fig. 5-14. The method for making the variable-position stop. The eyebolt is an efficient clamp. The T-nut must be installed before assembly.

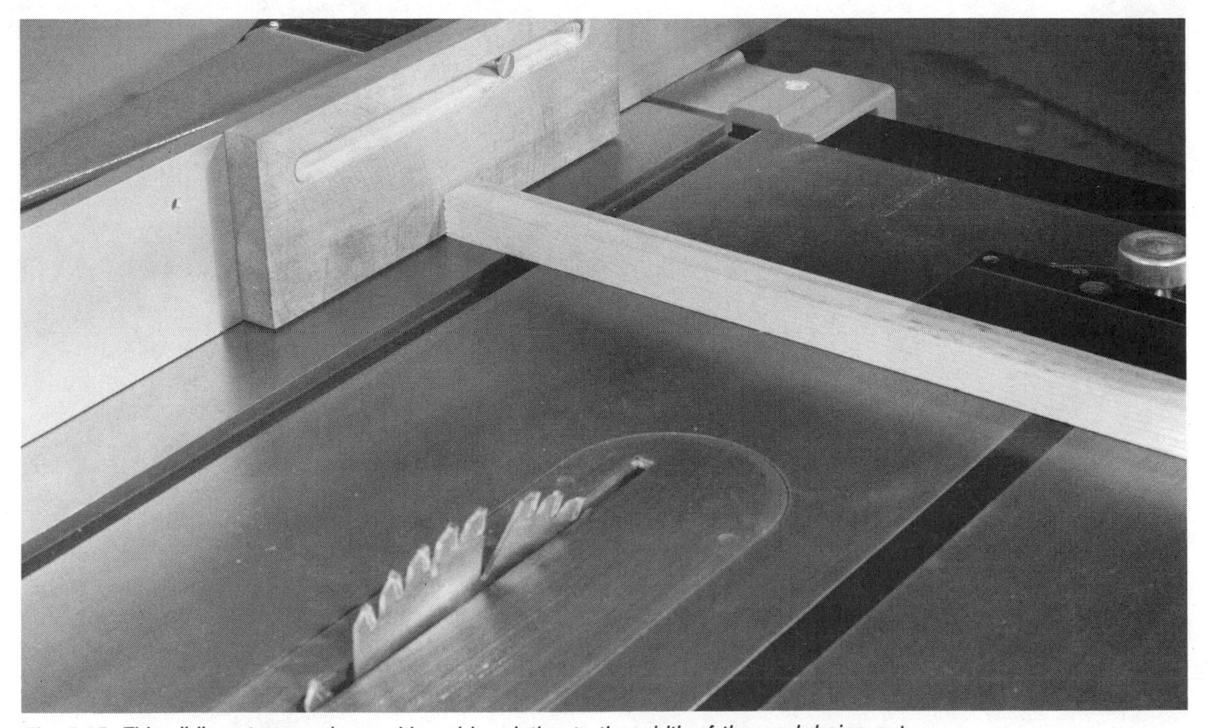

Fig. 5-15. This sliding stop can be positioned in relation to the width of the work being cut.

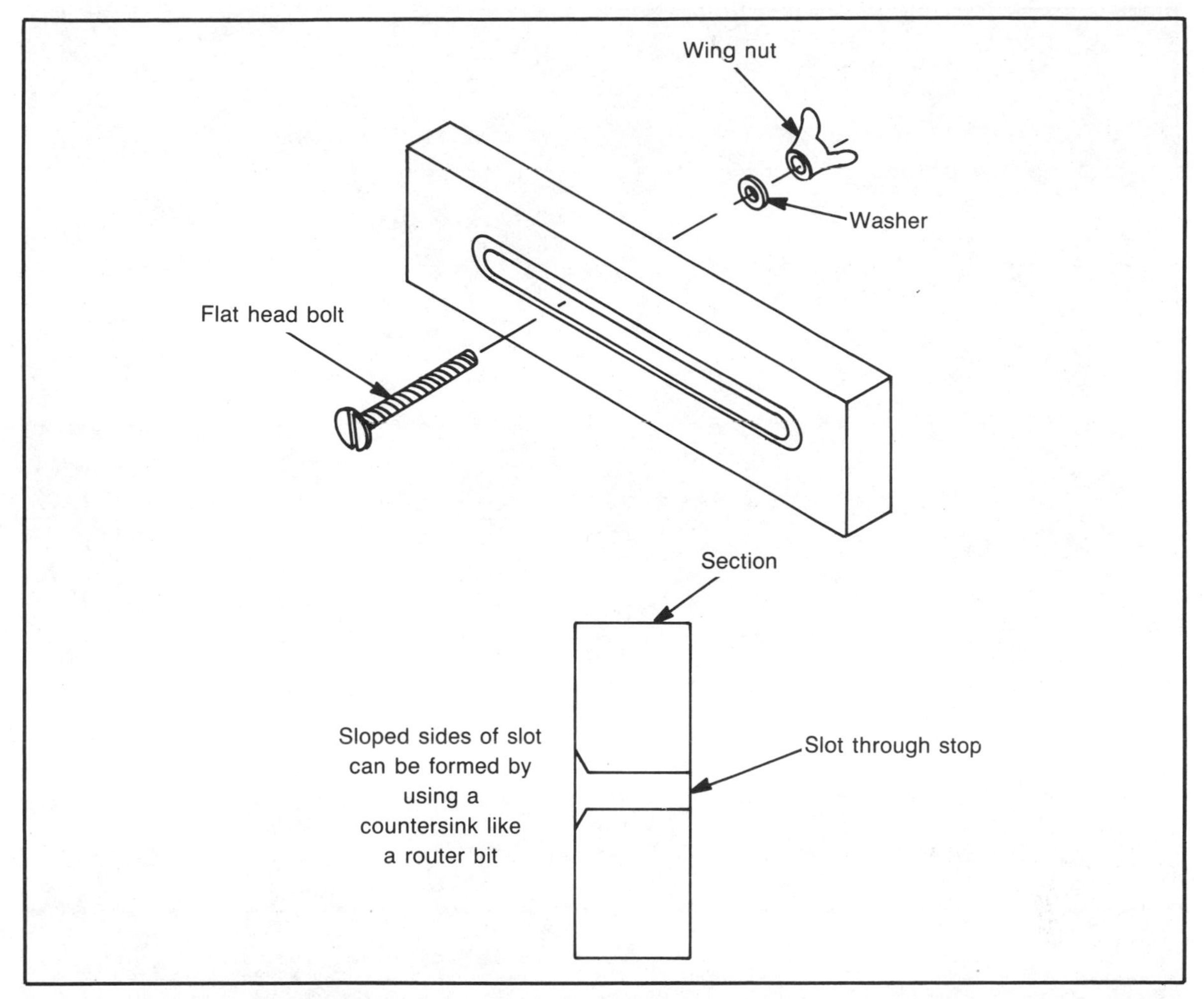

Fig. 5-16. Since the workpiece is rather small, it is best to form the slot in the sliding stop by drilling end holes and then cutting between them using a jigsaw or a coping saw.

SPECIAL EXTENSION WITH A SLIDING STOP

The project shown in Fig. 5-17 is essentially a miter-gauge extension, but it incorporates a sliding stop that can be reversed to gauge, within its capacities, the length of short or long cutoffs (Figs. 5-18 and 5-19). Projects like this become lifetime accessories, so you should construct them with care. You can form the limited slot in the extension by making repeat, stopped rip passes, a procedure that will be demonstrated in Chapter 6. You should size the block that is attached to the stop and that rides in the slot to slide smoothly, but without wobble. Construction details for the project are offered in Fig. 5-20.

MAKING A DOUBLE-BAR MITER GAUGE

One of the advantages of the homemade tool shown in Fig. 5-21 is that it supplies support on both sides of the saw blade. As for other extensions, a saw kerf through the head of the unit makes it easy to line up work for accurate crosscutting.

Fig. 5-17. This more sophisticated version of a miter-gauge extension employs a reversible sliding stop to provide more flexibility in workpiece lengths.

Fig. 5-18. Place the stop one way and move it to the front of the slot to gauge short workpieces.

Fig. 5-19. For longer work, reverse the stop and move it to the rear of the slot. Projects like this last a lifetime, so make them very carefully.

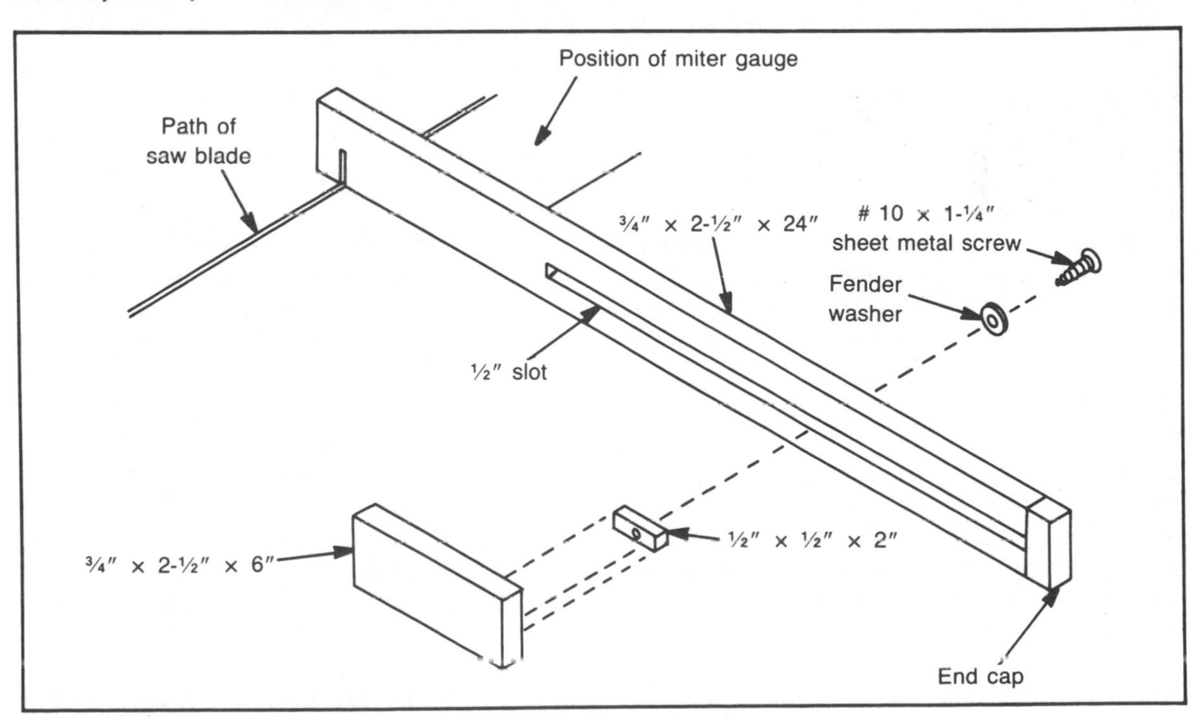

Fig. 5-20. The method for constructing the miter-gauge extension with a sliding stop.

Fig. 5-21. A double-bar miter gauge provides good support on both sides of the blade, whether you are crosscutting long or short pieces. A kerf through the head of the project makes it easy to position work for accurate sawing.

In order for the project to work as it should, you must be certain when you assemble its components that the angle between the head and the saw blade is 90 degrees (Fig. 5-22). Also be sure to size the bars riding in the table slots so they provide a nice sliding action without wobble.

An accessory you can make for the double-bar miter gauge is the sliding stop, which is shown being secured in Fig. 5-23. You can use this stop as a gauge when you need multiple pieces of similar length. Position the stop as needed by measuring between it and the saw blade (Fig. 5-24). As always, if the blade has set teeth, be sure to measure from one that points toward the stop. After you have established the position of the stop, you can crosscut as many pieces as you need, knowing that all will be exactly the same length (Fig. 5-25).

Construction details of the project and its accessory sliding stop are shown in Fig. 5-26 and 5-27.

THE SQUARING BOARD

The squaring board that is set up on the table saw in Fig. 5-28 might be considered a ripping accessory more than a crosscutting tool since its major purpose is to allow you to make a straight cut on stock whose edges are too irregular to bear against a rip fence. However, it is placed in this chapter since it is used in essentially a crosscutting environment.

The squaring board is a simple tool, consisting of a plywood platform attached to a wooden bar that rides in the table slot, and a fence that serves to brace workpieces. It is essential that you attach the fence so the angle between it and the saw blade is 90 degrees (Fig. 5-29).

Fig. 5-22. An accessory like the double-bar miter gauge isn't available commercially, but you can easily make one. When assembling, be sure the angle between the head of the project and the side of the blade is 90 degrees.

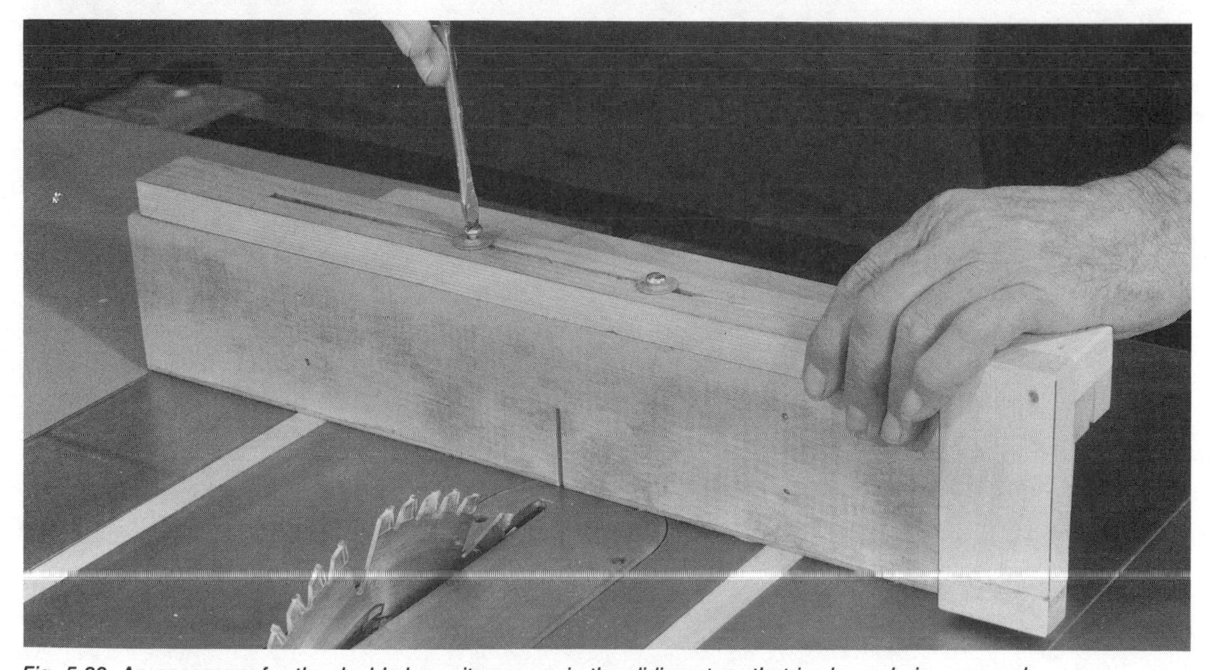

Fig. 5-23. An accessory for the double-bar miter gauge is the sliding stop, that is shown being secured.

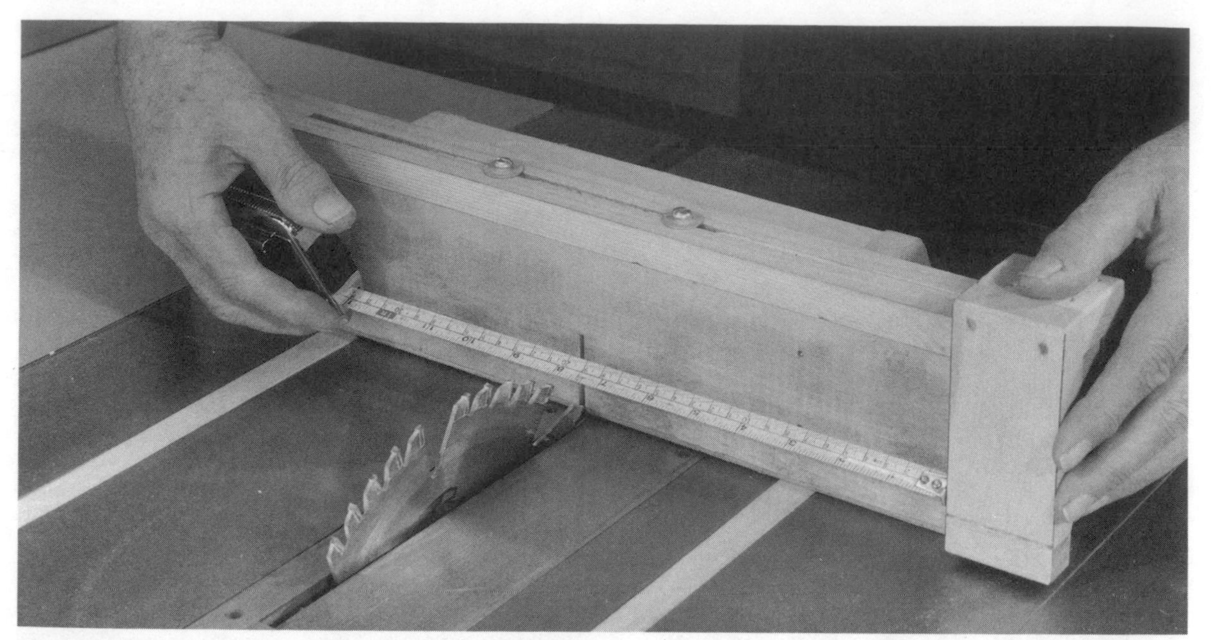

Fig. 5-24. As for other tools used for crosscutting to length, the distance between the stop and the saw blade determines the length of the workpieces.

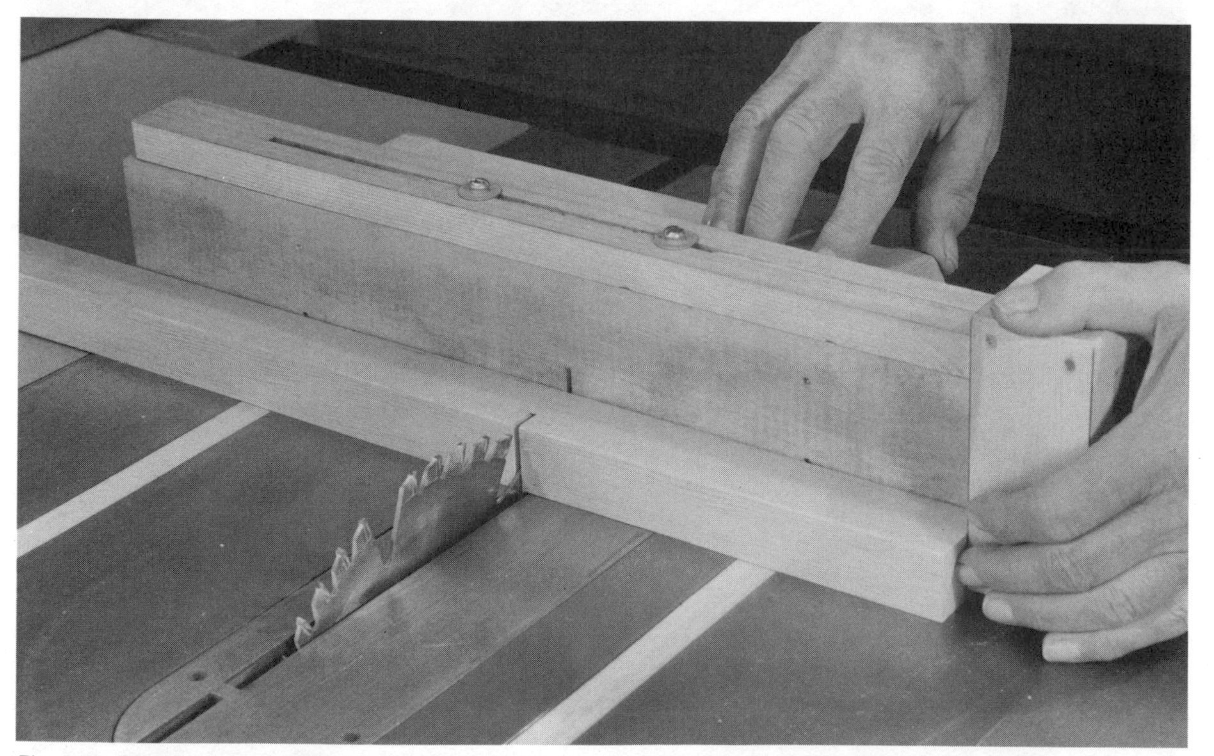

Fig. 5-25. Once the setup is established, you can cut any number of pieces of equal length. Do not advance the tool any farther than is needed to cut through the workpiece.

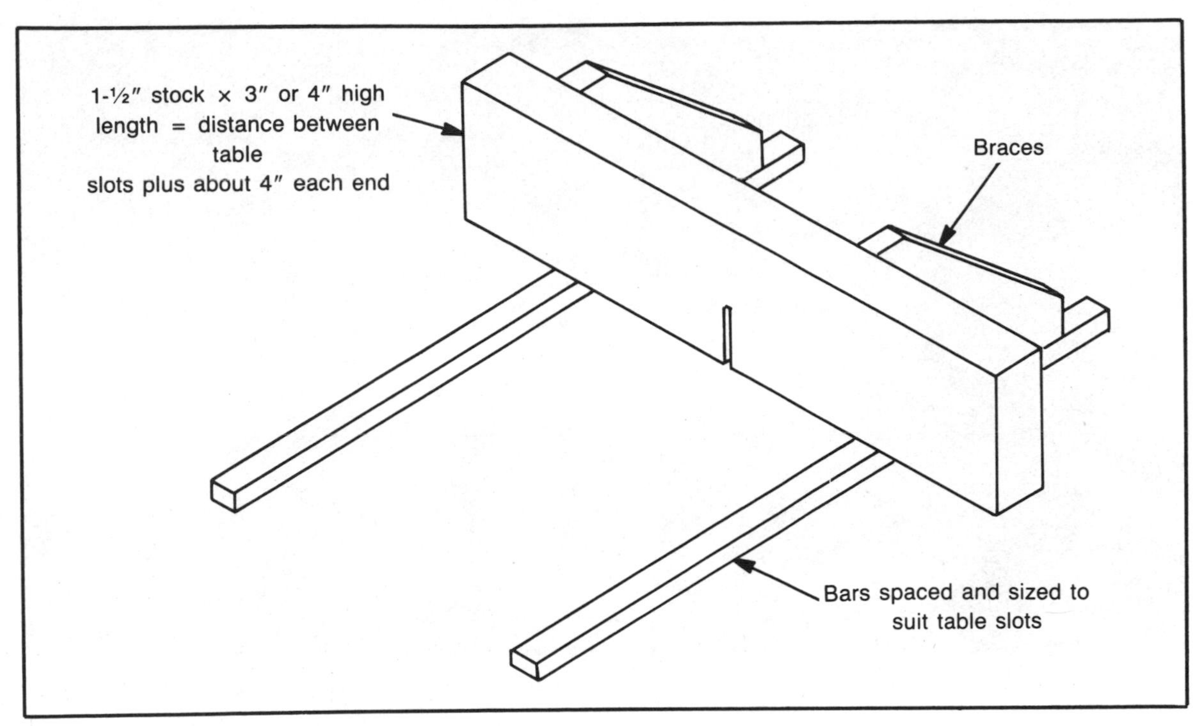

Fig. 5-26. You can make a double-bar miter gauge this way.

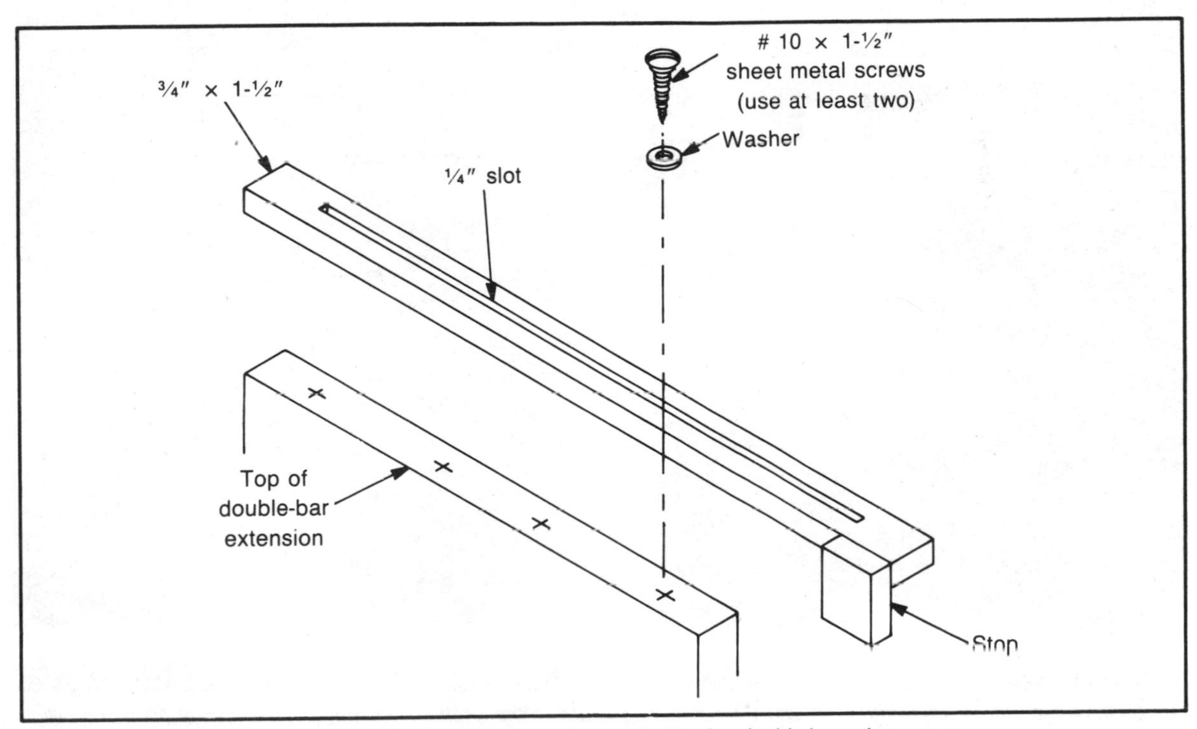

Fig. 5-27. Construction details of a sliding stop that can be used with the double-bar miter gauge.

Fig. 5-28. The squaring board is a simple project. It is simply a platform riding on a miter gauge bar, and a fence for positioning the work.

Fig. 5-29. Attach the fence to the platform of the squaring board so the angle between it and the side of the blade will be 90 degrees.

To use the tool, abut the workpiece against the fence and move the board forward, something like a miter gauge, to move the work past the saw blade (Fig. 5-30).

As for many homemade tools, use starts to suggest other practical applications. For example, there is no reason why you cannot use the squaring board for simple crosscutting (Fig. 5-31). By adding a triangular guide that will form a 45-degree angle with the saw blade, you can use the squaring board to make simple miter cuts (Figs. 5-32 and 5-33). Carry this a bit further by tack-nailing a stop to the platform, and you can control the length of miter-cut workpieces (Fig. 5-34). The miter guide is secured with screws so you can place it in position only when you need it.

You can make a basic squaring board by following the details in Fig. 5-35.

A FEW REPEAT-PASS IDEAS

When you set the projection of a saw blade so

Fig. 5-30. The squaring board enables you to make straight cuts on stock that has uneven edges.

Fig. 5-31. You can also use the squaring board for routine crosscutting.

Fig. 5-32. You can make the squaring board even more useful by adding a triangular gauge for miter cuts. The angle between the bearing edge of the guide and the side of the saw blade must be 45 degrees.

Fig. 5-33. The squaring board, with the 45-degree guide attached, becomes an accurate mitering tool. To keep the work-piece in firm position as you cut, face the bearing edge of the guide with a strip of fine sandpaper.

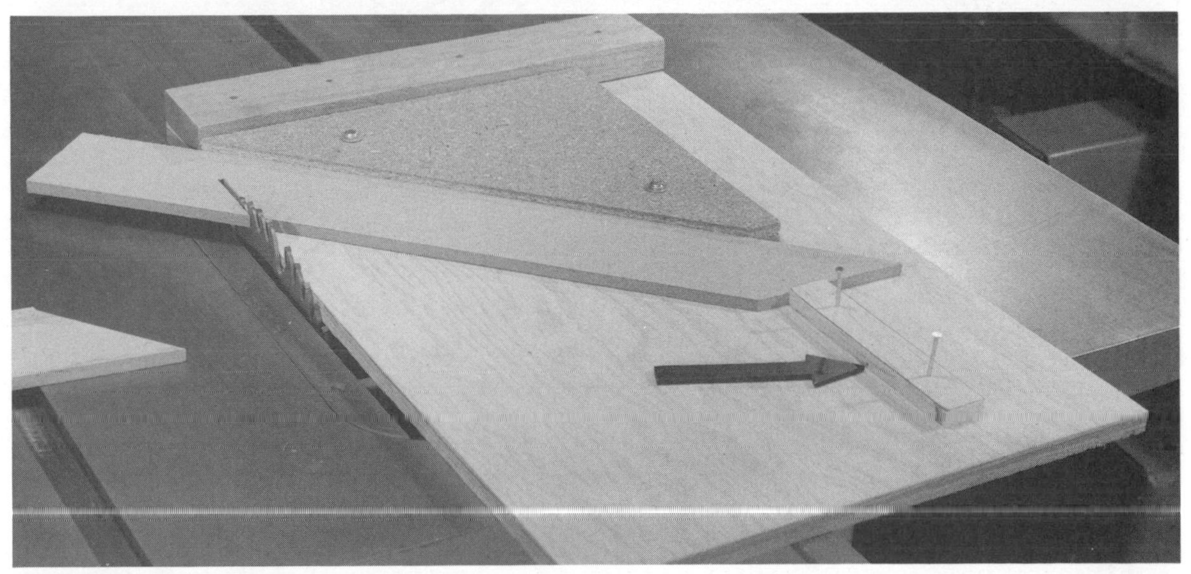

Fig. 5-34. If you tack-nail a stop to the platform (arrow), you will have a setup that will accurately gauge the length of mitered pieces.

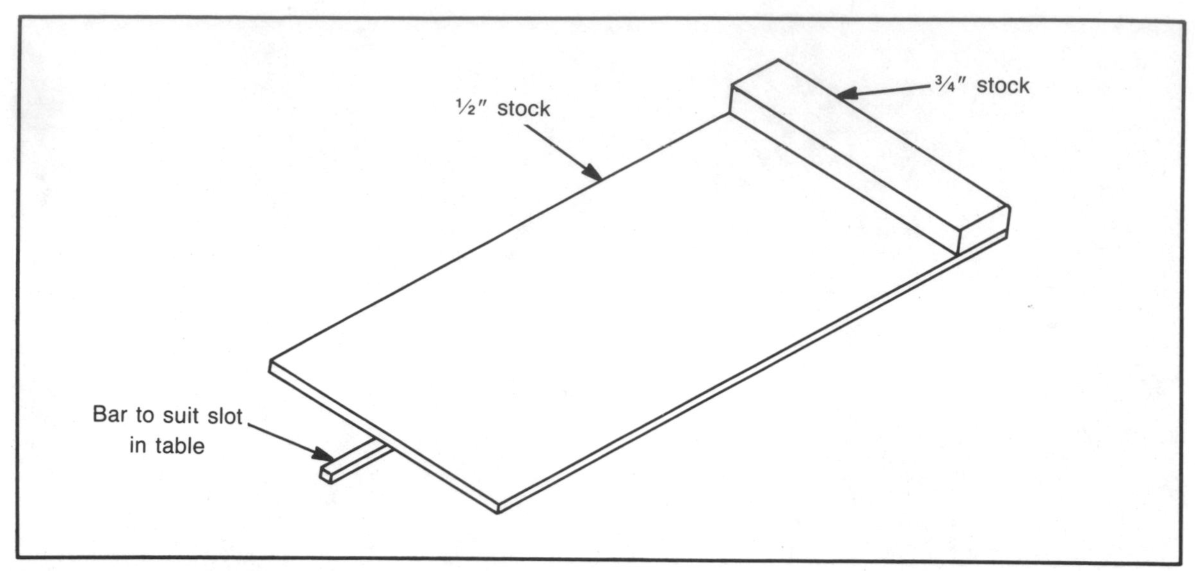

Fig. 5-35. The basic components of a squaring board.

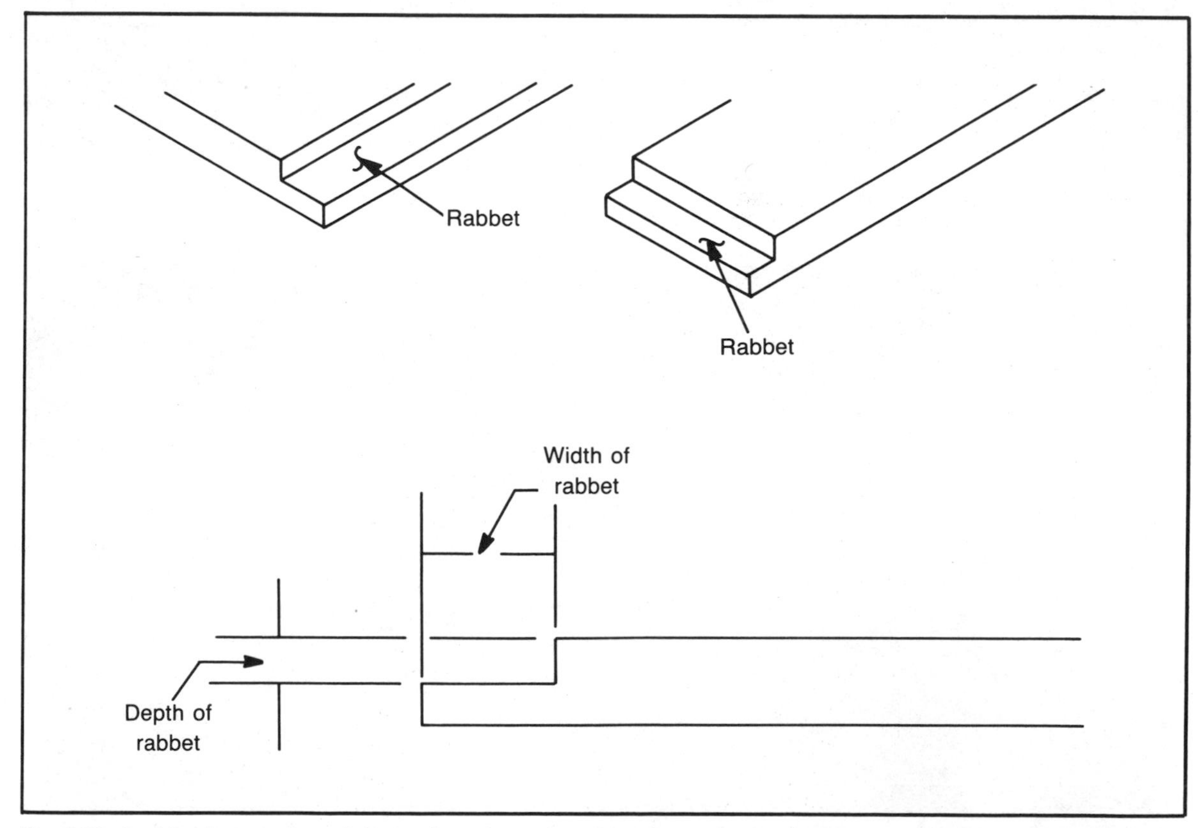

Fig. 5-36. A rabbet is an L-shaped cut made at the end or along the edge of a workpiece. Its width equals the thickness of the part to be inserted. Its depth is usually one-half to three-fourths the thickness of the stock in which it is cut.

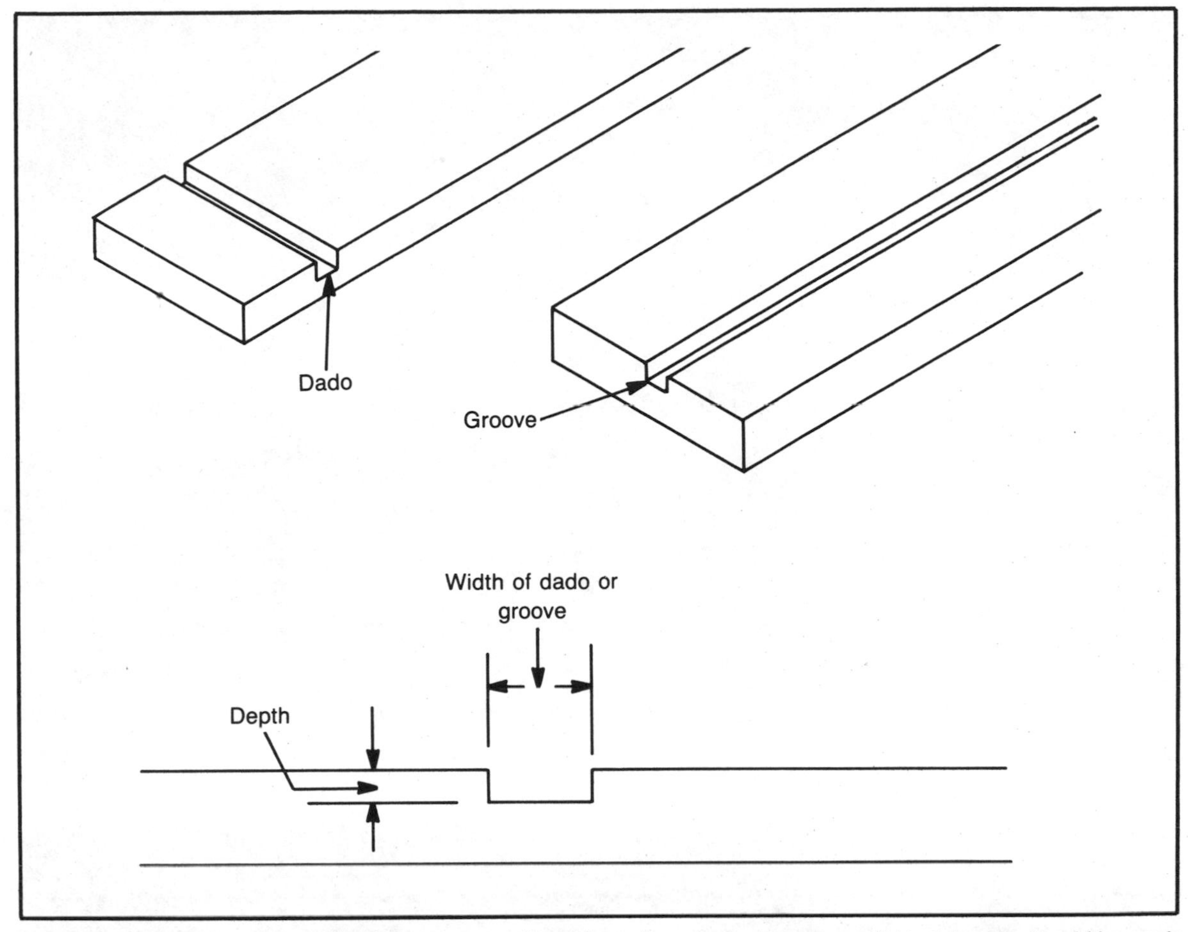

Fig. 5-37. A U-shaped cut is a dado when cut across the grain, and a groove when cut parallel to the grain. Its width equals the thickness of the insert piece. Its depth is rarely less than one-half the thickness of the part in which it is cut.

it is less than the thickness of the stock, and make cuts that overlap each other just a bit, you can use a single blade to shape, for example, joint forms like the dadoes and end rabbets in Fig. 5-36 and 5-37. The drawings also show grooves and edge rabbets that you can form with repeat passes during ripping operations, but this technique will be covered in Chapter 6.

An *end rabbet* is an L-shaped cut that is made across the end of a workpiece. To use the repeat-pass procedure, elevate the blade to provide the depth of the cut, and use the rip fence or a stop block to gauge the width of the cut (Fig. 5-38). You can use the rip fence as a stop when sawing this way since the part is not being cut off. Next, work as shown in Fig. 5-39. Repeat passes that overlap each other just a bit remove the remaining waste stock.

You can follow the same procedure and use a single blade to form a U-cut, which is the shape of a dado. Set blade elevation to provide the depth of the cut, and use the rip fence or a stop block to gauge the position of the shoulder cuts (Figs. 5-40 and -41). Then, make repeat passes to remove the waste between the first cuts (Fig. 5-42).

To form a stud tenon, first make shoulder cuts into opposite surfaces of the stock (Fig. 5-43), and then clean out the waste by making repeat passes.

Fig. 5-38. The first step of a repeat-pass end rabbet is to form the shoulder cut. Blade height equals cut depth; distance from the fence to the outside of the blade equals cut width. You can use the fence as a stop since the part is not being cut off.

Fig. 5-39. The second step is to make the repeat passes that remove the waste. The projection of the saw blade remains the same. Results are smoothest when you overlap the cuts just a bit.

Fig. 5-40. The dado requires two shoulder cuts. The height of the blade equals the depth of the cut. Make the first cut by measuring from the fence to the inside of the blade.

Fig. 5-41. Make the second shoulder cut for the dado by measuring from the fence to the outside of the blade.

Fig. 5-42. Remove the waste material between the shoulder cuts by making overlapping, repeat passes. Use the fence to gauge the accuracy of the starting cuts for the rabbet or dado when you have many similar cuts to make. If you need just one, you can be guided just by cut lines marked on the stock.

Fig. 5-43. You can use the same technique to shape a stud tenon, which is actually back-to-back rabbet cuts. First, make the shoulder cuts into opposite surfaces of the workpiece. The height of the saw blade determines how thick the tenon will be. To finish the job, make the repeat passes that remove the waste.

Table 5-1. Crosscutting Problems and Methods for Corrections.

PROBLEM	POSSIBLE CAUSE	CORRECTION
Cuts not square	Misalignment	Check miter gauge setting/Reset auto-stop if necessary
Work jams/Hard to feed	Miter gauge square to slots, but slots not parallel to saw blade	Check if table slots are parallel to saw blade
Inaccurate cuts	Misalignment	Check miter gauge setting/Check auto-stops/Make test cut and check with square
	Poor work handling	Keep work secure when making pass/Use hold-down if available
Cut has slight bevel	Misalignment	Blade must be square to table/Check blade-tilt 0° auto-stop
Miter gauge hard to move	Poor tool-keeping	Clean table slots and miter-gauge bar/Apply paste wax and rub to polish
Blade binds	Dull blade—Poor blade	Replace or sharpen
	Excessive work-overhang	Provide adequate support to keep work level
Blade stalls	Dull blade	Replace or sharpen
Burn marks on work	Forcing cut	Allow blade to cut at its own pace
	Tough wood	Feed more slowly/Make repeat passes when necessary
	Dirty blade	Clean to remove wood residue
	Projection of hollow-ground blade in correct	Hollow-ground blade requires more projection than conventional blade

You can make cuts like these much faster using a dadoing tool, which will be demonstrated in Chapter 8. Also, the art of joinery on the table saw will be covered in greater detail in Chapter 10. The repeat-pass idea is handy when you have only a few cuts to make and don't wish to bother changing cutting tools.

Possible problems that can occur when crosscutting, and suggestions for correcting them, are listed in Table 5-1.

Chapter 6

Ripping Operations

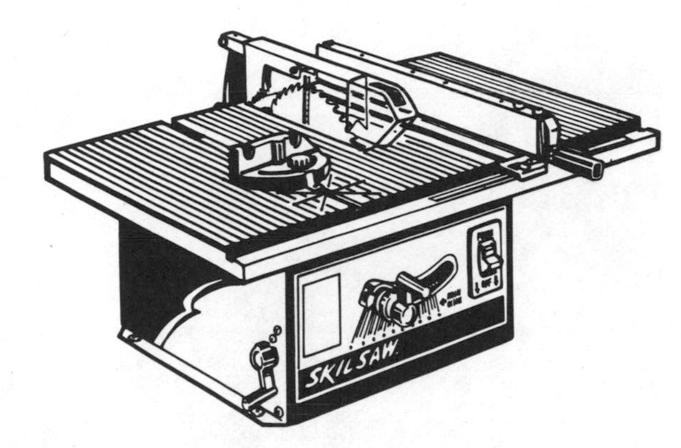

RIPPING IS THE PROCESS THAT IS USED, ESSENtially, to saw material to correct width. The application is often described as cutting parallel to, or with, the grain, which is apt when referring to lumber. Plywood, however, is often ripped across surface veneers, and materials like hardboard and particleboard are grainless. Therefore, viewing the operation simply as a means of sawing material to a specific width is more realistic.

In any event, in the ripping operation the workpiece is passed between the locked rip fence and the saw blade (Fig. 6-1). It is possible that the part being cut off is the component needed, but more often, it is the section between blade and fence that is needed.

The farthest distance the rip fence can be situated away from the saw blade dictates the machine's maximum rip cut. On small saws, this might not be more than the distance from the blade to the table's edge. On a multipurpose tool like the Shopsmith, which has a comparatively small table, rip capacity increases drastically through the use of an outboard extension table on which the rip fence can be mounted. When the Delta Unisaw is equipped with extension rails or with the Unifence (Fig. 6-2), rip capacity increases beyond the point where a 4- x -8-foot panel can be sawed in half across the small dimension.

Many saws have scales on the front rail so the fence can be positioned for the cut width, but some scales are considerably less accurate than others. Until you are positive that you can rely on the scale, it is best to use it for an approximate setting and make the final adjustment by measuring between the blade and the fence (Fig. 6-3). Take the measurement from the side of the blade facing the fence. If you measure from the outboard side of the blade, the width of the cut will be reduced by the width of the kerf. Two other factors to remember: If the blade has set teeth, be sure to measure from one that points toward the fence. If you have aligned the fence to provide some offset at the rear, measure from the front of the blade.

Fig. 6-1. Making a rip cut is essentially a matter of passing the workpiece between the rip fence and the saw blade. The distance between the two components determines the cut width.

Start a rip cut by placing the workpiece firmly down on the front edge of the table and snug against the rip fence. Most times, the left hand is used at the start to control the work while the right hand moves the work forward. Once the work is engaged with the blade, the left hand is free, except on very wide material where the hand can help without coming close to the blade. Feed the work steadily until it is past the saw blade. There is no return on a rip cut. The pass is complete when the work has passed the blade.

Many operators will hook the fingers of their feed hand over the rip fence as they move the stock forward. It is a wise precaution which guards against slipping, but it assumes that there is ample room between the fence and blade for the method to be used without danger. Other procedures you can adopt to contribute to safety for ripping applications, especially when cuts are narrow, will be demonstrated shortly.

WORK SUPPORT

Ripping is often required along the long dimension of a man-made panel or on extra-long boards. Trying to use muscle at such times to control the cut and support the work can be dangerous and will not contribute to accuracy. You can ask for other hands as long as the helper has been educated to provide sensible assistance, but it is better to equip the shop with outboard support stands, which will always be available and will do your bidding exactly.

Commercial extension stands that have roller tops and are height-adjustable are

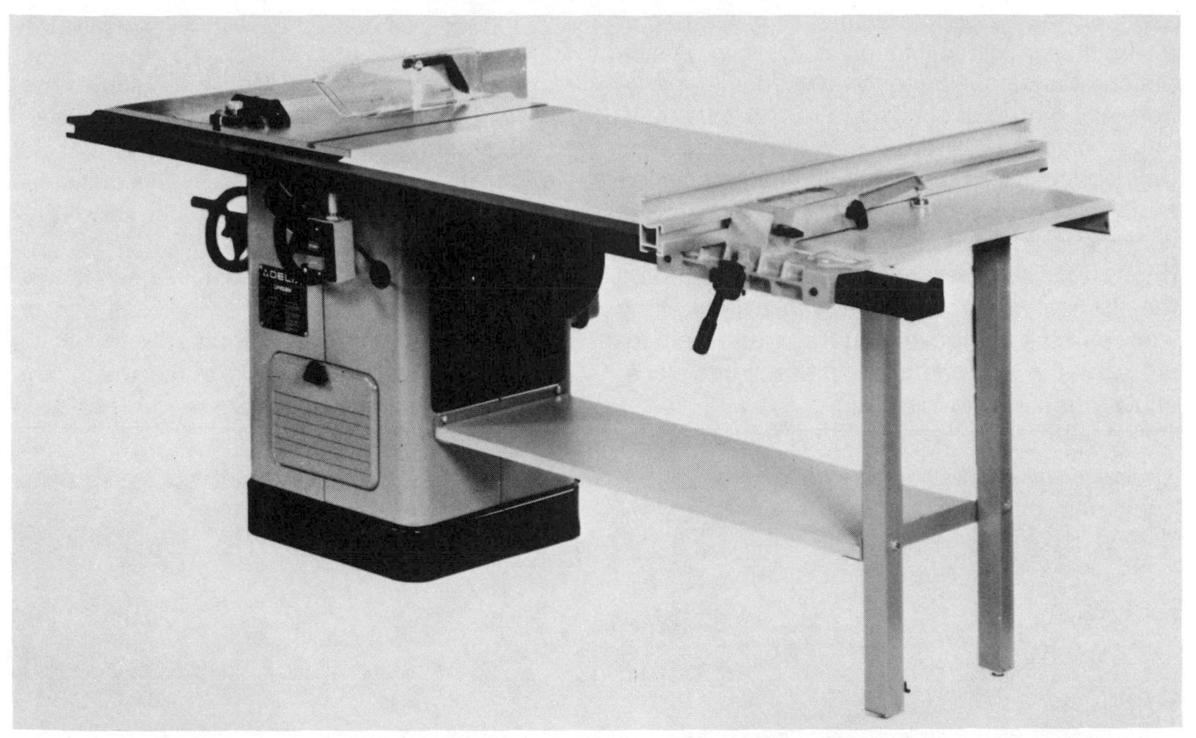

Fig. 6-2. Many table saws, like this Delta unit, provide for much wider cuts than would be possible if the fence could not be moved off the main table.

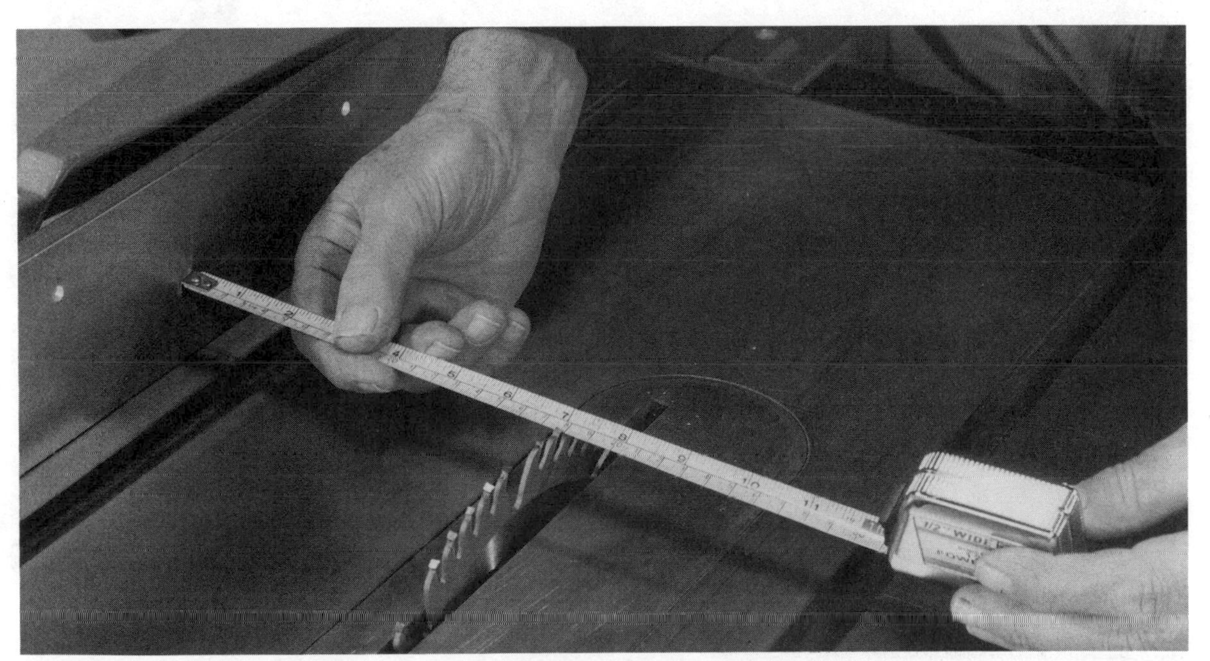

Fig. 6-3. To determine the width of the cut, measure between the fence and the saw blade. Always measure from the inside surface of the blade and, if the blade has set teeth, from one that points toward the fence.

available—an example was shown in Chapter 2—but it's not difficult to make your own. A simple one that is designed to be used with a sawhorse as a base is shown in Fig. 6-4. When the unit is placed behind the saw and the roller is adjusted so its top is on the same plane as the surface of the saw's table, long or heavy work will receive adequate support when being ripped (Fig. 6-5). The roller on my unit is a steel tube that is sealed at each end with a tight-fitting wooden disk. A conventional nut and bolt, installed in the disk before it is pressed into place, allows the roller to turn (Fig. 6-6).

Figure 6-7 shows how to make the unit, and suggests that you use a length of large-diameter closet pole as the roller. Figure 6-8 shows how to seal the ends of a metal tube, whether steel or aluminum. The important factor is that the wooden disk, which you can form on a drill press with a hole saw or a fly cutter, must fit very tightly in the tube.

Since the roller top I use is designed for use on a sawhorse, and since sawhorses can be used for many other workshop applications, it's only fair to suggest ways to make them. The brackets shown in Fig. 6-9 are readily available

Fig. 6-4. A roller of this type, mounted on a board that can be clamped to a sawhorse, will provide good outboard support when ripping long or heavy pieces of material. It's much better and safer to work this way than trying to maintain correct work position with muscle alone.

in hardware stores or home supply centers. With the addition of five pieces of standard 2- × -4 lumber, they make sturdy support units.

Another type of sawhorse bracket is shown in Fig. 6-10. It doesn't even require nails. The prongs that are part of the brackets dig into the wood as the wing nut is tightened. A Black & Decker Workmate bench (Fig. 6-11), since it has a built-in clamping device, can efficiently support a roller-top support system.

Another thought for an outboard support stand is the independent, height-adjustable unit detailed in Fig. 6-12. Since the height of the unit is easily adjustable, you can use it to provide additional work support not only when using a table saw, but when working with a band, jig, or radial arm saw. You can design the top unit on stands of this type as shown in Fig. 6-13. The rollers are standard, plate-mounted swivel-type casters. Since they can turn in any direction, they can supply outboard support for crosscutting jobs, as well as for ripping operations.

RIPPING NARROW PIECES SAFELY

A fundamental safety rule when doing table-saw work is to never allow your hands to come close

Fig. 6-5. When the top of the roller is on the same plane as the table's surface, long or heavy workpieces will be supported after the cut and will move easily with minimum effort on the part of the operator.

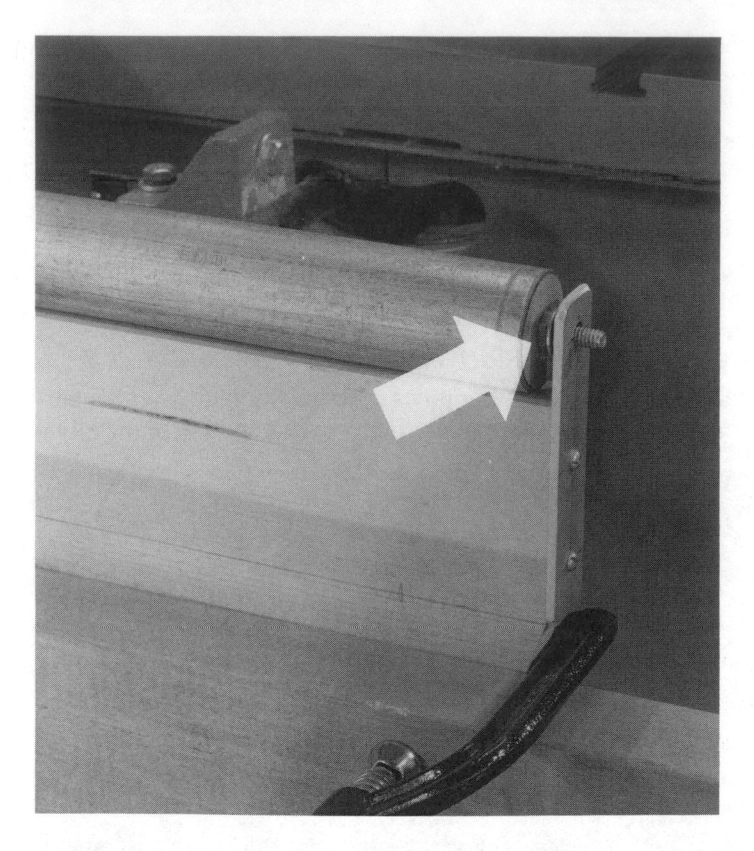

Fig. 6-6. The roller rotates on a common bolt secured with a nut in the center of the wooden disks that are used to seal the tube. The hole through the mending plate on which the roller rides should be just a fraction larger than the diameter of the bolt.

to the saw blade. You must follow this rule even more carefully whenever you are involved with ripping narrow strips. When there isn't ample room to use your hand to feed the stock, the wise alternative is to substitute something for your fingers. The "something" is called a *push stick.* Too often, too many workers will pick up a piece of scrap to use as a pusher. It is better than fingers, but won't have the shape to do the job efficiently or with maximum safety. It is better to make special push sticks that you can keep as permanent accessories.

The most simple type of pusher is shown in Fig. 6-14 and detailed in Fig. 6-15. The tool is long enough to keep your hand in a safe position, and is shaped at the base so it will move the work forward while providing a degree of hold-down action. It is a popular design, but there is room for improvement.

A pusher design that takes a step toward a more sophisticated version is shown in Fig. 6-16. Here, the project is designed with a long base to help keep the work flat on the table while the stick moves it forward. A horizontal piece, attached to the back of the pusher, rides on top of the rip fence and helps to keep you from tilting the pusher as you make the pass.

My personal preference—and much use has justified my feeling—is for a pusher that is U-shaped so that it can straddle the rip fence. Two fence-straddling units that I use in my shop are shown in Figs. 6-17 and 6-18. You can make one or both of them by following the construction details in Figs. 6-19 and 6-20. Bear in mind that the thickness of the component which actually does the pushing—the part which bears against the front surface of the rip fence—is variable. If it were ¾ inch thick you wouldn't use it for ripping strips ½ inch wide since you would be moving it into the saw blade. You should make several projects, each with a pusher component of different thickness.

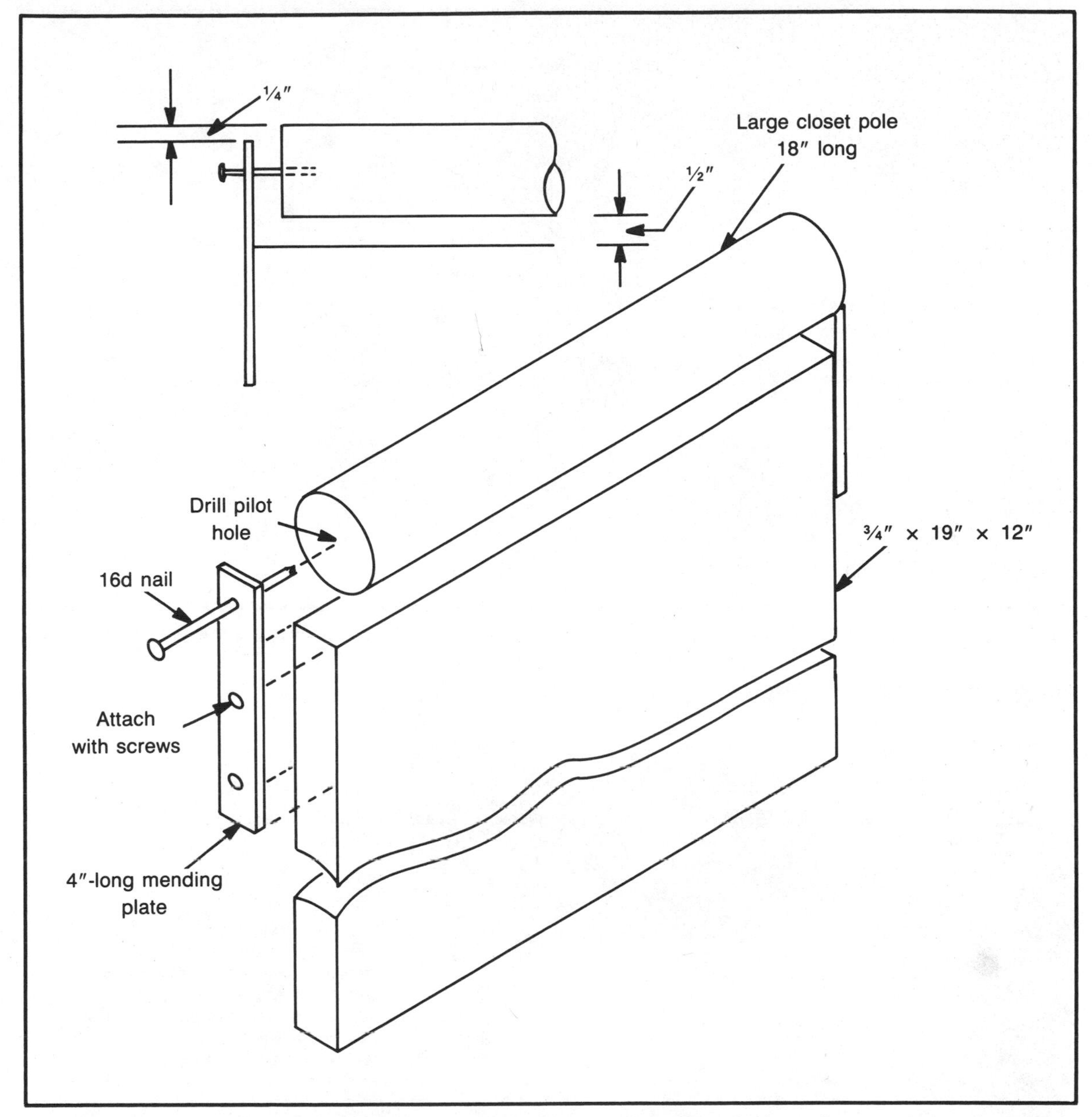

Fig. 6-7. Method for making a roller support for use with a sawhorse. As suggested, a large-diameter closet pole, or something you might choose to turn in a lathe, can be substituted for a metal tube.

Another favorite homemade tool for ripping operations is the combination pusher/hold-down shown in Fig. 6-21. It is not designed for use when sawing narrow strips, but can improve accuracy and safety on many routine ripping operations. You can make one like it by following the details in Fig. 6-22.

Those times when you need many similar narrow strips, it is better to provide a special jig instead of working in routine fashion with the rip

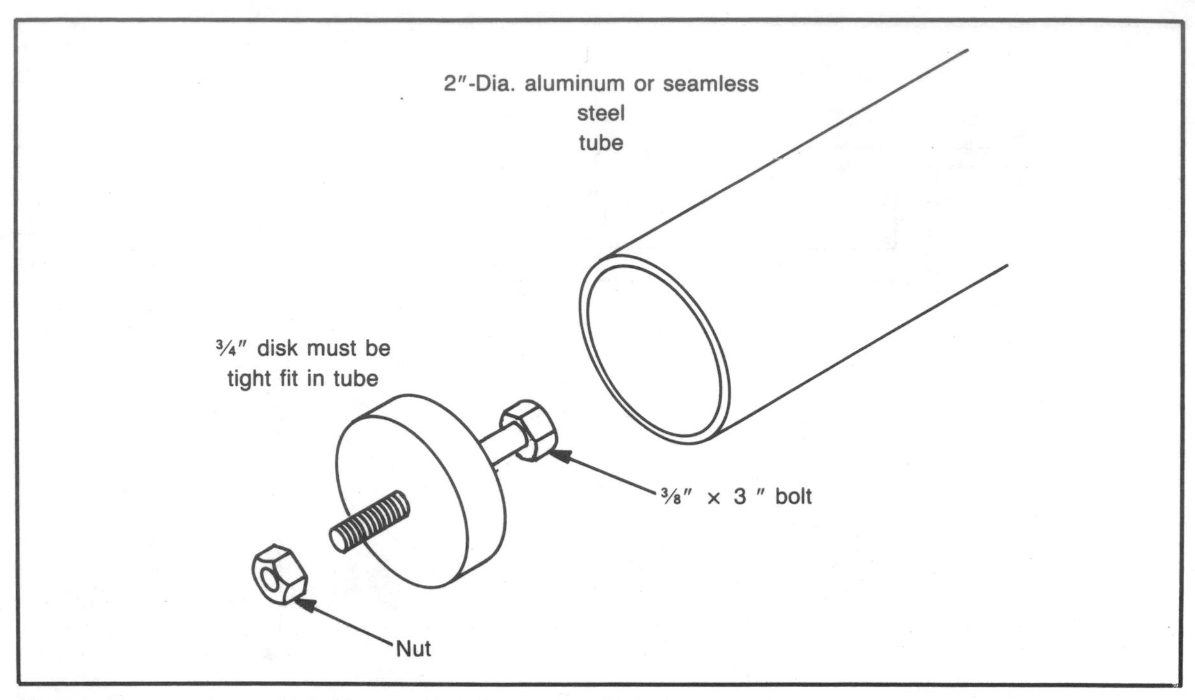

Fig. 6-8. If you use a metal tube, be sure that the wooden disks used to seal the tube ends are a very tight fit.

Fig. 6-9. Brackets like this, which are generally available, make it easy to put together a sturdy sawhorse. If you add a shelf at a lower point between the slanted legs, the project will be even more stable.

fence. The *jig* (Fig. 6-23), is a platform that is wide enough to provide ample room between fence and blade. The working edge of the jig is notched a limited distance so a stop area is left against which the work can be braced for sawing. The width of the notch suits the width of the pieces that are needed.

Since the piece being cut will be trapped between the jig and the blade with the possibility that it might be thrown up, it is a good idea to tack-nail a pair of hold-downs to the jig. The hold-downs are just small pieces of wood that overlap the notch in the jig. Another idea is to add a handle to the jig so it can be moved more easily. A typical jig, with hold-downs and handle added, is shown ready for use in Fig. 6-24.

To do the cutting, start with the jig at the front of the table. Place the piece to be cut so it is snug in the notch and then move jig and work past the saw blade (Fig. 6-25). Don't make a return. When the jig is past the blade, lift it from the table and return it to the starting position for the next cut. Incidentally, there's no reason why you cannot use the guard on operations of this nature.

Often, you can do a woodworking chore faster, more accurately, and even more safely by devising a special jig. Once I had a quantity of long strips that I wished to slice in half lengthwise. Instead of doing the ripping in routine fashion, I set up the jig shown in Fig. 6-26. This jig is a block of wood with cutouts at each end so it can be clamped to the rip fence. I cut a groove close to the block's outer edge, sizing it to suit the strips. I then positioned the jig so the saw blade was centered in the groove, and I raised the blade so it barely poked through the top of the block. Next, I fed the strips into the front of the jig; each succeeding strip acted as a pusher for the one being sawed. I sawed the final strip

Fig. 6-10. This is another type of standard sawhorse bracket. Legs and crosspiece are secured by the prongs that dig into them as the wing nut is tightened.

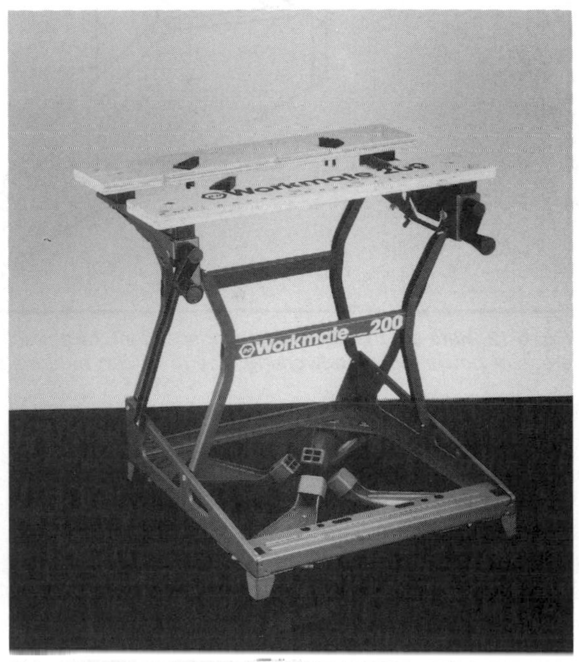

Fig. 6-11. Black & Decker's Workmate bench has a built-in clamping device. If one is available for your saw, you can use it to grip a roller-top support system.

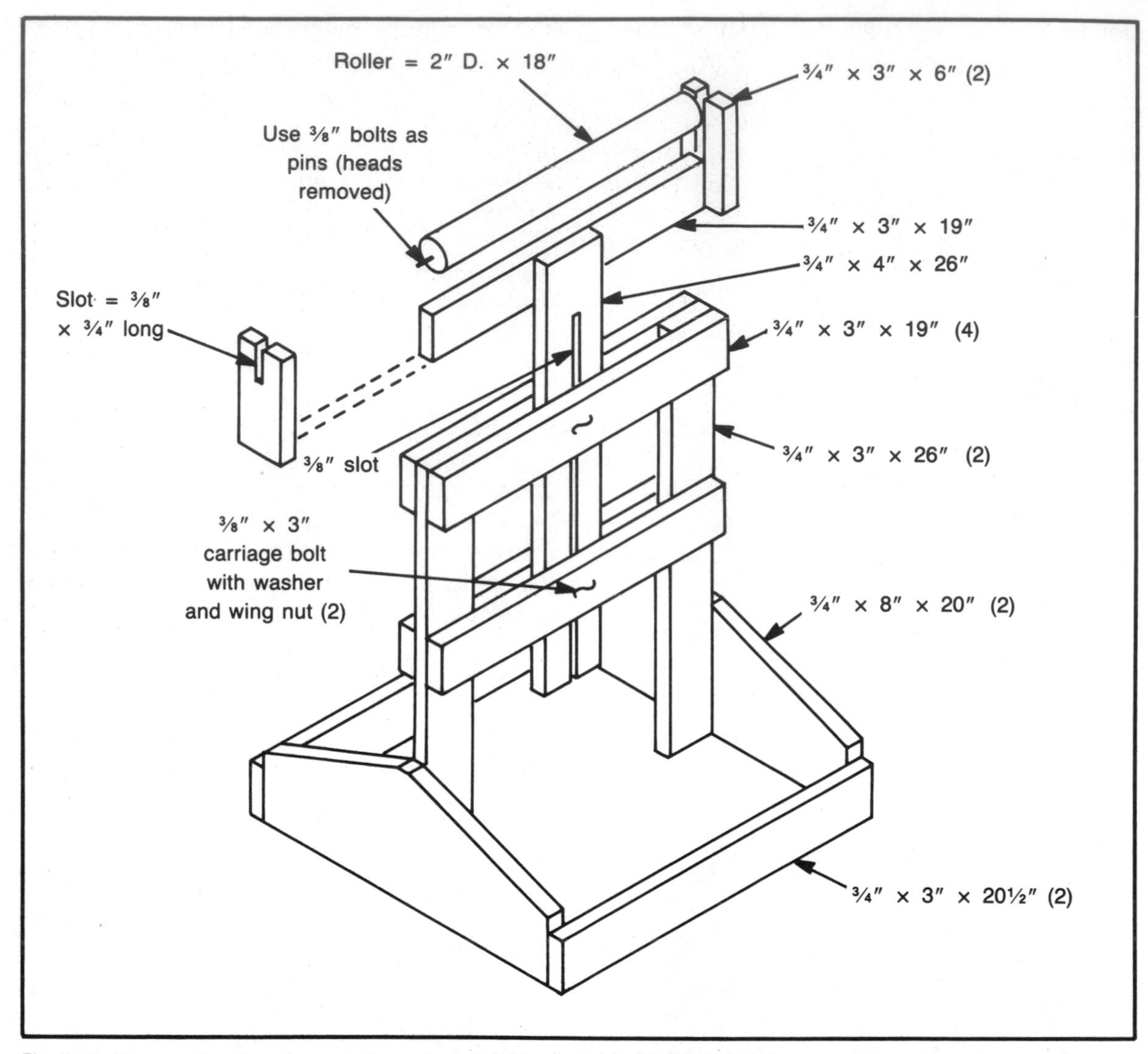

Fig. 6-12. Here are the plans for an independent, height-adjustable floor stand that can provide outboard support in many areas of power-tool woodworking. The roller can be a metal tube or a cylinder of solid wood.

by pulling it from the back. The work went quickly and all the strips were equal in thickness.

This might seem like a lot of explanation for a setup you might never use, but it is offered as an example of how to become knowledgeable on table-saw operations.

RIPPING EXTRA-THICK STOCK

The two-pass procedure used in Chapter 5 to crosscut stock that is thicker than the maximum height of the saw blade also applies when ripping (Fig. 6-27). Make the first pass with the blade elevated to a bit more than half the stock's thickness. Then turn the work over so the same surface is against the fence, and make the second pass. There will be some drag here since the blade is buried, so feed only as fast as the blade can cut, but feed steadily to avoid burn marks.

Fig. 6-13. The top of support stands can be made this way. Plate-mounted, swivel type casters, which are attached with screws, can turn in any direction so the concept is as functional for crosscutting as it is for ripping.

Fig. 6-14. An elementary pusher looks like a miniature hockey stick with a short blade. The blade is notched so the tool can bear against the back edge of the workpiece.

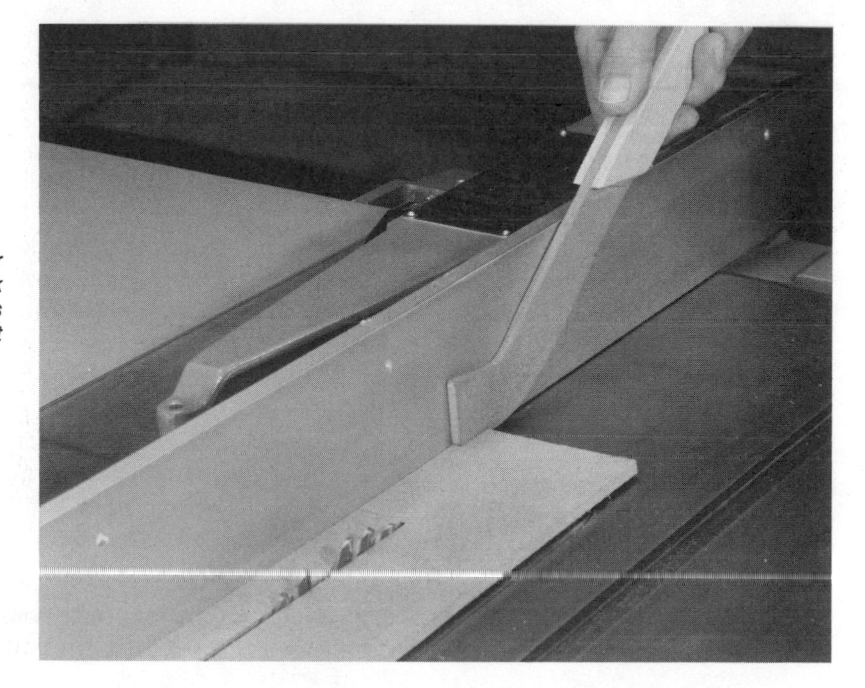

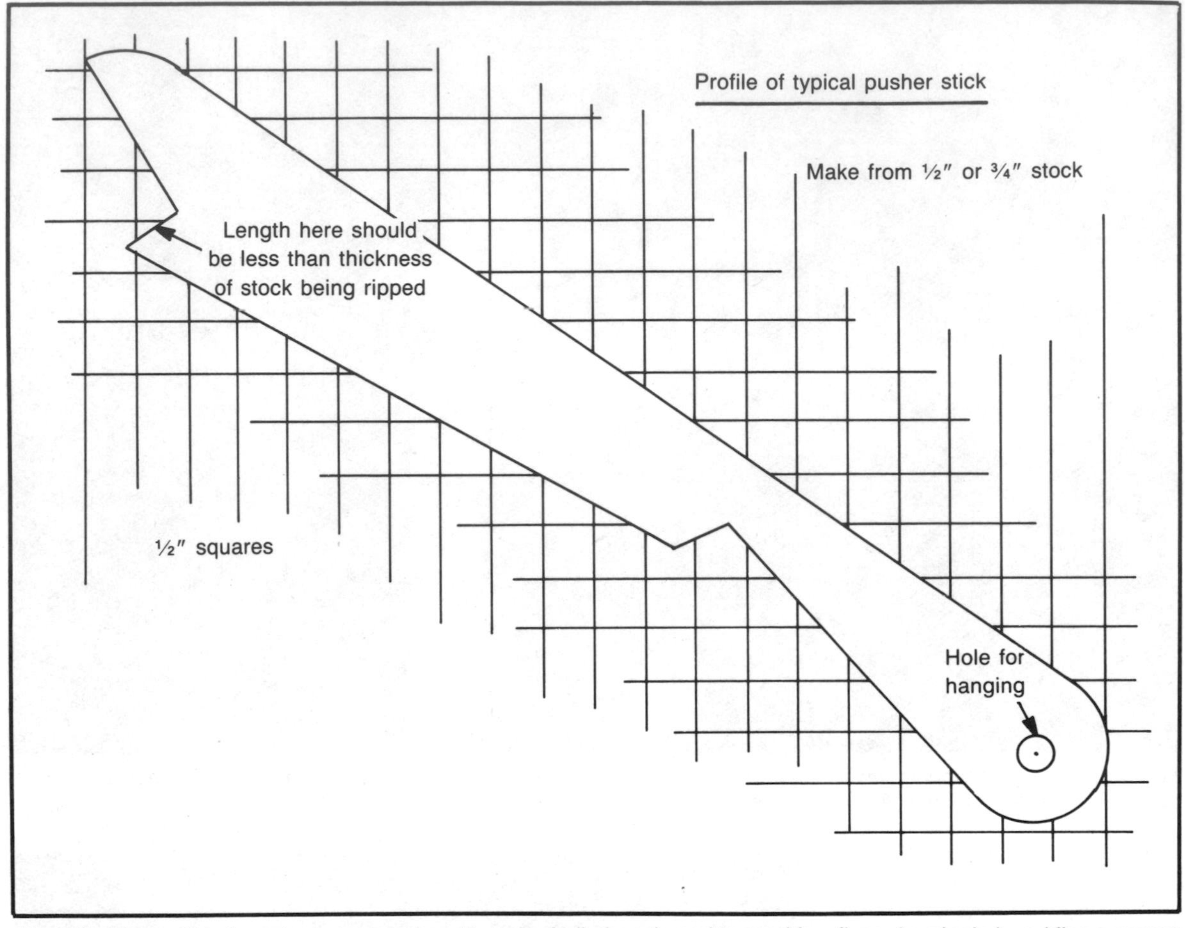

Fig. 6-15. The profile of an elementary pusher. You can bulk the grip end to provide a firmer hand grip by adding a narrow piece to each side of it.

This operation can be viewed as *resawing*; that is, turning thick stock into thinner pieces. Results will be good if the rip fence is correctly aligned and the angle between the blade and the table is 90 degrees.

RIPPING VERY THIN MATERIAL

Thin material, like a plastic laminate, has a tendency to climb the blade and to undulate during the pass. Also, the material might slip into the gap that exists on most table saws between the underside of the rip fence and the table. So it is necessary to take the precautions shown in Fig. 6-28 for the work to be done accurately and safely. Attach to the rip fence an auxiliary wood facing that bears down firmly on the table, and clamp a hold-down to the fence to keep the material from vibrating.

You can use the homemade springsticks that were demonstrated in Chapter 2 and that can be shaped like those in Fig. 6-29 instead of the wood block for the hold-down function. Other uses for the springsticks will be demonstrated as we move along.

RIPPING ODD ANGLES

Often, you can use a special ripping operation to make a cut that is inconvenient to do with a

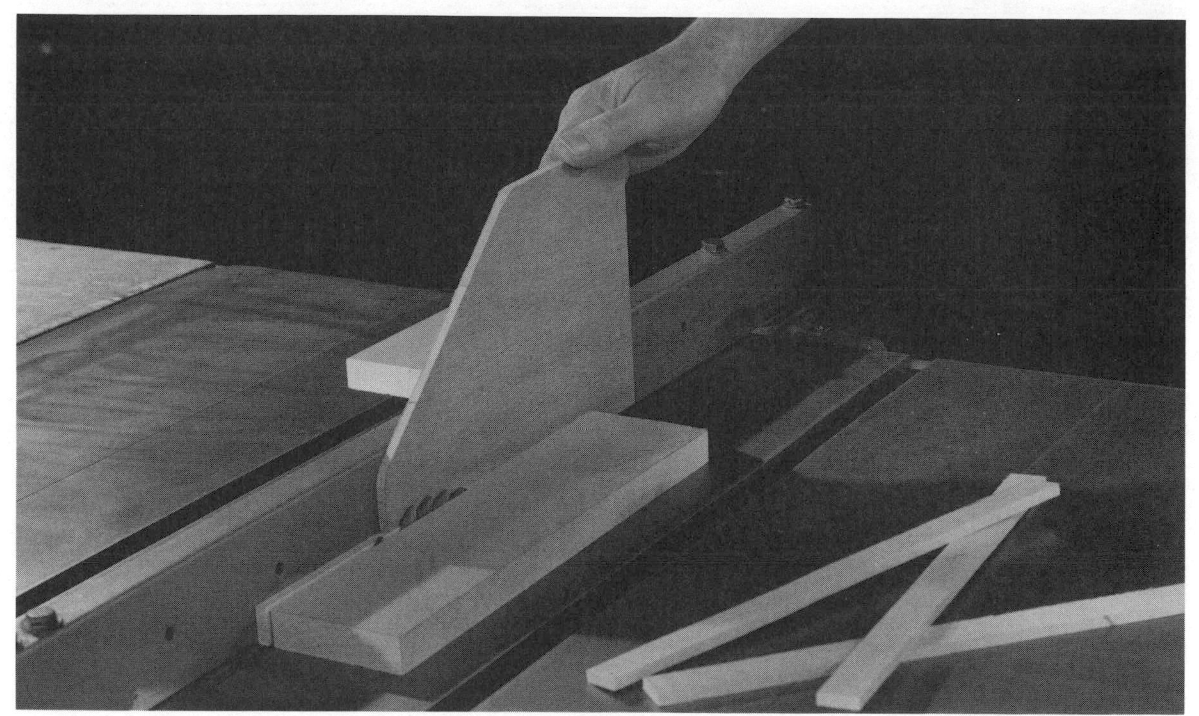

Fig. 6-16. A step up in terms of pusher design is one with a longer base so it can provide more hold-down action. Add a horizontal piece at the rear. This piece rides on top of the rip fence and helps guard against tilting the tool as you make the cut.

Fig. 6-17. A fence-straddling push stick is a better design than most. This one has a slanted, front-bearing edge.

Fig. 6-18. This fence straddler has a notched pusher component. The notch that suits the thickness of the work is used to move the work forward.

Fig. 6-19. This is how the project with the angled pusher component is made. The angled edge will grip better if it is faced with fine sandpaper.

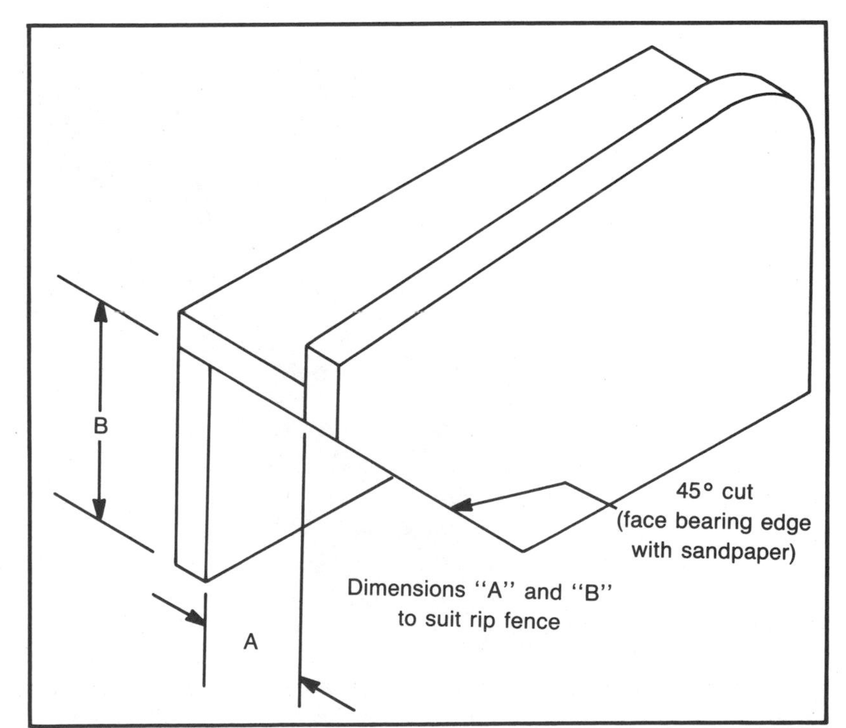

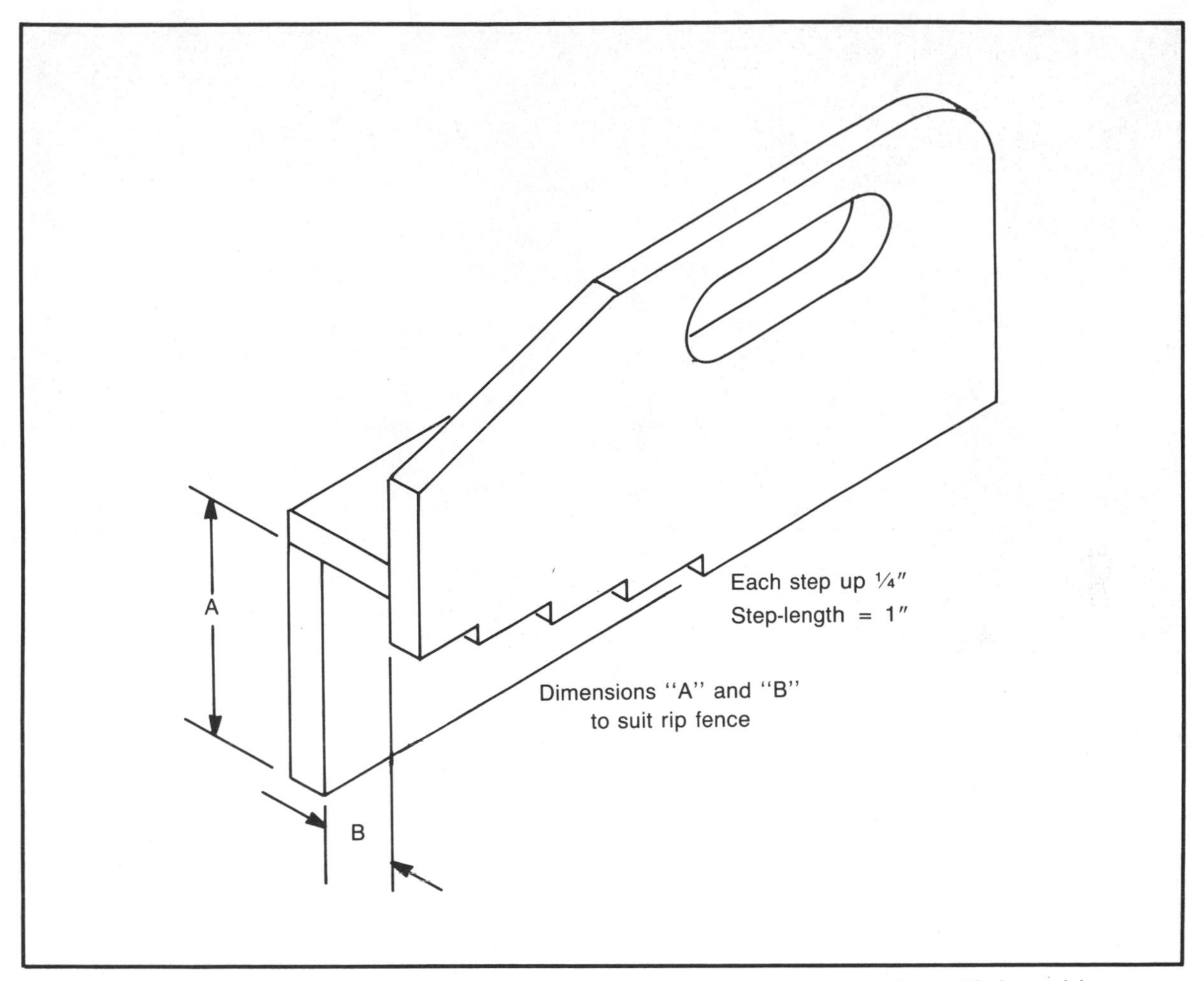

Fig. 6-20. Construction details for the stepped pusher. You can form the hand-grip slot by boring end holes and then removing the waste between them with a coping saw.

miter gauge. Another use for the technique is to make a straight cut on material whose edges aren't even enough to bear against the rip fence. The idea is to tack-nail a wide piece of stock to the surface of the workpiece to serve as a guide for the cut (Fig. 6-30). Since the guide overhangs the work, it is a good idea to attach a strip of wood whose thickness matches that of the workpiece to the underside of the guide.

When cutting an odd angle, situate the edge of the guide on the cut line and locate the fence so the blade and the edge of the guide will be in line.

STOPPED CUTS

Stopped cuts are saw cuts that do not separate pieces. They might start at one end of the stock and terminate before reaching the opposite end, or they might start and stop short of both ends. A simple way to control the length of a cut is to use one of the rip-fence stop blocks shown in Chapter 5 (Fig. 6-31). When the cut is complete, you can very slowly retract the workpiece. You might be more comfortable by turning off the machine and waiting for the blade to stop turning before removing the work.

One problem with a rip-fence stop block is

Fig. 6-21. A combination pusher/hold-down is useful on many ripping operations. A cleat at the rear bears against the work to move it forward. The bulk of the tool bears down on the work so it will be flat on the table throughout the pass.

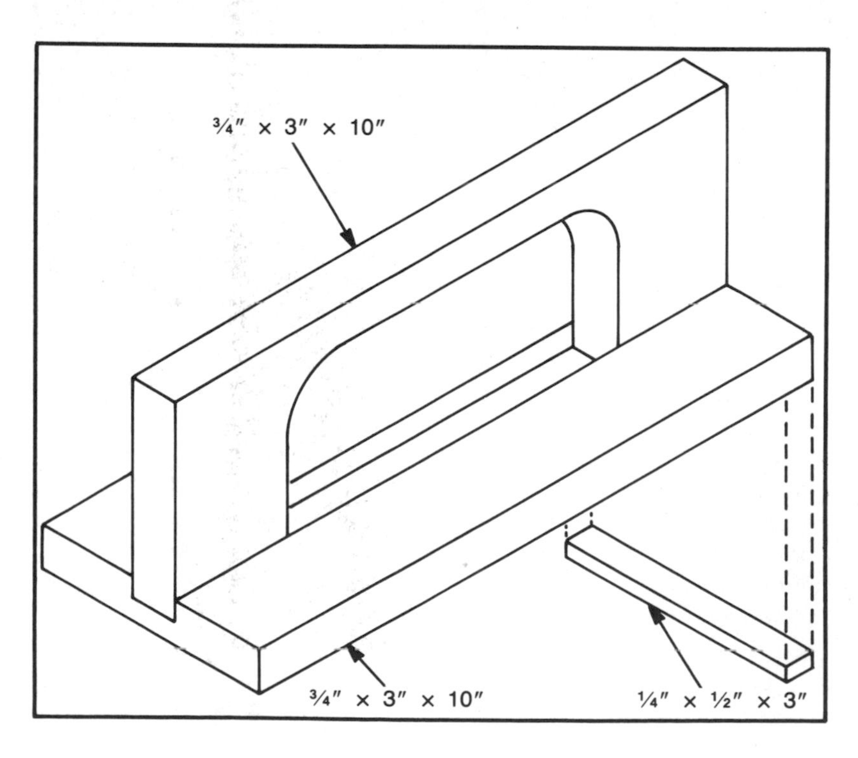

Fig. 6-22. The method for making a combination pusher/hold-down. The ends of the gripping component should be rounded off.

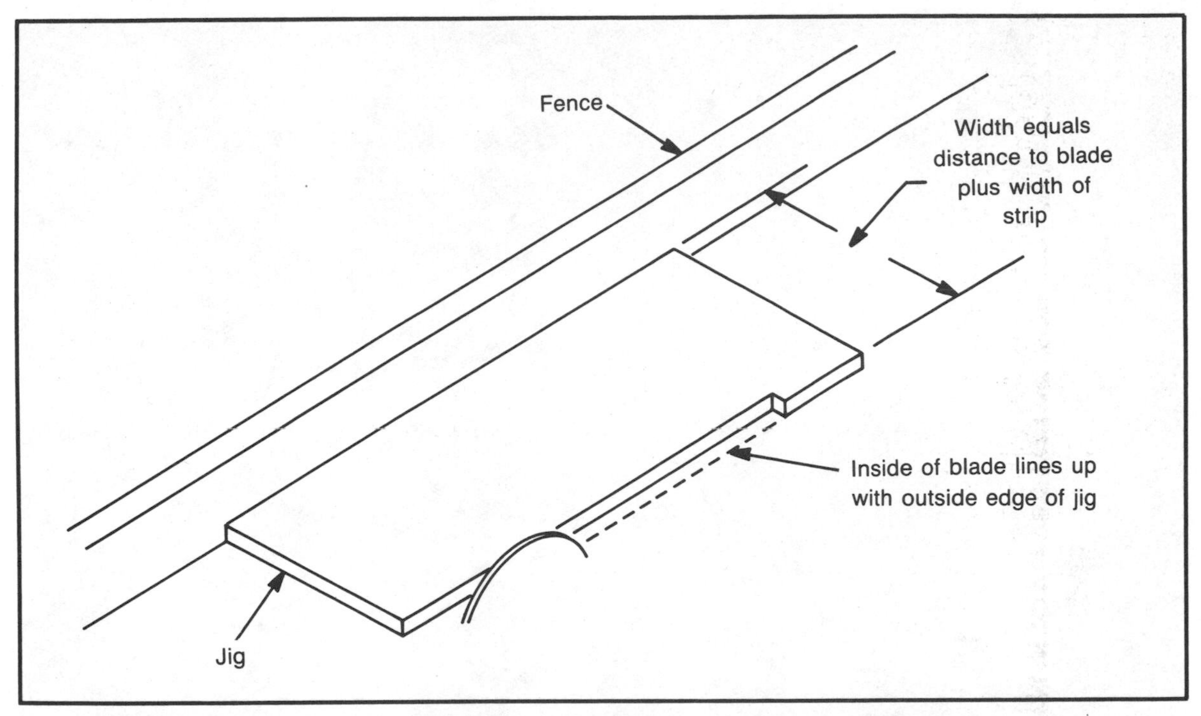

Fig. 6-23. A jig that can be used for production ripping of slim strips.

Fig. 6-24. To make the strip-cutting jig safer and more convenient to use, tack-nail small pieces of wood so they overhang the notch, and add a handle so you can move the jig more easily.

Fig. 6-25. Place the stock firmly in the notch. The jig moves it forward. The cut is complete when the jig has passed the saw blade.

that the length of the cuts you can make is determined by the length of the fence. You can solve this problem by making a rip-fence extension with an attached stop block like the one in Figs. 6-32 and 6-33. The sawing procedure, of course, is the same. Place the work at the front of the table and against the fence just as if you were making a routine rip cut, and then advance the work until it contacts the stop block (Fig. 6-34).

The stop-block technique is a good way to form the kerfs that are needed, for example when fabricating springsticks (Fig. 6-35). When you are doing jobs like this, start kerfing at the outer edge of the stock so that the remaining bulk will be between the saw blade and the fence. If you work the other way, the springy fingers will not provide adequate support.

A more sophisticated type of rip-fence extension, one with twin sliding stops, is shown in Figs. 6-36 and 6-37. One of the advantages of the unit is that the stops can be set to control the length of cuts which stop short of both ends of the material. An example is the groove

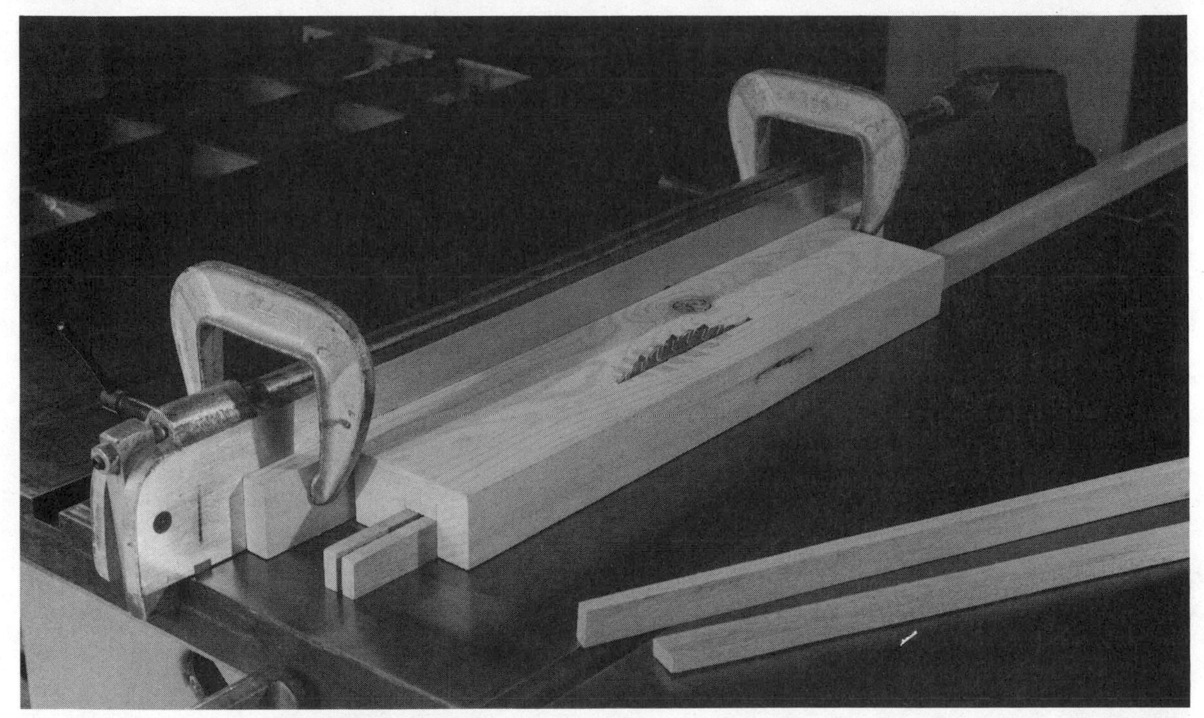

Fig. 6-26. This jig was a big help when I needed to rip long strips of wood in half. It's an example of how being jig-wise can help to get jobs done faster, safer, and more accurately.

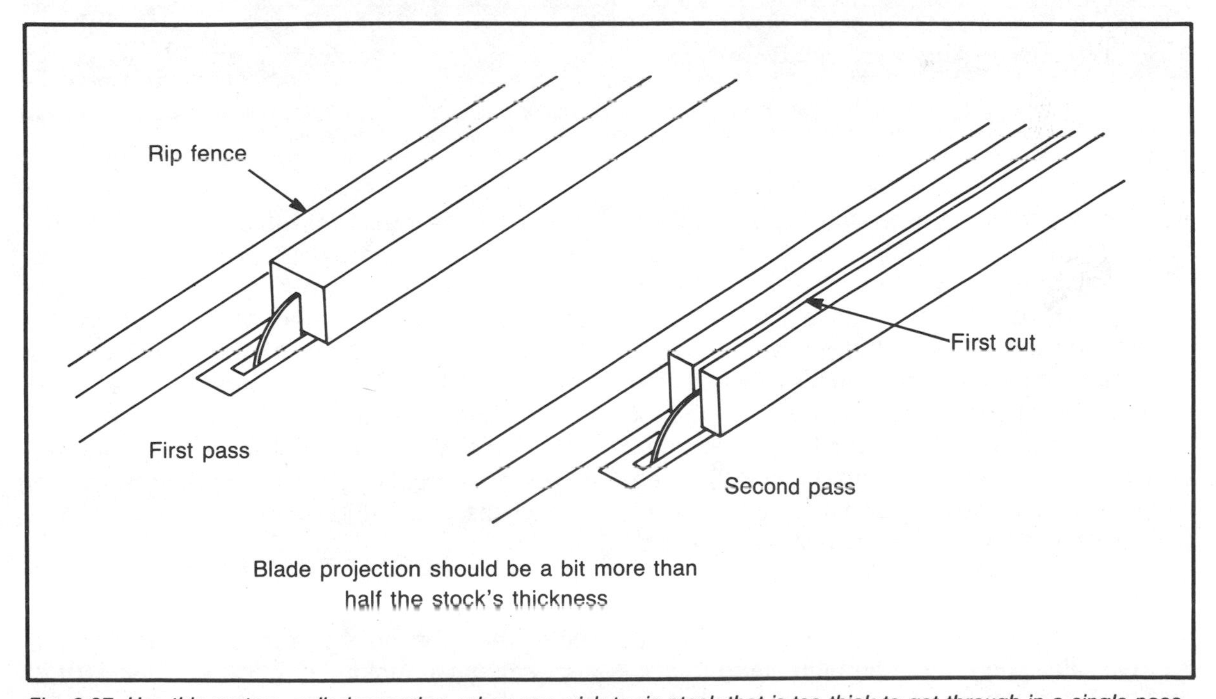

Fig. 6-27. Use this system, called resawing, when you wish to rip stock that is too thick to get through in a single pass.

Fig. 6-28. A setup like this will keep thin material from sliding under the rip fence and from undulating during the pass. Don't let the hold-down bear down so firmly that you will have difficulty moving the material.

needed when you are reinforcing an edge-to-edge joint with a blind spline. Locate the fence, with jig attached, so the cut will be in the center of the stock's edge. The height of the saw blade determines the depth of the groove.

Start the job by turning on the saw and very carefully placing the workpiece so it is clear of the blade but is firmly against the fence and braced against the front stop (Fig. 6-38). Then tilt the workpiece down until it rests firmly on the table. Advance the workpiece until it hits the second stop (Fig. 6-39).

Some operators prefer to start with the blade below the table and with the workpiece flat on the table. Then they raise the blade to correct projection and move the work forward to complete the cut. I feel this method puts the worker in an awkward position since he must use one hand to hold the work, and the other to turn the blade-elevation wheel. Anyone using this procedure should clamp the workpiece while he raises the blade. Then he can remove the clamp to complete the cut.

You can use the double-stop project for routine stopped cuts by simply working with only one of the stops (Fig. 6-40).

REPEAT-PASS WORK

You can use a conventional saw blade for cuts wider than the normal kerf by employing the repeat-pass technique (Fig. 6-41). This method is simply a matter of resetting the rip fence for

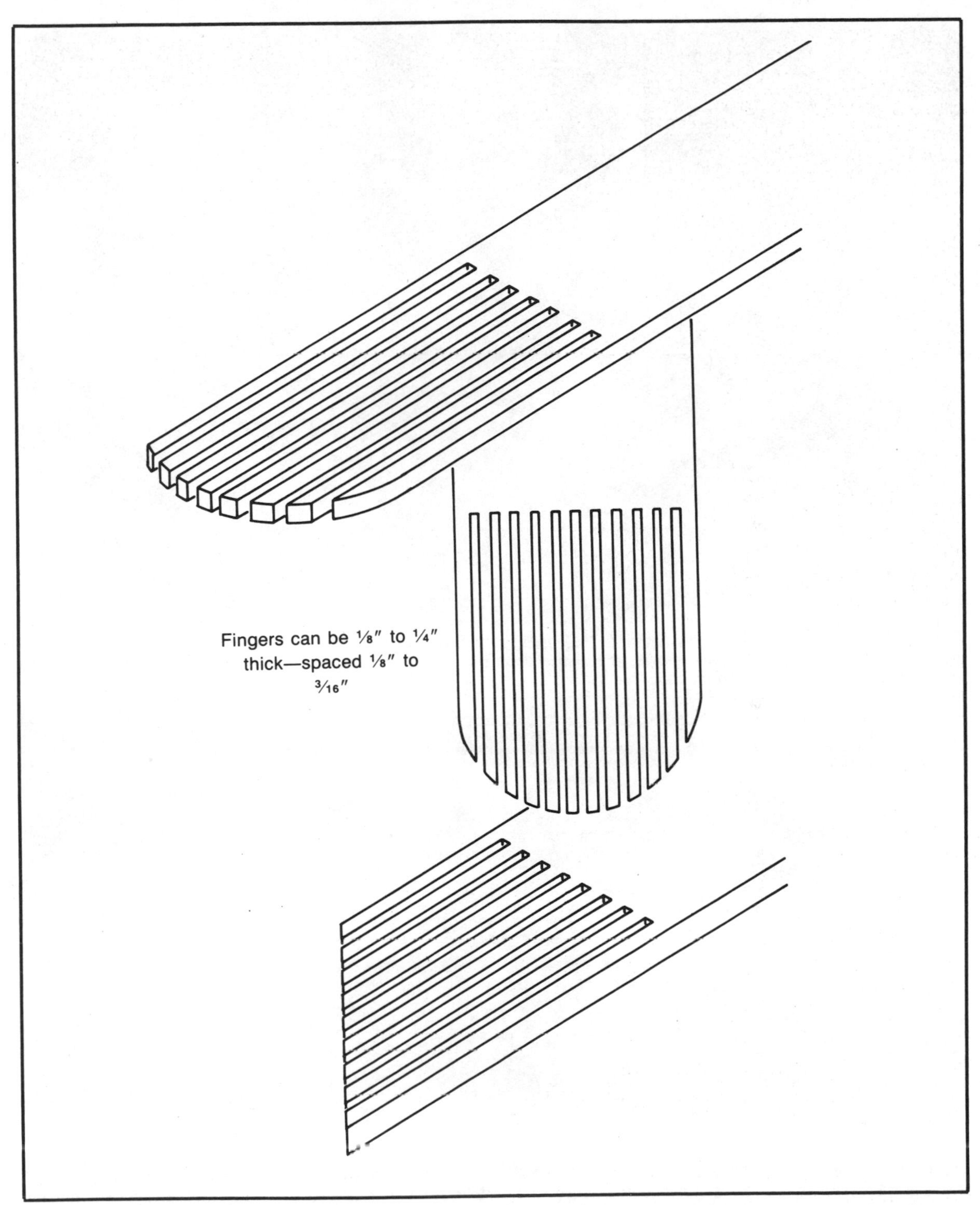

Fig. 6-29. Designs for springsticks are variable. They can be wide or narrow, long or short, and have ends of various shapes.

Fig. 6-30. You can use a special guide that is tack-nailed to the work at one edge, and whose opposite edge rides the rip fence to move work for odd-angle cuts. It's also a good way to work when you need to rip stock with edges that are not even enough to ride the rip fence.

Fig. 6-31. You can control the length of cuts by using a stop block on the rip fence. The problem with this arrangement is that maximum cut length is controlled by the length of the fence.

Fig. 6-32. You can secure a homemade extension with an attached stop block anywhere on the rip fence, thus giving you more control over cut lengths.

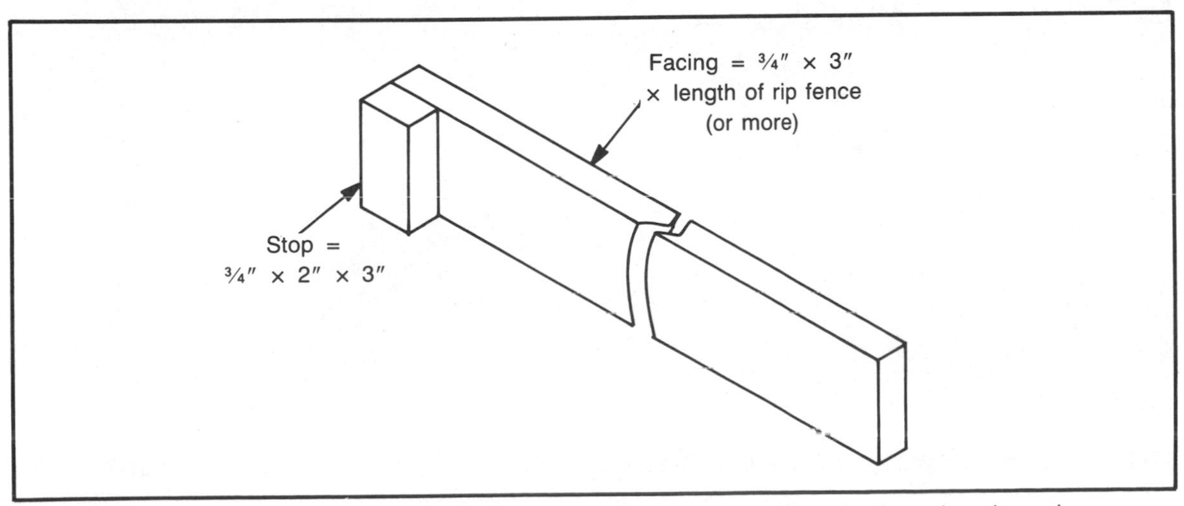

Fig. 6-33. The extension, or fence facing, doesn't need to be more than a straight strip of wood or plywood.

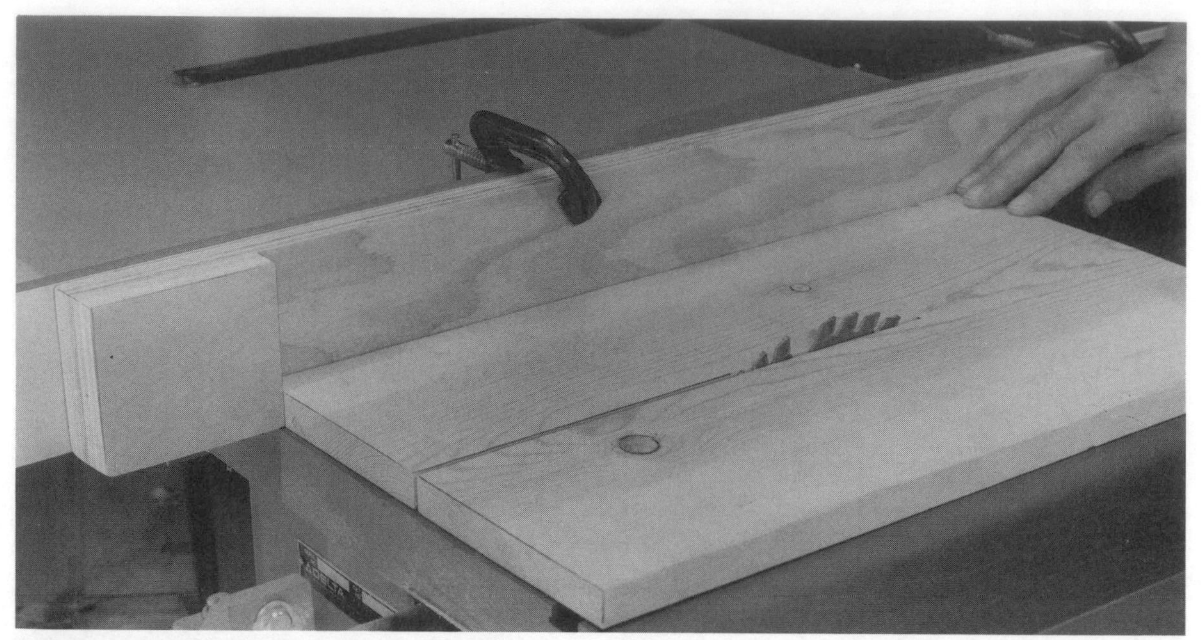

Fig. 6-34. The sawing procedure is the same, whether you work with a stop block on the rip fence or make the extension. Start as if you were making a routine rip. The cut is complete when the work contacts the stop block.

Fig. 6-35. The stop block procedure is the ideal way to make similar cuts of equal length; an application you must follow, for example, when making springsticks.

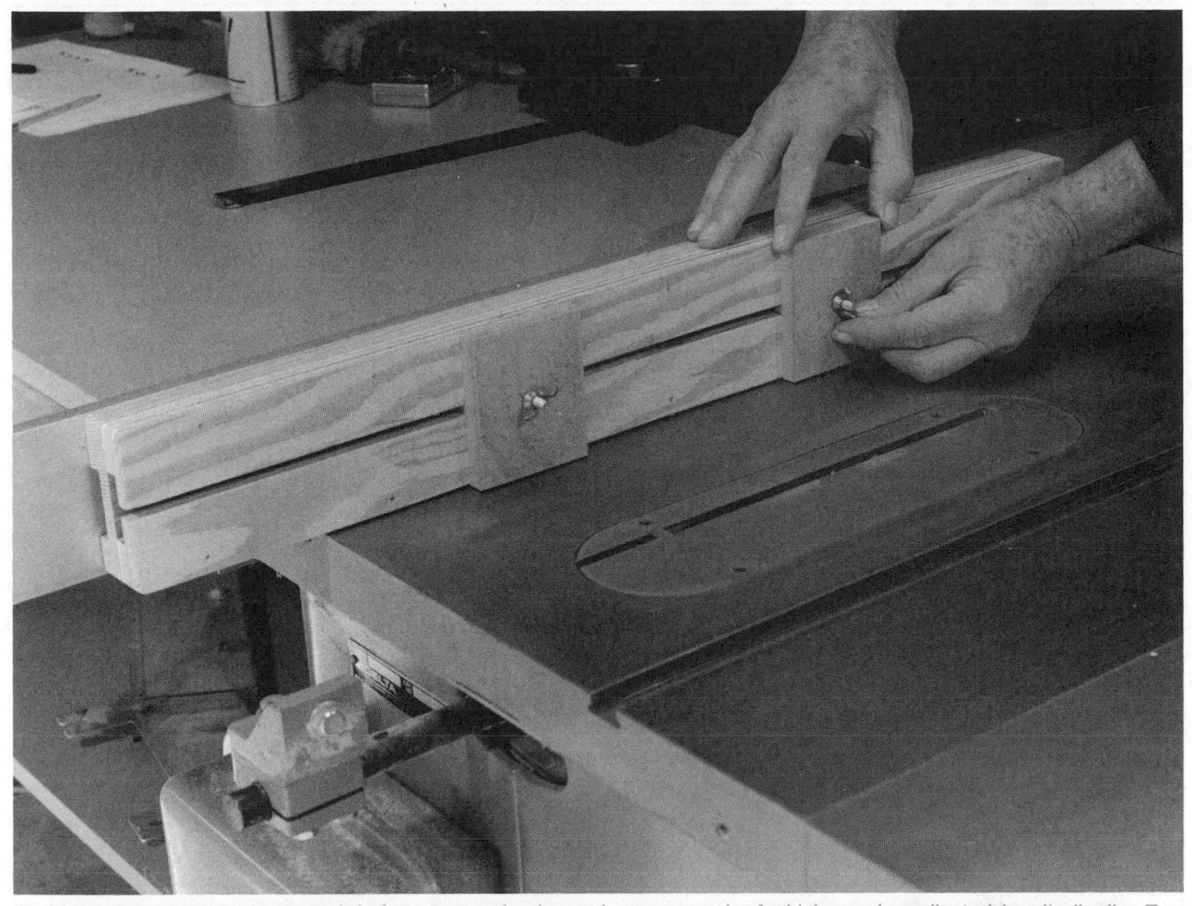

Fig. 6-36. This more sophisticated rip-fence extension has twin stops, each of which can be adjusted longitudinally. Two stops make it possible to automatically control the length of a cut that is confined between stock edges.

each of the cuts. The cuts should overlap just a bit for smoothest results.

You can use this method, for example, to cut a groove in a single component. It is best to work with a dadoing tool when you need the same cut in many pieces. The use of a dadoing tool will be covered in Chapter 8.

TWO-PASS OPERATIONS

The two-pass idea allows the use of a regular saw blade to shape forms that usually require other tools. An *edge rabbet*, which is an L-shaped cut along the edge of stock, will serve to demonstrate the procedure. The first step, as shown in Fig. 6-42, is to set the blade's elevation for the depth of the rabbet, and to position the rip fence for the width of the rabbet. Be sure to measure from the fence to the outside of the blade.

Use the first cut as a gauge to set the blade's elevation for the cut that follows (Fig. 6-43). Then locate the rip fence for the cut that will remove the waste piece (Fig. 6-44). This critical dimension is from the fence to the *inside* of the blade. If you have been careful when making the setups, the corner of the cut will be clean and the planes of the cut will form a 90-degree angle (Fig. 6-45).

Check Table 6-1 when you feel that a ripping operation is not going as smoothly as it should.

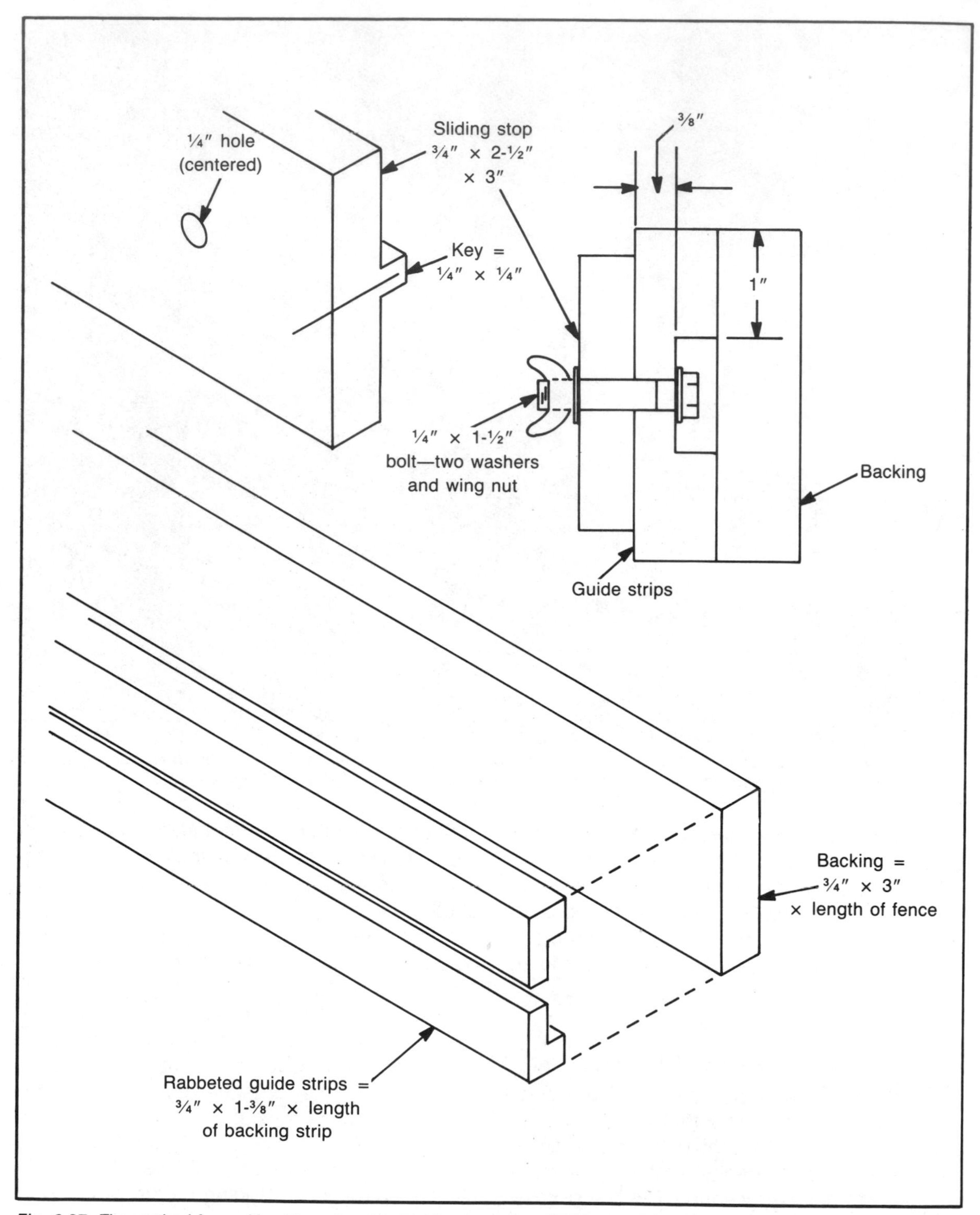

Fig. 6-37. The method for making the extension that incorporates sliding stops.

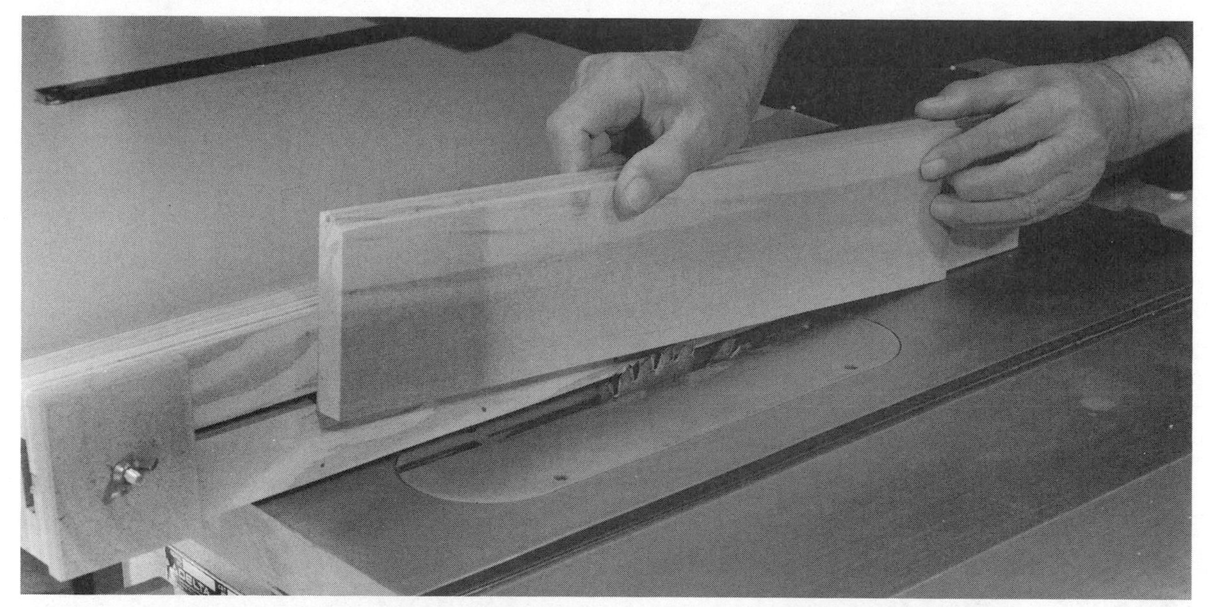

Fig. 6-38. To start the cut for a groove for a blind spline, hold the work clear of the blade. Keep it firmly against the fence and braced against the front stop. Lower the work very slowly over the turning blade until it sits firmly on the table.

Fig. 6-39. The next step is to move the work forward until it contacts the rear stop. At this point, you can carefully lift the work free of the blade, or you can hold the work still, turn off the machine, and wait for the blade to stop before removing the work.

Fig. 6-40. You can use the two-stop extension for routine stopped cuts by just sliding off one of the stops.

Fig. 6-41. You can form grooves that are wider than the normal saw kerf simply by making repeat passes. If you have similar pieces to do, use a dadoing tool to make the work go much faster and be more accurate.

Fig. 6-42. The first step when using the two-pass technique to form an edge rabbet. Set blade height for the depth of the rabbet. The distance from the fence to the outside of the blade determines the width of the rabbet.

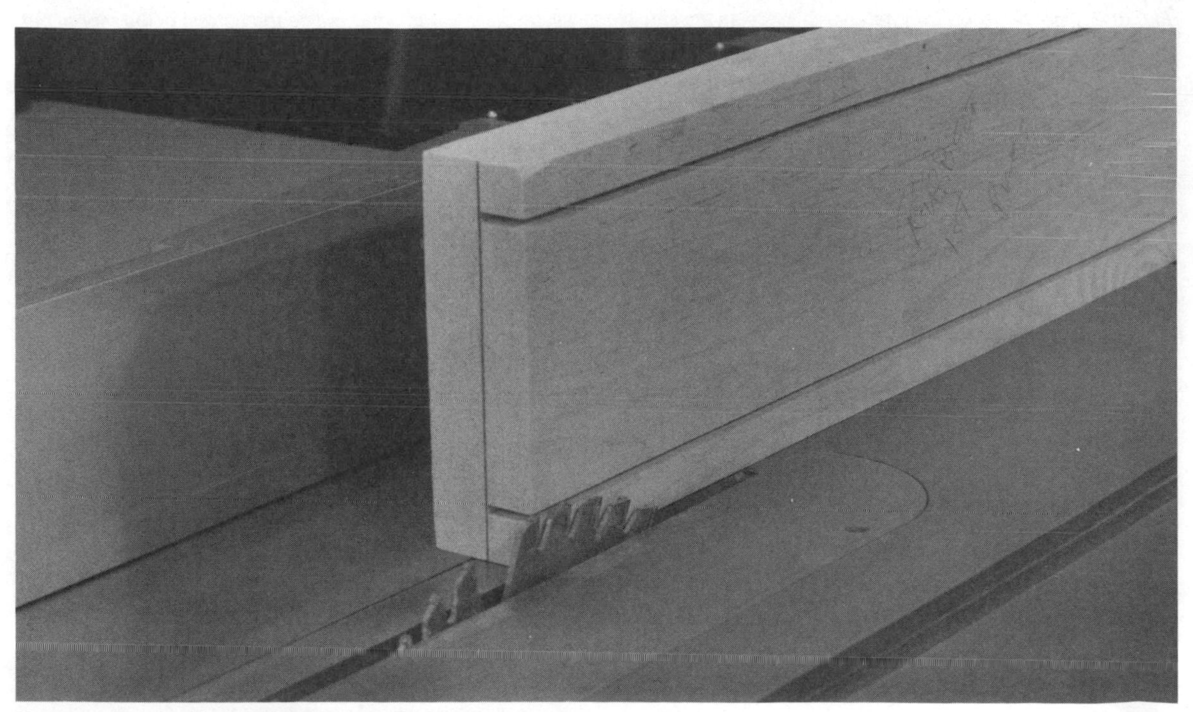

Fig. 6-43. Set the blade height accurately by using the first cut as a gauge. The top of the teeth should be on line with the top plane of the kerf.

Fig. 6-44. The third step is to set the rip fence so the saw blade will be on line with what will be the bottom of the rabbet cut.

Fig. 6-45. The final step is to make the cuts that remove the waste. Be sure to keep the workpiece snugly against the fence throughout the pass.

Table 6-1. Troubleshooting Chart for Ripping Operations.

Problem	Possible cause	Correction
Work too narrow	Wrong fence setting/Measuring from wrong side of blade	Careful measuring/Measure from side of blade nearest fence and from tooth that points toward fence
Edge not square	Angle between blade and table incorrect	Check alignment/reset auto-stop if necessary
	Distorted stock	Material may need surfacing before ripping
Sides gouged/Excessive tooth marks	Blade chatter	Feed more slowly/Blade may be dull/Poor blade
	Work fed incorrectly	Work must be flat on table and snug against fence throughout the pass
	Bad blade—Dull blade	Replace or sharpen
Blade stalls/Burn marks on work	Dull blade	Replace or sharpen
	Forcing the cut	Allow blade to cut at its own pace
	Tough wood	Feed more slowly/Make repeat passes for deep cuts
	Dirty blade	Clean frequently to remove wood residue
	Projection of hollow-ground blade incorrect	Hollow-ground blades require more projection than conventional blades
Kerf closes and binds blade	Moisture content in wood	Use splitter/Use wedge in kerf if necessary
Work moves away from fence/ work jams between fence and blade	Misalignment	Table slots not parallel to blade/Fence not parallel to table slots
	Poor work-handling	Keep work against fence throughout pass
	Distorted stock	Edge that rides rip fence must be true
Work won't pass splitter	Misalignment	Be sure splitter lines up with saw blade
Kickback	Misalignment	Adjust anti-kickback fingers/Check fence alignment
	Pass not completed	Move work completely past the saw blade
	Dull blade	Have sharpened

Chapter 7

Angular Sawing

ANGULAR SAWING IS THE TERM I USE TO DESCRIBE saw cuts that are made with the miter gauge pivoted to an angle other than 90 degrees, or with the saw blade tilted, or with both components adjusted to particular settings for compound angle joints. Terminology can be confusing. For example, the assemblies shown in Fig. 7-1 are called miter joints, so the cuts required are often called, simply, *miters*. To talk the same language, let's identify them as shown.

The *simple miter* is cut at an angle across the surface of the stock. The *cross miter* is also cut across the stock, but at an angle through its thickness. The *rip miter*, or *bevel*, is an angle cut through the stock's thickness, but made parallel to the grain.

There are some frustrating chores in woodworking, and cutting a good miter joint is one of them. It's not that the sawing itself is any different from other jobs; it's that you won't be excused for the least bit of inaccuracy. If you are off slightly when making a 45-degree miter cut, the parts will come together nicely, but the angle between them will not be 90 degrees. The single, little error is multiplied eight times when you have cut pieces, say, for a box or a picture frame. The total error becomes frustratingly evident at assembly time.

Don't be discouraged, though. The secret to perfect miter cuts consists simply of being very careful when you are adjusting the miter gauge or blade and of making and checking a test cut on scrap stock before you saw good material. Bear in mind that the angle at which sawing is done is always one-half of the joint angle (Fig. 7-2).

BASIC MITERING

To make simple miter cuts (frame miters), pivot the head of the miter gauge and lock it at the angle required. Then, as in crosscutting, hold the work firmly in position against the gauge as you make the pass. It's always a good idea to mark the cut line on the work with a protractor or triangle so you can judge accuracy as you cut (Fig. 7-3).

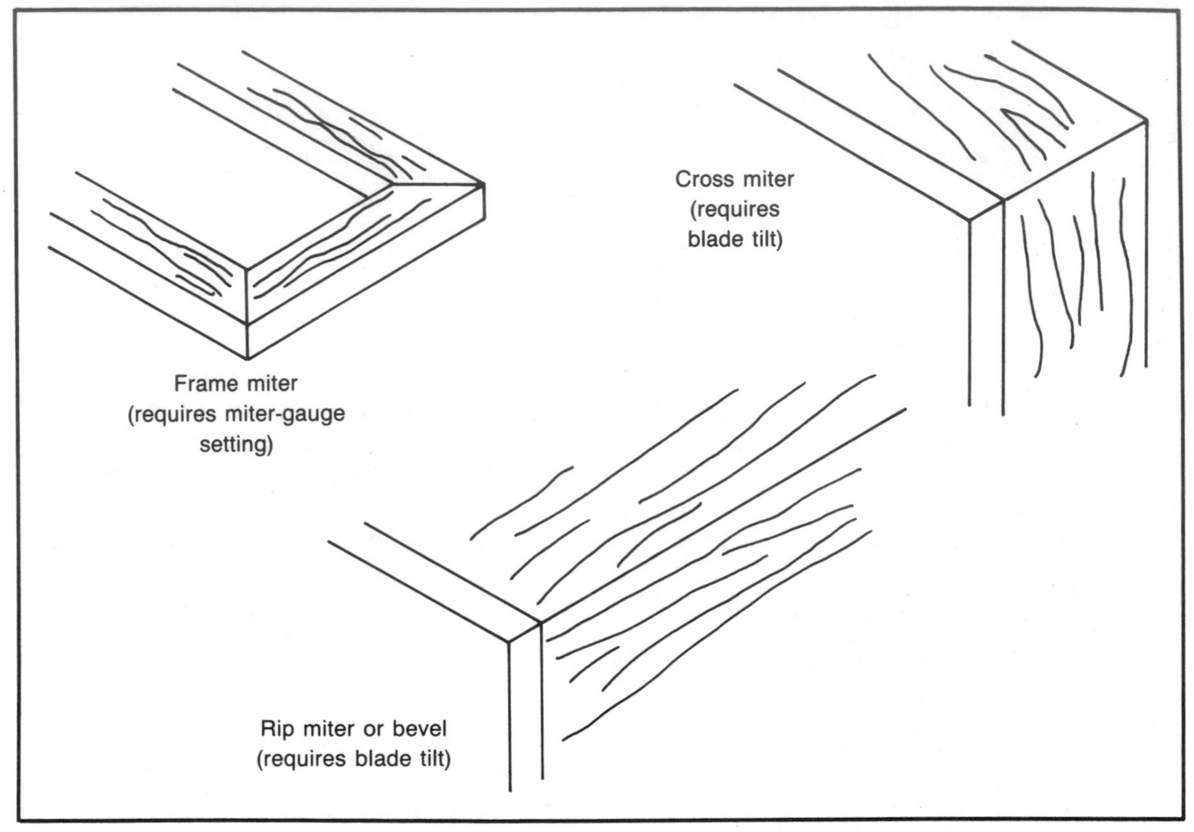

Fig. 7-1. Miter joints are popular because mating pieces can be joined without exposing unattractive end grain. These are the basic wood connections that involve angular cuts.

Two factors to be aware of that can spoil accuracy even when you have been careful with setups are: the action of the saw blade will tend to pivot the workpiece about the forward edge of the miter gauge (Fig. 7-4), and also will try to move it along the miter gauge, which is a negative action commonly called *creep*. To avoid the spoilers, be certain to keep the work firmly in position as you cut and to make the pass a bit more slowly than you normally would.

A miter-gauge extension, especially if it is faced with sandpaper, will provide a great assist for keeping the workpiece where it should be as you saw (Fig. 7-5). You also can rely on a miter gauge/hold-down for extra help (Fig. 7-6).

The shape of the material will often dictate how you must use the miter gauge. You can make some mating cuts with a single miter gauge setting. Others might require that you pivot the miter gauge left for one cut and right for another. Another possibility is that you might need to use the gauge on each side of the blade. The latter two possibilities point up how critical it is to be precise with adjustments.

You can miter stock that is flat at each end without needing to change the setting on the miter gauge. After the first cut, simply flip the stock and turn it end for end to position it for the second cut (Figs. 7-7 and 7-8). If the stock is not flat, molding for example, make one cut with the miter gauge on one side of the blade. Make the second cut, on the opposite end of the stock, by readjusting the head of the miter gauge and using it in the slot on the other side of the blade (Figs. 7-9 and 7-10).

Many operators find that mitering moldings

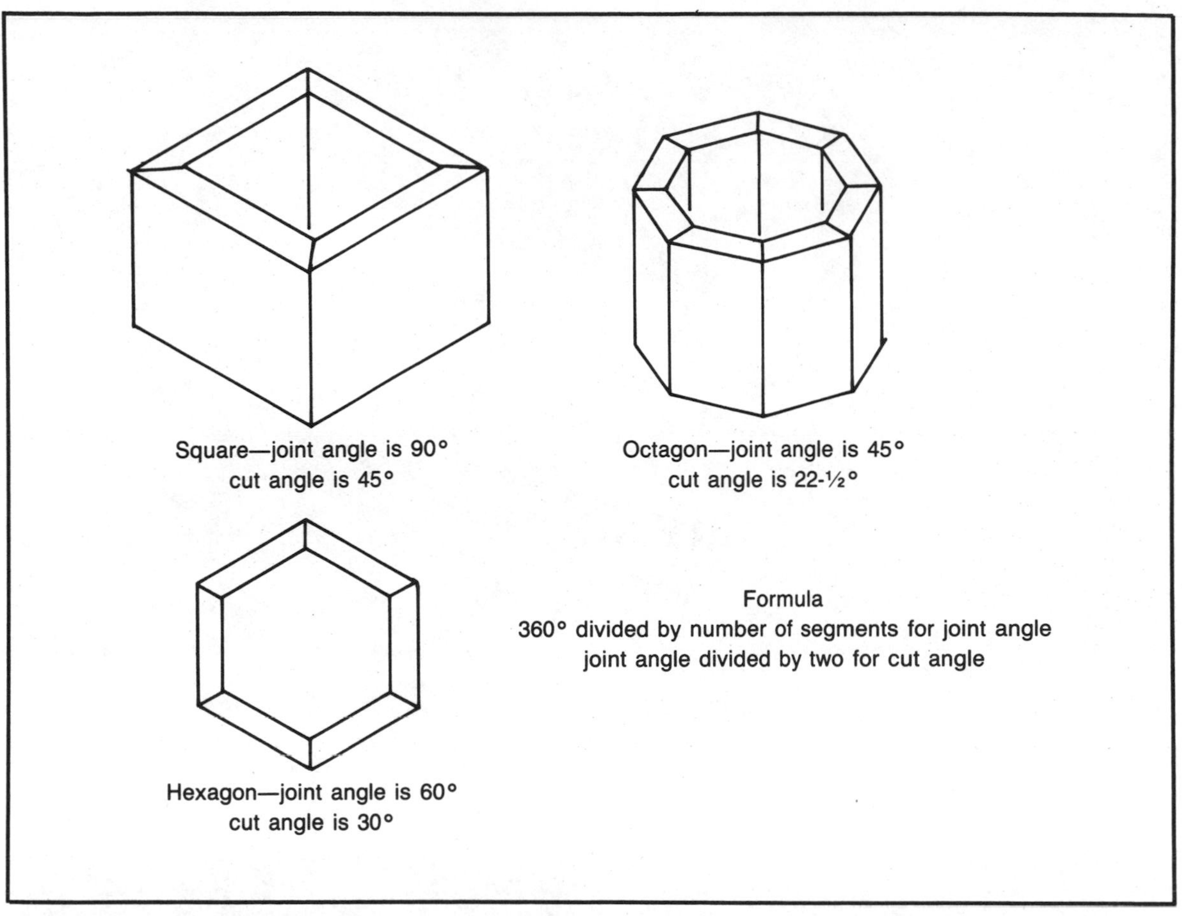

Fig. 7-2. Be aware of the difference between the angle formed by two connected pieces and the angle at which the cut should be made.

is easier to do, and can be done more accurately, with two miter gauges. A miter gauge is used in each table slot with heads pivoted so the angle between them is 90 degrees. Chapter 11 contains more information on this technique.

CUTTING TO LENGTH

Many projects that are assembled with miter joints require that components be of equal length, for example, a square or rectangular picture frame. To cut the pieces consecutively from one length of stock, first make a miter cut at one end of the material and then use a miter gauge/stop rod to control the work for the following cuts (Fig. 7-11). Flip over the material for each of the cuts. You can use a homemade miter-gauge extension with a clamped-on stop block in the same way.

Another system—one that is a bit wasteful of material but that ensures correct work lengths—is to cut the frame pieces to correct length at the beginning. Miter the ends of the parts by using a stop block on the rip fence to gauge the cutoff (Fig. 7-12).

MITERS ON UNEQUAL PIECES

Figure 7-13 shows a method you can use to establish the correct cut angle when it is necessary to use a frame miter joint to assemble parts of unequal width. Use one part as a gauge to

Fig. 7-3. To make the simple frame miter, pivot the head of the miter gauge to the required angle and then make the pass as you would for a basic crosscut. If the blade doesn't follow the cut line you have marked, you'll know that an adjustment is needed.

Fig. 7-4. Moving the work against the saw blade creates a force that tends to pivot the work about the front edge of the miter gauge. It's also possible for the work to move along the front of the miter gauge. Holding the work firmly against the miter gauge is essential for accurate cutting.

Fig. 7-5. A miter-gauge extension, especially if faced with sandpaper, will contribute considerably to accurate mitering. You can use the kerf in the extension to position the workpiece for the cut. Remember that the cut must be on the waste side of the stock.

mark its width on the mating pieces. Draw a line from the mark to the outside corner of the second piece. This line is the angle at which you should make the cut on both components.

Table 7-1 lists possible causes of inaccuracies when sawing simple miters and suggests methods of correction.

HOMEMADE AIDS

One way to saw miters accurately without needing to worry about miter-gauge settings is to make the sliding table jig shown in Fig. 7-14 and detailed in Fig. 7-15. The jig is a plywood platform that rides on bars which slide in the table slots. It supports stationary guides that position the workpieces exactly right for the cut. Because the work is moved by the jig, problems that are characteristic when using a miter gauge are eliminated.

When you are making a sliding table, cut the platform first in a size that suits your equip-

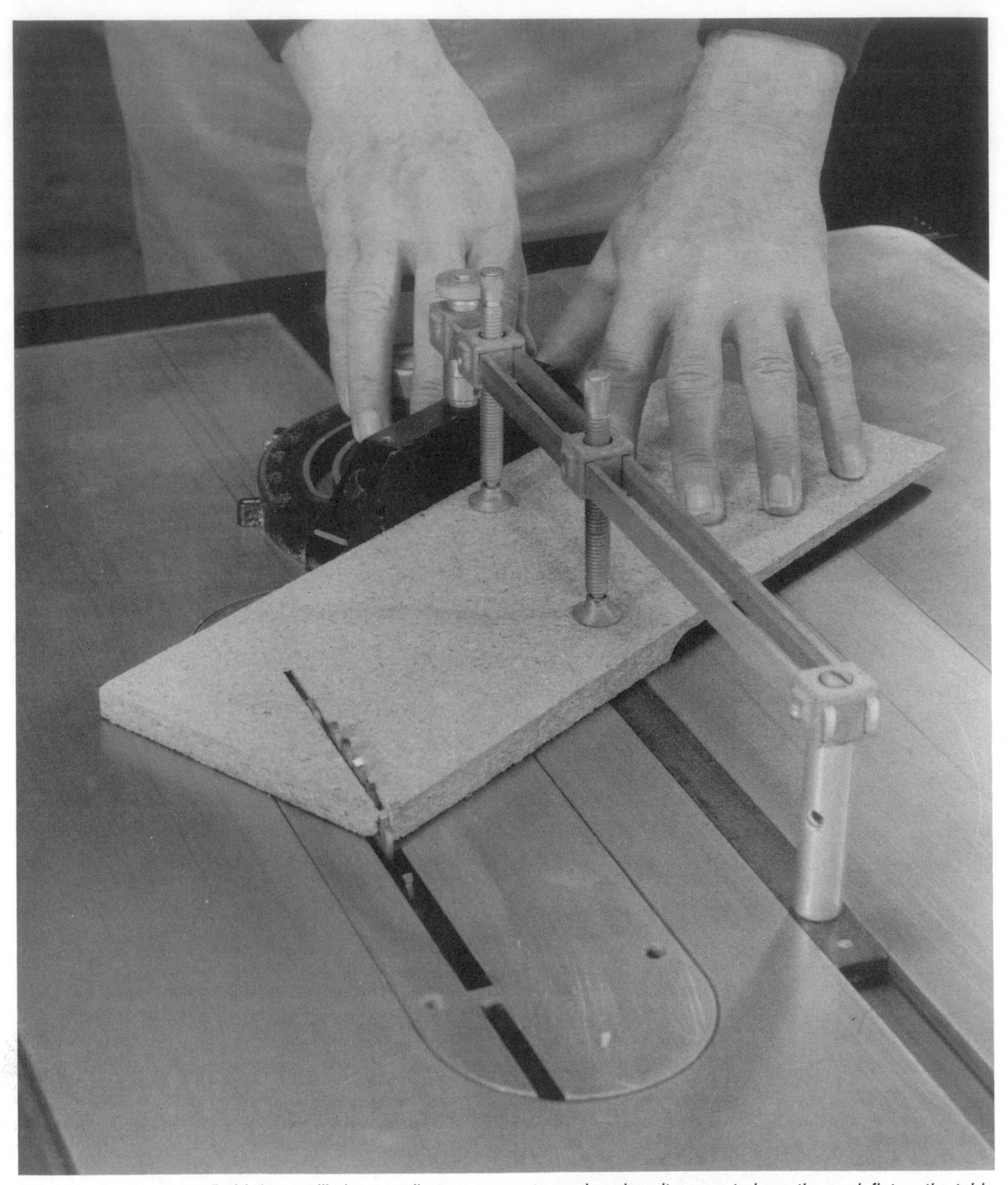

Fig. 7-6. A miter gauge/hold-down will also contribute to accurate sawing since it serves to keep the work flat on the table and snugly against the miter gauge throughout the pass.

Fig. 7-7. You can miter pieces of flat stock at each end without having to change the location or the pivot setting of the miter gauge. Make the first cut in routine fashion.

Fig. 7-8. Before you make the second cut, flip the stock over and turn it end for end. Be careful when you position the stock for the second cut or you might end up with miter cuts that parallel each other.

Fig. 7-9. Stock that is shaped, like molding, can't be treated like stock that is flat. Make the first cut with the miter gauge on one side of the blade.

Fig. 7-10. The cut at the opposite end of the molding must be made after the head of the miter gauge has been pivoted to the opposing 45-degree position and with the gauge used on the other side of the blade. This points up the necessity of being precise with adjustments.

Fig. 7-11. You can work with a miter gauge/stop rod or an extension that has a stop block to automatically control the length of mitered workpieces. This method is better than trying to control work length by sawing to marks on the work.

Fig. 7-12. Another way to work when mitered pieces must be of equal length is to cut the parts to length first. Then work as shown to saw the miters. The position of the stop block on the rip fence gauges the cutoff.

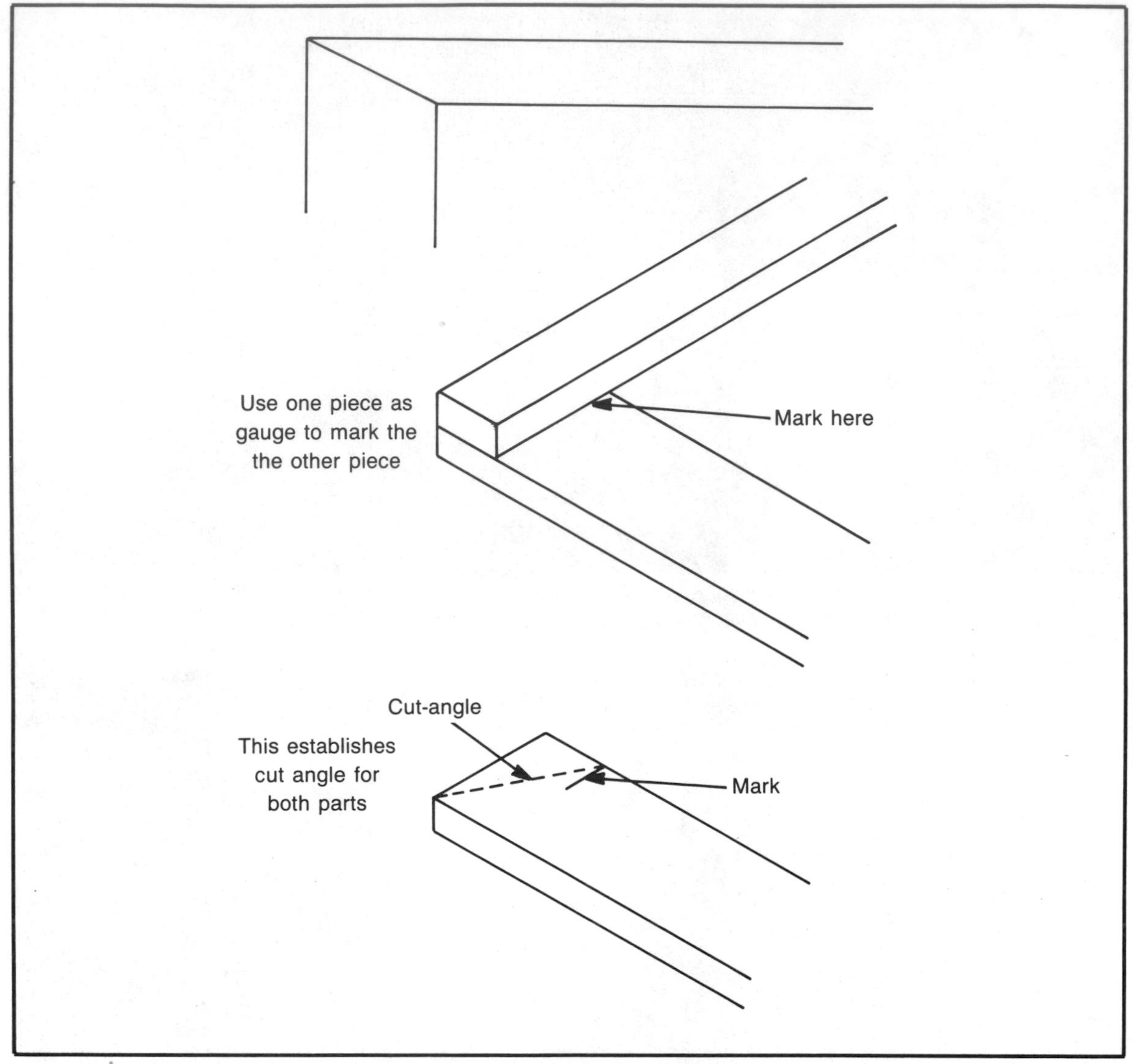

Fig. 7-13. This is one way to determine the angle for an accurate miter cut when the parts that must be joined are not equal in width.

ment. Next, cut the bars so they will be a nice sliding fit in the table slots. With the bars in place, lower the saw blade so it is under the table, and then place the platform so it is centered over the blade and lines up with the forward edge of the saw's table. Tack-nail through the platform at each end of each bar and then raise the blade while holding down the platform so the blade will cut through. Elongate the slot by moving the platform, and then shut off the machine and remove the platform. Turn the platform over and attach the bars permanently by driving small flathead screws up through the bottom surfaces.

Establish the position of the guides, or fences, by accurate layout using the kerf in the platform as a centerline. Be sure to check the relationship of the fences with a square before you attach them permanently (Fig. 7-16). If you

Table 7-1. Some Common Causes of Inaccurate Miters and Correction Methods.

PROBLEM	POSSIBLE CAUSE	CORRECTION
Inaccurate cuts	Misalignment	Check miter-gauge setting/Check auto-stop if used/Check test cut with protractor
	Poor work-handling	Keep work firmly against miter gauge/Use hold-down if available
	Creep	Hold work firmly throughout pass/Use hold-down
	Work support	Use miter gauge extension
Cut has slight bevel	Misalignment	Be sure angle between table and blade is 90°

Fig. 7-14. A sliding table jig for miters will eliminate the factors contributing to inaccuracies. The guides, or fences, will position the work correctly. Since work and jig move together, the chances of the work moving during the cut are nil.

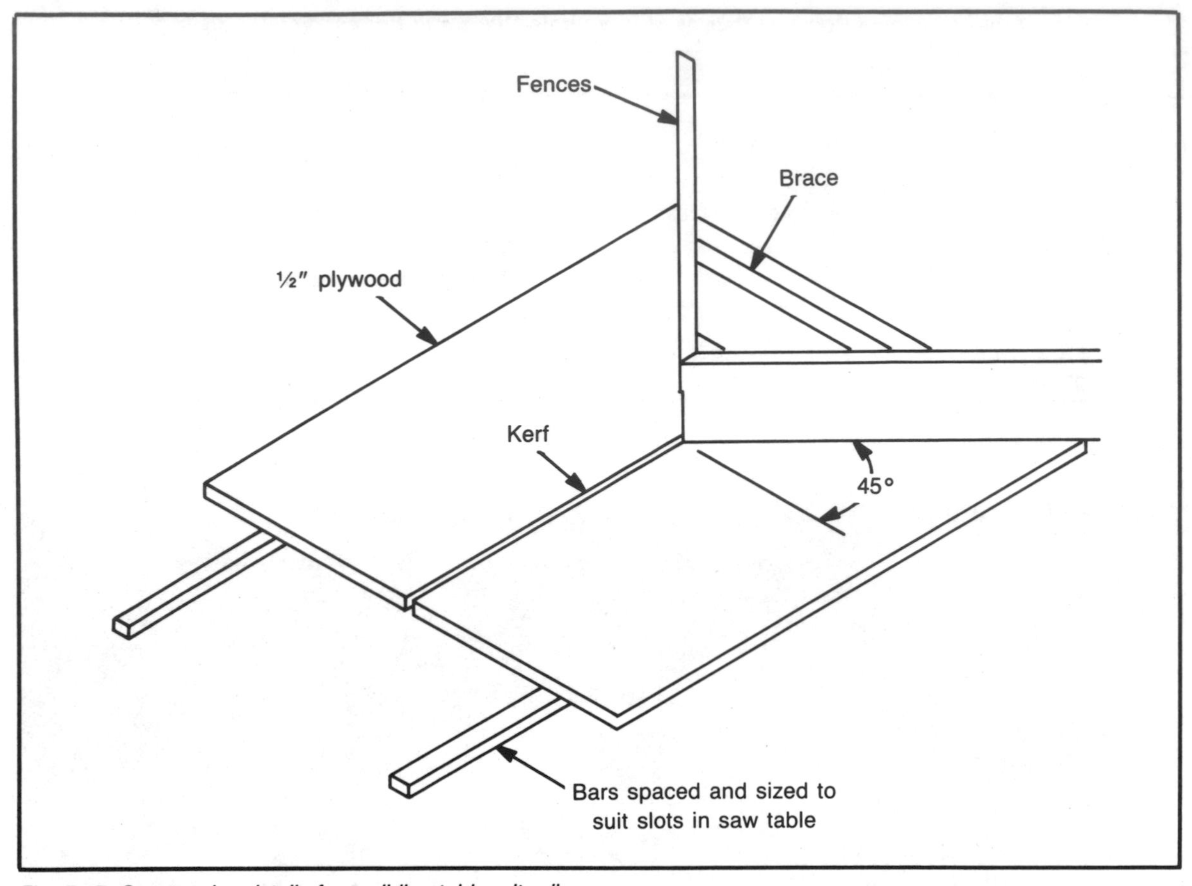

Fig. 7-15. Construction details for a sliding-table miter jig.

make the jig carefully and care for it as you would any precision tool, it will serve you faithfully for as long as you care to cut miters.

You will really appreciate the jig when you are mitering shaped pieces: moldings that normally require the miter gauge to be used on both sides of the blade and the head to be readjusted between moves. With the jig, accurate cutting of moldings is simply a matter of using one fence or the other, depending on which end of the molding you are sawing (Figs. 7-17 and 7-18).

It doesn't take much extra work to make an L-shaped stop block as an accessory for the jig. Then you can organize the jig to control the length of miter-cut pieces (Fig. 7-19). You also can establish the length of cutoffs by marking the platform with a gauge line. The system of making parallel miter cuts (Fig. 7-20), which results in pieces that can be assembled as chevrons (Fig. 7-21), is an example of ways to extend the utility of the project.

MAKE A MITER-GAUGE "CLAMP"

I find the homemade miter-gauge accessory displayed in Fig. 7-22 to be very useful for various mitering applications. For one thing, work gripped between the clamp and the head of the miter gauge has little opportunity to move during the cut, especially if the bearing edge of the clamp is faced with a strip of sandpaper (Fig. 7-23). Since the clamp can be inverted, it can be used for left- or right-hand miter cuts.

You can make the slot in the clamp by using the stopped-cuts technique explained in Chapter 6. You will need to drill and tap the miter-gauge bar for the bolts used to secure the clamp's position (Fig. 9-24).

CROSS MITERS

You can make a *cross miter* by tilting the saw blade to the required angle and then proceeding as you would for a simple crosscut (Fig. 7-25). Testing the blade tilt for accuracy and keeping the work firmly in position throughout the pass are critical steps. Since the blade is at an angle, it will actually be sawing more wood than it does when at 90 degrees, so slow up on the feed just a bit, especially on thick material.

Cuts of this type are usually required at both ends of the material, and often on components of equal length. Gauging the length of parts mechanically is better than sawing to a line marked on the work. You can be sure of accuracy by using a miter-gauge extension with a stop block or a miter-gauge stop rod (Fig. 7-26).

A cross miter is a cut through the full thickness of the stock. A cross *chamfer*, which is always done along edges, cuts away only part of the stock's thickness. Chamfers are usually made to provide a decorative detail. The sawing operation (Fig. 7-27) is the same as for a cross miter.

Fig. 7-16. The jig, of course, must be made accurately. Locate the position of the fences by making a layout using the kerf as a centerline. Check the relationship of the fences with a square before you assemble the parts.

Fig. 7-17. The sliding table makes it easy to cut miters on shaped pieces which, unlike flat stock, can't be flipped over for opposing cuts. Make the first cut with the work against one fence.

Fig. 7-18. Cut the miter at the opposite end of the part by positioning the material against the second fence.

Fig. 7-19. Make an L-shaped stop block as an accessory for the jig so you can automatically gauge the length of similar pieces. The block is L-shaped so a small C-clamp can be used to secure it.

Fig. 7-20. You can also gauge work lengths by working to a mark on the jig's base. Making parallel miter cuts on short pieces is a sample application.

Fig. 7-21. The small pieces with matching end cuts can be assembled as chevrons for decorative purposes.

Fig. 7-22. Secure the homemade miter gauge clamp to the gauge's bar with short bolts. Use large washers under the head of the fasteners.

Fig. 7-23. The purpose of the clamp is to hold work securely between itself and the head of the miter gauge. Face the bearing edge of the clamp with a slim strip of sandpaper.

A CROSS-MITER JIG

I find the fence-straddling jig displayed in Fig. 7-28 very helpful when I cross-miter many similar pieces. You can obtain accurate 45-degree cuts, with the blade in normal 90-degree position, by clamping the workpiece in place and moving the jig past the saw blade (Fig. 7-30).

Another practical application for the project is sawing grooves for slim strips of material called *splines*, which are often used to reinforce cross-miter joints (Fig. 7-29). The groove for the spline must be perpendicular to the plane of the cross-miter cut. This position is easy to accomplish if you set up the work and jig as shown in Fig. 7-31. You can make the jig by following the plans in Fig. 7-32, but don't be casual when constructing it. It will do accurate work only if it is made accurately.

Another way to cut a spline groove is demonstrated in Fig. 7-33. The blade is tilted to 45 degrees so the cut will be square to the mitered surface. Chapter 10 contains more detailed information on what splines are and how they can be used.

BEVELS

The major difference between a routine rip cut and a *bevel* cut is the latter is sawed with the blade tilted to a specific angle (Fig. 7-34). The side of the blade on which the fence should be used will depend on the direction the blade is designed to tilt. It is always best to work so that the workpiece is on the open-angle side of the blade. Avoid the possibility of kickbacks by not making setups that will cause slim cutoffs to be trapped between the blade and fence.

Obey all the safety rules that apply to routine ripping operations. Use push sticks when necessary, and keep the guard in place. Keep the work snug against the fence throughout the pass by using a springstick as shown in Fig. 7-35. Clamp the springstick so that its fingers will bear against the work in *front* of the blade.

You can make chamfers on long edges the

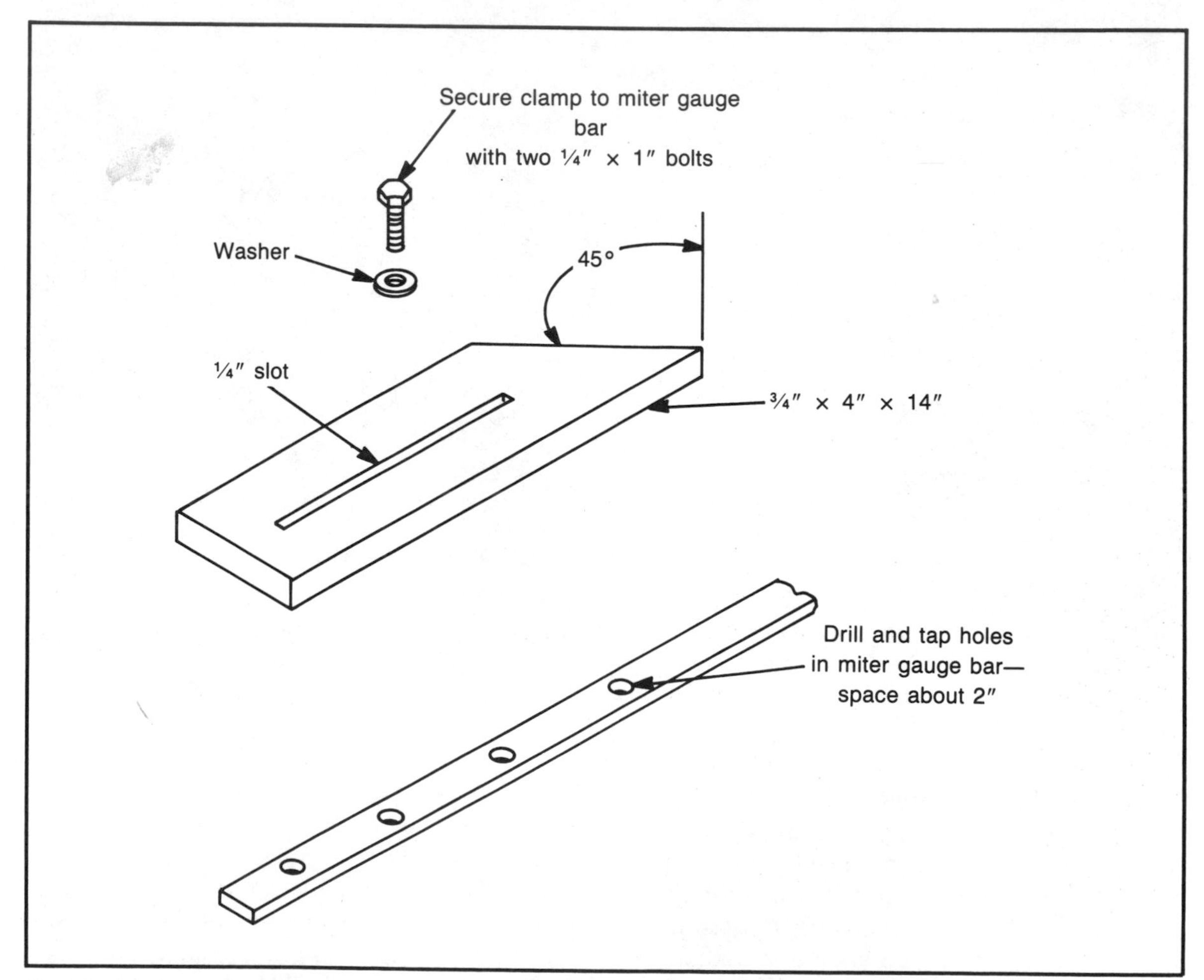

Fig. 7-24. The bar of the miter gauge will need to be drilled and tapped for the bolts that secure the clamp. Making several clamps with different front angles allows using them for miters at angles other than 45 degrees.

same way. The only difference is that the cut removes only part of the work's edge (Fig. 7-36). When a chamfer or a bevel is required on all four edges of a workpiece or on two adjacent edges, make the cross-grain cuts first, since feathering is more likely to occur at the end of these cuts. The final cuts, made with the grain, will eliminate the imperfections.

Problems that cause inaccuracies when bevel cutting and ways to to correct them are listed in Table 7-2.

SEGMENT CUTTING

The technique of bevel-cutting a certain number of pieces at a specific angle can be used to construct hollow columns, lamp bases, and other projects like the barrel furniture displayed in Fig. 7-37. The concept is based on simple arithmetic. When the included angles of the number of pieces you bevel-cut total 360 degrees, the pieces can be joined edge to edge to form a circle (Fig. 7-38). This idea is useful for making parts that will turn a corner and for half-round

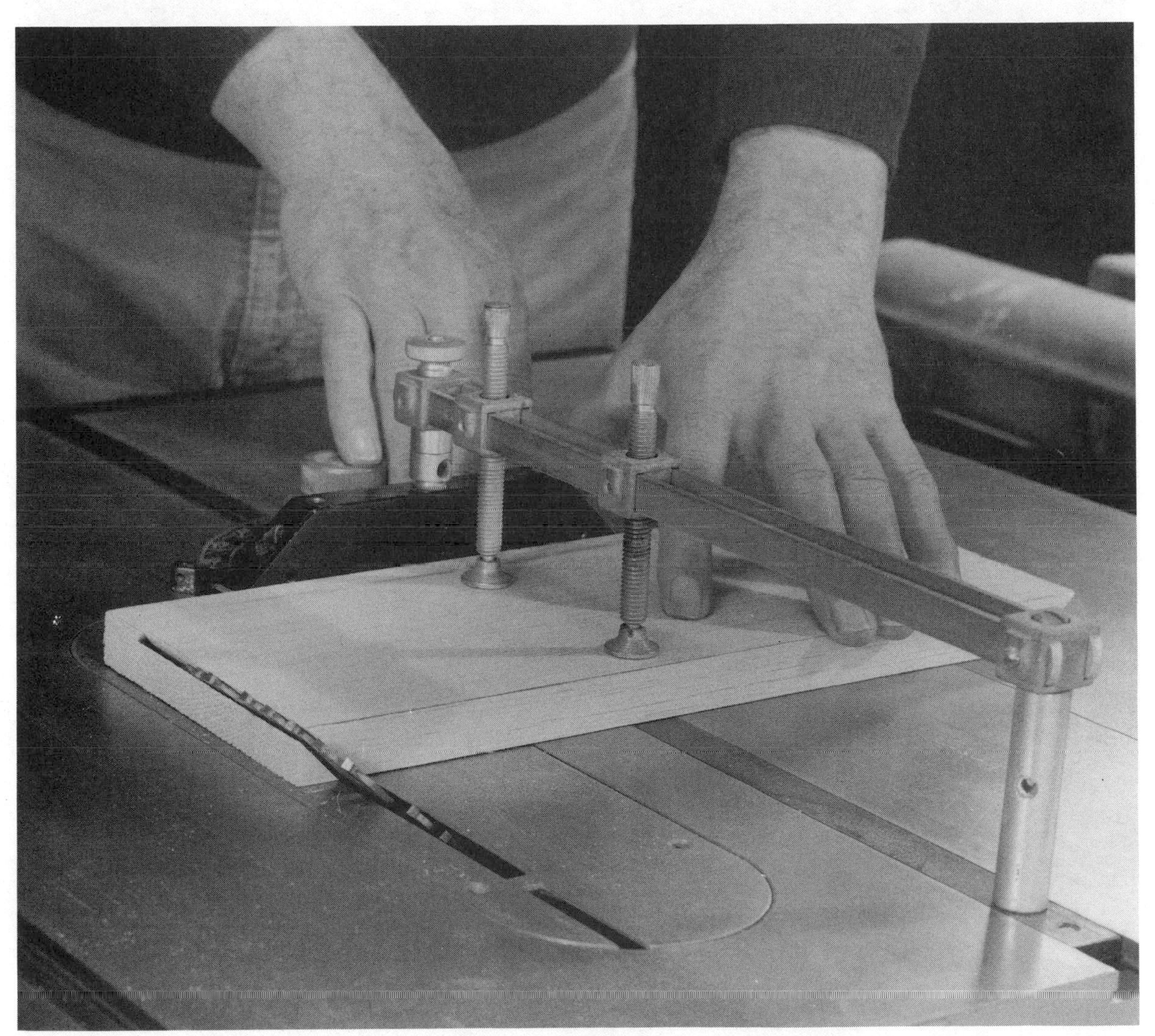

Fig. 7-25. A cross miter is cut like a simple crosscut but with the blade tilted to the required angle. Results are best when you use a smooth cutting blade. Slow up at the end of the cut to minimize feathering.

Fig. 7-26. You can gauge the length of cross-mitered components by using a miter gauge/stop rod or an extension with a clamped-on stop block.

Fig. 7-27. A chamfer is done like a cross miter, except that the cut removes only part of the work's edge. Chamfers are usually made to provide decorative details.

Fig. 7-28. You must make the fence-straddling cross miter jig very carefully, but it will serve forever to help you do accurate cross miter work without having to tilt the blade.

shapes. You can even change the direction of a curve by how you assemble the pieces (Fig. 7-39).

It's very important to remember that the *cut* angle that is required is one-half of the *joint* angle. This rule applies whether you are constructing a cylinder-type project or a flat assembly that can, for example, be used as a tray or a tabletop (Figs. 7-40 and 7-41).

Following is the formula to use when bevel-cutting parts that can be assembled as a circle. Divide 360 degrees by the number of pieces you plan to use to get the total angle of each part. Divide the total angle in half to find the bevel angle that is required on each edge of the parts. The greater the number of pieces, the truer the circle they will form.

Accuracy is very critical. A tiny part of a de-

Fig. 7-29. Use a clamp to secure the work's position against the guide strip and then move the jig forward for the cut. The jig must travel smoothly and without wobble.

gree doesn't seem like much to worry about, but multiply the error by 10, 15, or 20. Can you imagine what you will face when you are ready to install the final part?

MAKING COMPOUND-ANGLE CUTS

Many projects require compound-angle cuts. Picture frames that slop toward or away from a wall, a simple box with sloping sides, truncated pyramid forms, and peaked figures all require that this particular sawing technique be used (Figs. 7-42 and 7-43). To make the cuts so the parts will slope as you want them to, and so adjacent components will make a precise turn—90 degrees if, as in a picture frame, only four parts are involved—is a test of how carefully you can work.

The simple miter requires a miter-gauge setting. A bevel or cross miter is done by tilting the blade. The compound angle requires that both settings, each at a specific angle, be used at the same time. This type of cutting often requires that you use the miter gauge alternately

Fig. 7-30. Splines are used to reinforce joints like the cross miter. Using material whose thickness matches the width of the saw kerf simplifies the job.

Fig. 7-31. You can use the cross miter jig to cut the groove for the spline. Blade projection must be one-half the total width of the groove when parts are joined. Make the cut so it favors what will be the inside corner of the joint.

in the table slots, and after changing the gauge's setting. It's obvious that there is much opportunity here for human error. Take heart, though. These aspects of the operation are not intended to discourage, but to impress you with the importance of being precise. Also, there are ways to work and jigs to make that will help you achieve professional results with minimum fuss.

Normally, a compound-angle cut is made by adjusting the head of the miter gauge and the tilt of the saw blade to the specific settings that produce a particular work slope (Fig. 7-44). Table 7-3 lists what these setting must be for various slope angle on four-, six-, and eight-sided projects. Some of the settings are to ¼, ½, or ¾ degree, which makes it obvious that you can't approach this aspect of table saw work too casually. Use the table when it is necessary, but don't overlook the fact that the slope angle on many projects, so far as appearance is

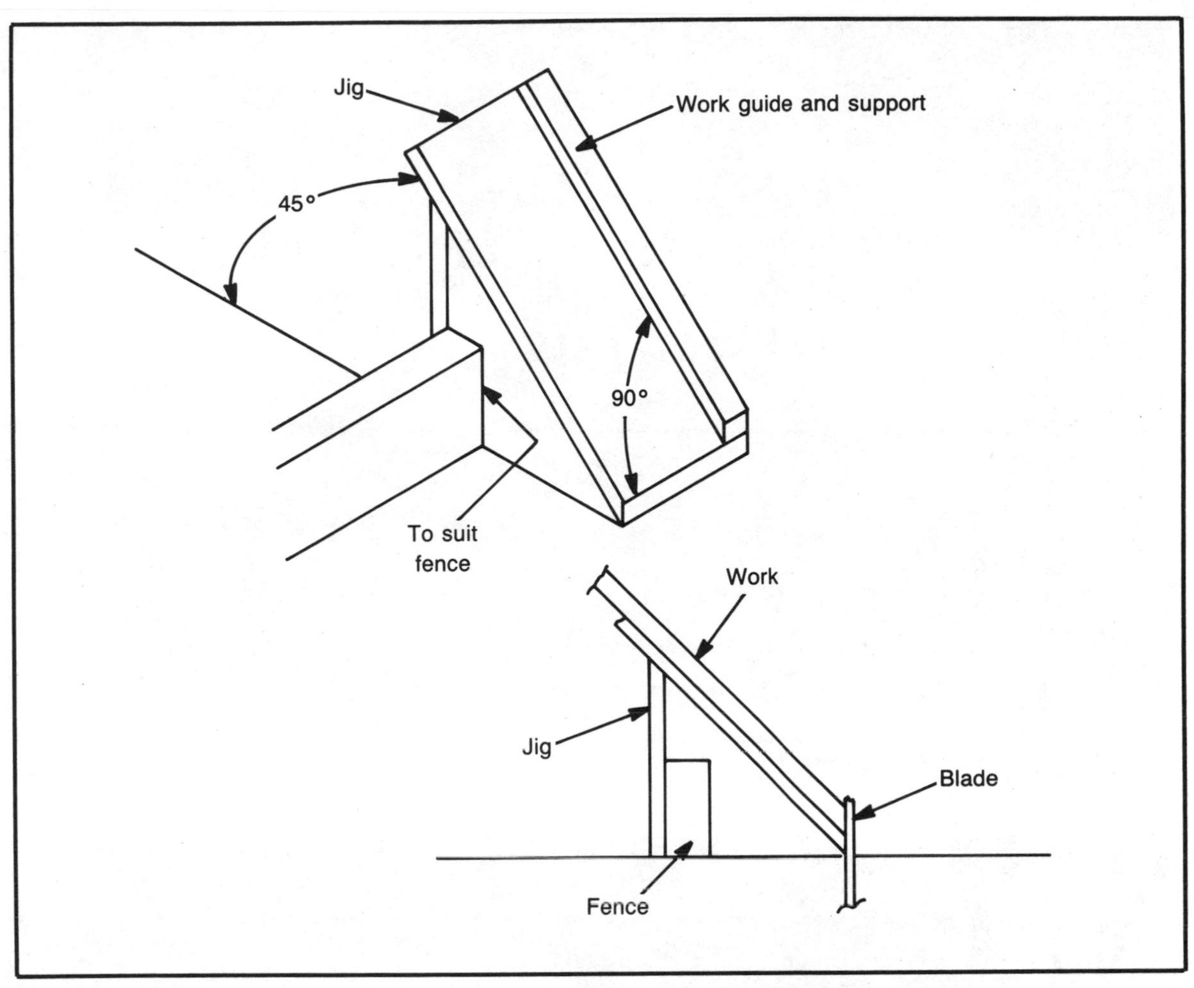

Fig. 7-32. The method for making the cross miter jig. Careful cutting and assembly are required.

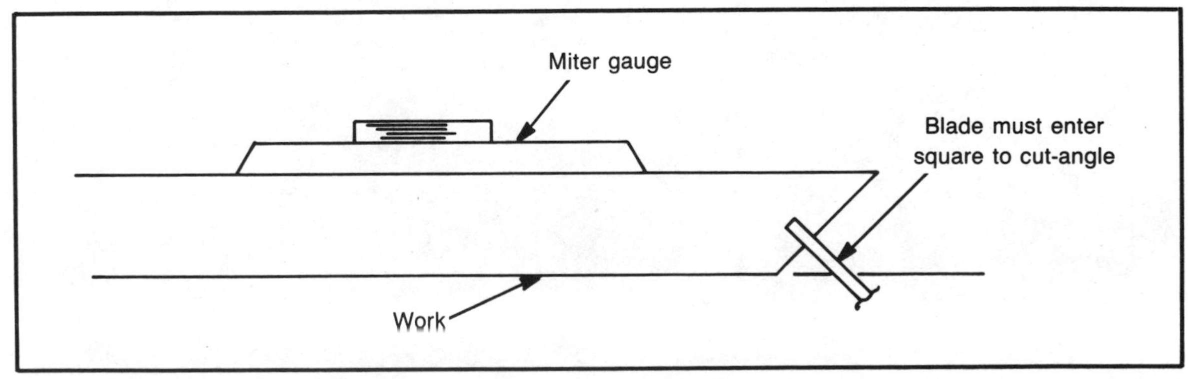

Fig. 7-33. You also can form grooves for splines in cross miters by using the miter gauge to advance the work, but care is required so the cuts will match. Clamp a stop block on the open angle side of the blade to help position the workpieces.

Fig. 7-34. To form a bevel, work as you would for routine ripping but with the blade tilted at the necessary angle. Be sure the workpiece bears against the fence throughout the pass.

Fig. 7-35. Use a springstick to help keep the work in correct position as you cut. The fingers should bear against the work in front of the saw blade. A pusher/hold-down also helps to achieve accurate results.

Fig. 7-36. To make a chamfer on long edges, work as you would for ripping. Always make the setup so cutoffs can't be trapped and thrown back. Use a smooth cutting blade for best results.

Table 7-2. Troubleshooting Chart for Cutting Bevels.

PROBLEM	POSSIBLE CAUSE	CORRECTION
Inaccurate cuts	Misalignment	Check auto-stop if used
	Wrong setting	Recheck setting of saw blade/Make test cut and check with protractor
	Poor work-handling	Keep work flat on table and snugly against fence throughout pass
	Cut not uniform	Do not allow work to move away from fence/Use spring-stick or other aid/Check fence alignment

Fig. 7-37. A specific formula is followed for bevel-cutting segments that can be assembled for projects like this barrel furniture.

concerned, is not always critical. It's a persnickety critic who would judge a project like a shadowbox picture frame or a patio planter box on the basis of whether the slope of the components should have been a few degrees more or less.

If you make an arrangement that tilts the workpiece at a visually pleasing slope angle and then make the cut with the blade at 90 degrees, as you would for a simple miter, the result will be a compound-angle cut. This result occurs regardless of the slope you have established for the work. That's where a special jig enters the picture. The platform, the bars, the triangular part, and the handle are permanently assembled. The two forward wood strips are height blocks which are tack-nailed in positions that suit the width of the work and the slope angle you wish to produce. Workpieces are held in a tilted position during the cut (Fig. 7-45). The tilt is the slope angle; the cut is a compound angle even though the blade is in its normal 90-degree position.

Another advantage of the jig is that it can be used to make cuts on moldings that normally require the miter gauge to be shifted from one side of the blade to the other (Fig. 7-46). Opposite cuts are made by alternating the position of the work from one bearing edge of the triangular guide to the other. There is no reason why the same jig can't be used for simple mitering. Just don't mount the height blocks (Fig. 7-47). Construction details for the jig are shown in Fig. 7-48.

Naturally, there are some limitations that are imposed by the size of the workpiece and the capacities of the saw. Work width is a consideration, but it becomes less of a factor as the work slope decreases. Anyway, Table 7-3 is always available when you can't or choose not to work with the jig.

A MITER-GAUGE MITER BOX

The project shown in Figs. 7-49 and 7-50 is attached to the miter gauge like an extension, but

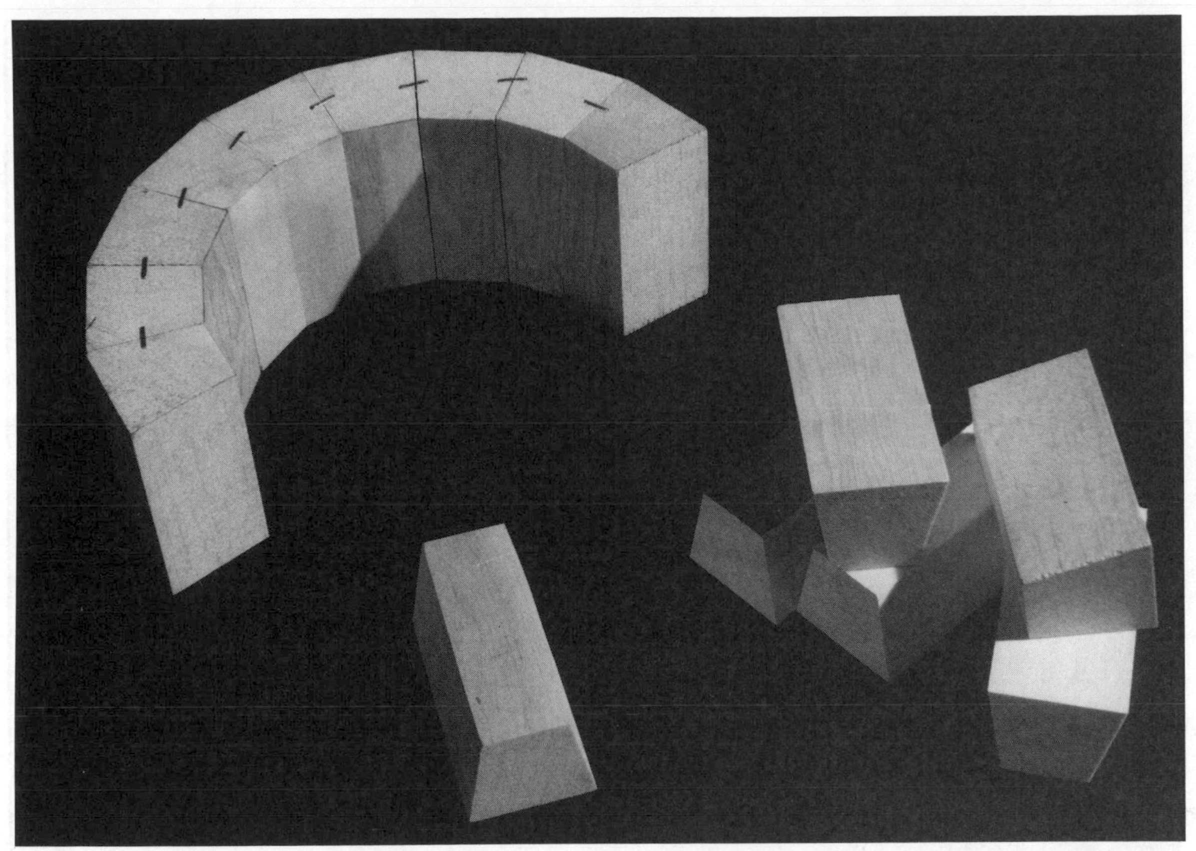

Fig. 7-38. When the bevel angle is correct and the work is done accurately, you can join beveled segments to form a circle.

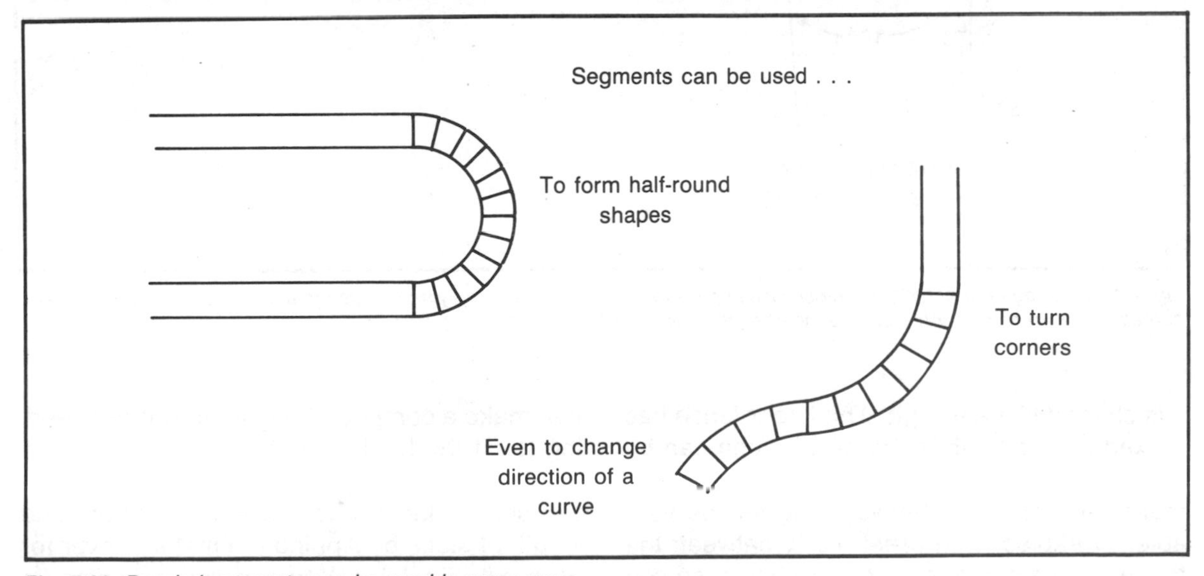

Fig. 7-39. Beveled segments can be used in many ways.

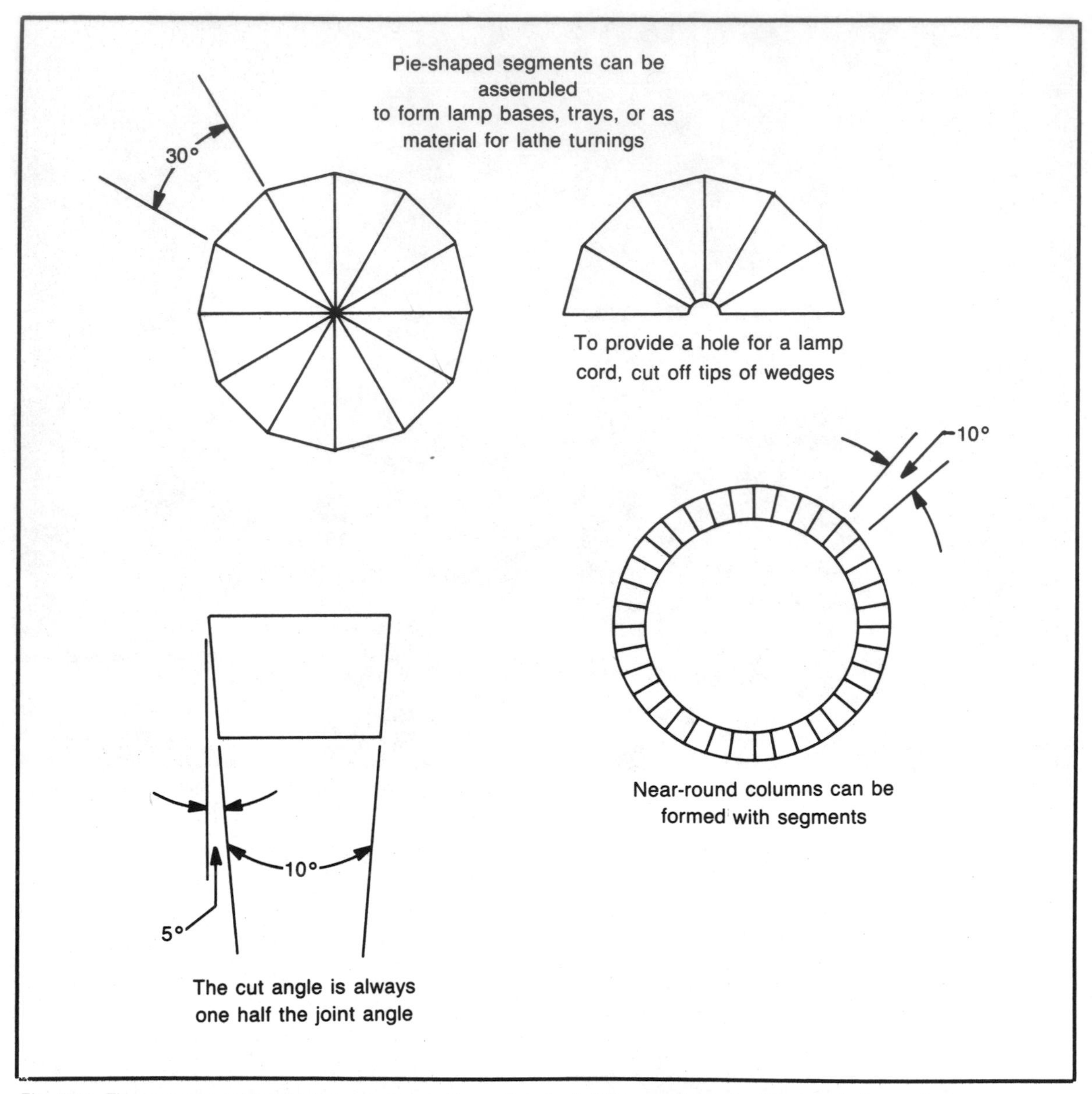

Fig. 7-40. The segment-cutting technique also applies to flat stock. When applying the idea to a hollow cylinder, the project comes closer to a true circle as you increase the number of segments.

it is shaped like a trough. The larger fence has a fixed position, while the smaller one can be tack-nailed anywhere on the tool's base, which makes the distance between the fences variable. Workpieces can rest nicely between the fences at a slope angle of your choice, so you can make a compound-angle cut without needing to tilt the blade (Fig 7-51).

When you are working with flat stock, you can cut similar pieces consecutively from one length of stock by flipping the material over for each cut (Fig 7-52). You can use the same jig

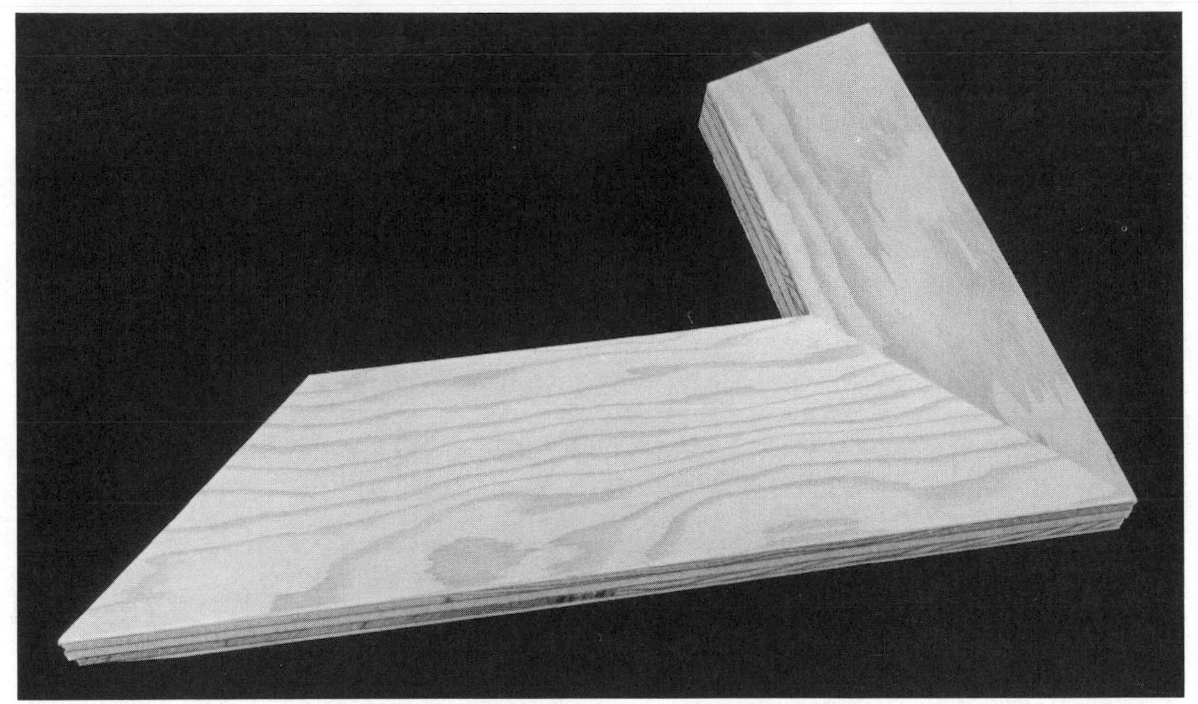

Fig. 7-41. A compound-angle joint is often found on picture frames. The frame will slope toward or away from the wall, depending on how the joint angles are sawed.

for sawing simple miters. Make the pass with the workpiece placed flat on the jig's base (Fig. 7-53).

You can also use the homemade miter-gauge clamp shown previously in this chapter to hold work at a slope angle (Fig. 7-54). You can brace the workpiece against the head of the miter gauge or add a miter-gauge extension to provide more work support.

Be very persnickety when you are sawing compound miters, whether you are using a jig or working to miter-gauge and saw-blade settings taken from Table 7-3. Use a blade that will produce smooth cuts; one with set teeth won't produce optimum results. Keep a firm grip on the work and make the pass slowly.

MAKING TAPER CUTS

Taper cuts are saw cuts that are made at an oblique angle across the work. Such cuts can be done with a miter gauge if the width of the work allows safe and convenient use of the gauge, but since tapers are often quite long, it is usually best to view the operation somewhat like a rip. A normal rip cut produces workpieces with parallel sides, but since the taper is an angular cut, an arrangement must be made to hold the work in correct position while it is sawed. The solution is to use a taper jig. This accessory provides a straight side, which can ride against the rip fence, and an adjustable side, which can be set for the degree of taper that is needed.

You can buy a taper jig (Fig. 7-55), or you can make your own (Fig. 7-56). Plans for a homemade unit are offered in Fig. 7-57, as are two methods of attaching the hinge to allow for taper adjustment. Each will work efficiently, but the surface-mounted version is easier to do. If you use this design, have two sides of the jig clamped together as you attach the hinge. The pivoting, slotted crosspiece that is used to secure settings can be made of sheet metal or a piece of ¼-inch-thick hardwood.

Figure 7-58 suggests how you can modify the jig so it can straddle the rip fence. The two additional pieces that are needed, assembled as shown to the fence leg of the jig, form a *U* which cradles the fence. The added feature ensures that the jig will not move away from the rip fence when you are cutting. The jig's stop block (Fig. 7-59) is what makes the workpiece move as a unit with the jig.

After you have assembled the jig, hold the two legs in a closed position and make a mark across them 12 inches from the hinged end. To set the jig for a particular cut, measure between the marks on the legs (Fig. 7-60). If, for example, you lock the jig so the distance between the marks is 1 inch, you will make a cut that tapers 1 inch per foot.

After you have adjusted the jig, place its straight leg against the rip fence and position the workpiece so it is flush against the opposite leg and so its back edge is abutted to the stop block. Then advance the jig and work past the saw blade. If the same taper is required on both edges of the stock, open the jig to twice the original setting before placing the work for the second cut.

You also can use the taper jig with the saw

Fig. 7-42. You can use segment cutting to make components that are too large to cut from a board, or to create inlaid effects by cutting the segments from contrasting woods.

blade tilted (Fig. 7-61). This is a good way to make compound-angle cuts when the size of the workpiece makes it inconvenient to use a standard procedure. In a sense, the jig substitutes for the miter gauge.

OTHER METHODS FOR TAPER CUTS

You also can make taper cuts by using a notched jig like the one shown in Fig. 7-62. The jig is a wide piece of wood with parallel sides and a recess on the saw-blade side that matches the size and shape of either the part you wish to remove from the workpiece or the part you need. Set the rip fence so the distance from it to the *inside* of the blade equals the width of the notched jig. Then move the jig and work simultaneously past the saw blade.

It is wise to make a special notched jig when the cut that is required might be too extreme for the safe use of a taper jig or other normal methods. Notched jigs are useful in many other phases of table-saw work. More of them will be demonstrated in Chapter 11.

Figure 7-63 shows yet another method you can use to saw tapers or to form wedge-shaped pieces. Tack-nail the guide, which has parallel sides, to the surface of the workpiece so its outboard edge is on the cut line you want. Position the rip fence so the distance from it to the saw blade equals the width of the guide.

FORMING A RAISED PANEL

The term *raised panel* describes an insert piece

Fig. 7-43. The joints for projects like this require compound-angle cuts.

Fig. 7-44. A compound angle-cut requires a blade tilt and miter-gauge settings. The settings must be precise when a particular slope on the project is desired.

Table 7-3. Table Saw Settings for Compound-Angle Joints.

WORK SLOPE	BUTT JOINT (4 Sides)		MITER JOINT 4 Sides		MITER JOINT 6 Sides		MITER JOINT 8 Sides	
	A	B	A	B	A	B	A	B
5°	½	85	44¾	85	29¾	87½	22¼	88
10°	1½	80¼	44¼	80¼	29½	84½	22	86
15°	3¾	75½	43¼	75½	29	81¾	21½	84
20°	6¼	71¼	41¾	71¼	28¼	79	21	82
25°	10	67	40	67	27¼	76½	20¼	80
30°	14½	63½	37¾	63½	26	74	19½	78¼
35°	19½	60¼	35½	60¼	24½	71¾	18¼	76¾
40°	24½	57¼	32½	57¼	22¾	69¾	17	75
45°	30	54¾	30	54¾	21	67¾	15¾	73¾
50°	36	52½	27	52½	19	66¼	14½	72½
55°	42	50¾	24	50¾	16¾	64¾	12½	71¼
60°	48	49	21	49	14½	63½	11	70¼

"A" = saw-blade tilt

Fig. 7-45. Position the height blocks, by tack-nailing, so the workpiece is tilted to provide a slope angle that you find visually pleasing. The cut is a compound angle even though the blade is at 90 degrees.

Fig. 7-46. The jig can be used for compound cuts on shaped pieces, like moldings. To make opposing cuts, use both sides of the jig. When the height blocks have been set for a particular work width, you can use them for cuts on narrower work, without moving them, by using a strip of wood between the work and the triangular guide.

Fig. 7-47. The sliding-table jig can be used without the height blocks for simple 45-degree miters. Always stop moving the table forward as soon as the workpiece has been cut through.

that is set in a frame. It might be the panel in a door or the side of a chest. The required cuts are done on all four edges of the workpiece and result in a decorative, three-dimensional effect, as well as a *tongue* that fits in grooves formed in the edges of the frame (Fig. 7-64). This type of work is usually done with a single cutter on a shaper or with a portable router. Because of the cutter's profile, the shaped edge looks like molding (Fig. 7-65). The same work can be done on a table saw and with a conventional saw blade, but the planes of the raised edge will all be flat (Fig 7-66).

Figure 7-67 shows the procedure to form a raised panel. You must be extremely careful when you are making the setups. Cuts must mate correctly for professional results. A good, sharp saw blade, one without set teeth, is also essential.

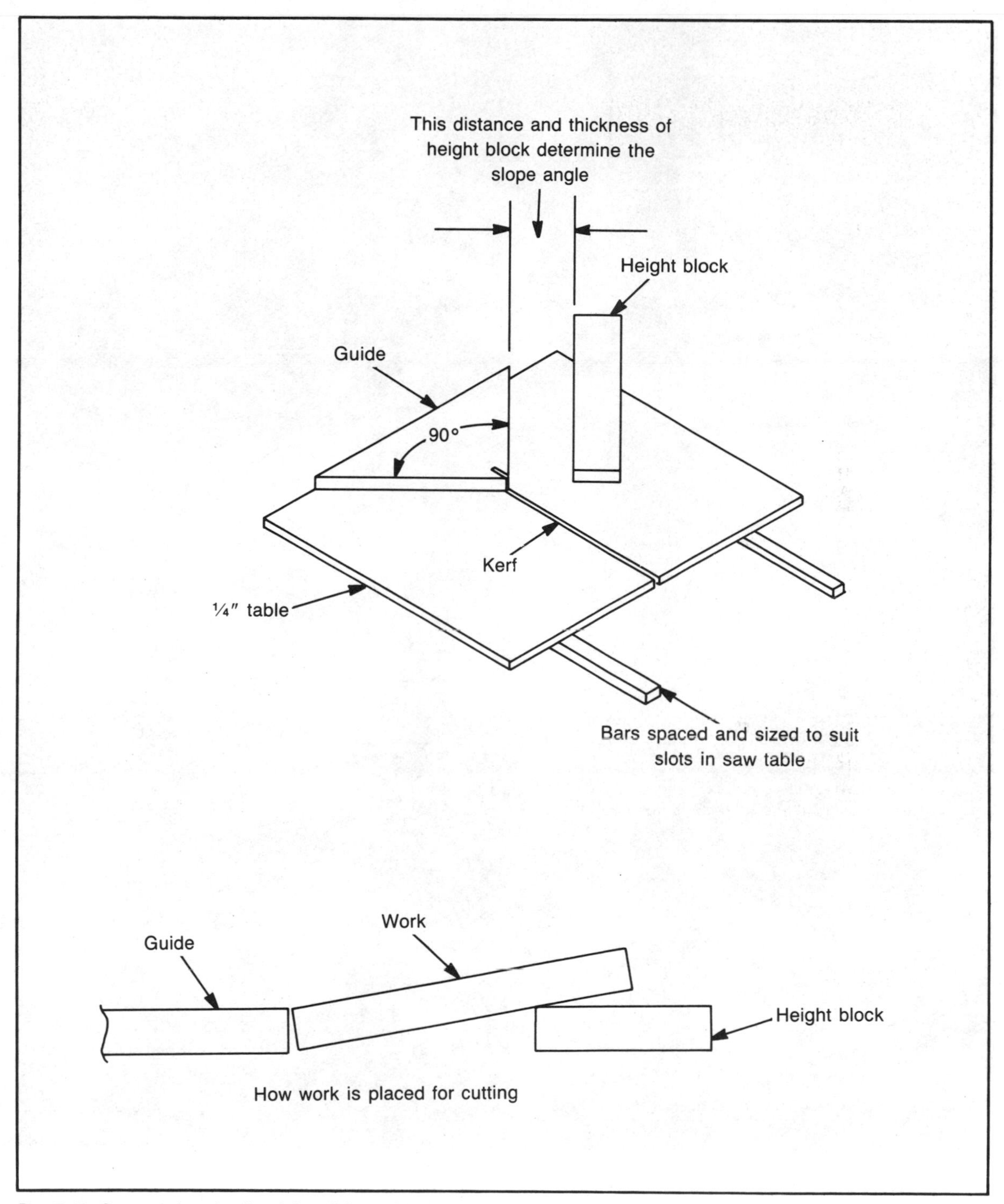

Fig. 7-48. Construction details of the sliding-table jig.

Fig. 7-49. The miter-gauge miter box is designed like a trough and is secured to the miter gauge like an extension.

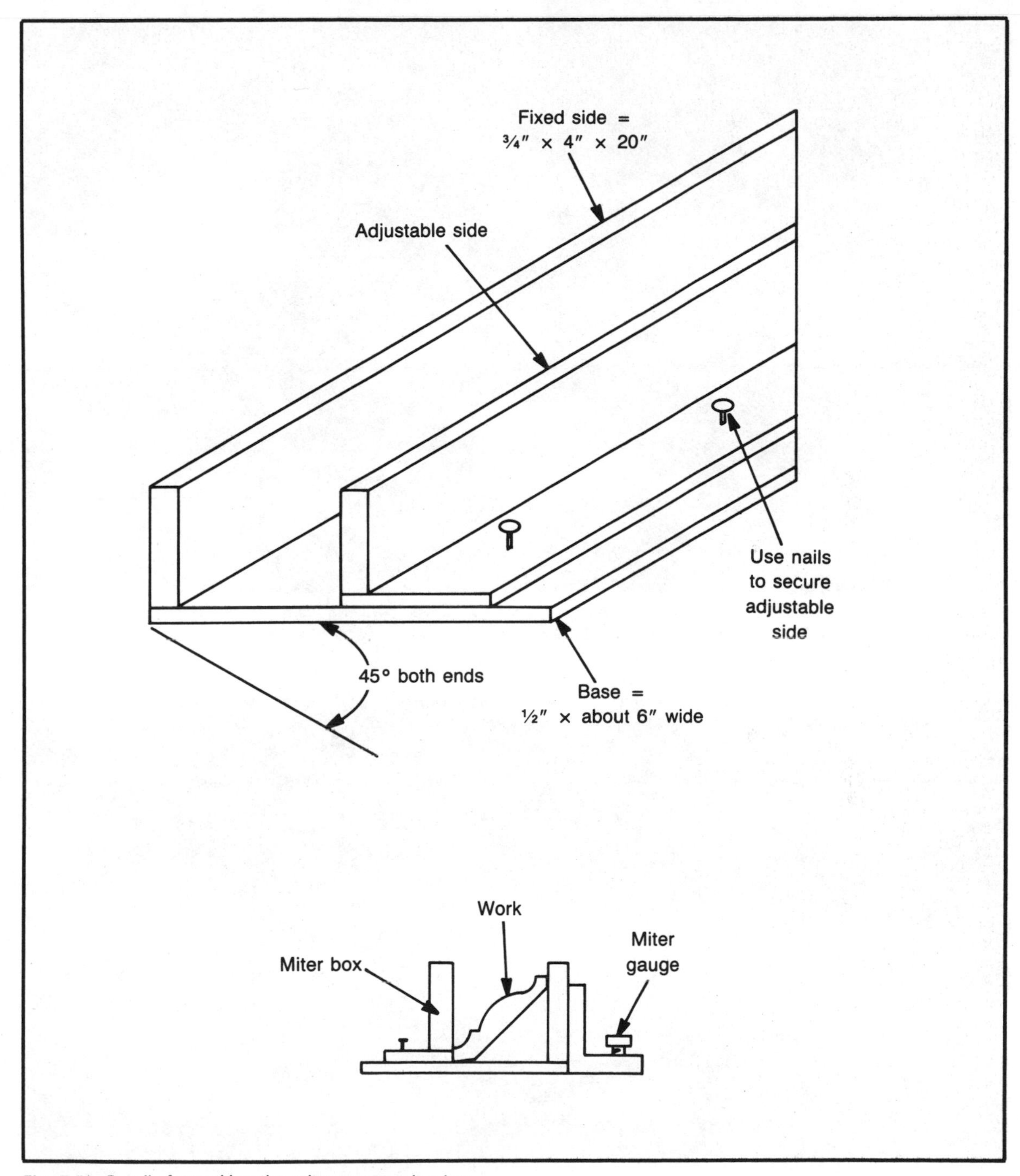

Fig. 7-50. Details for making the miter-gauge miter box.

Fig. 7-51. Position the adjustable fence of the miter box so the work placed between the fences will be tilted to the desired slope. The blade is at 90 degrees, but the cut will be a compound angle.

Fig. 7-52. You can cut similar pieces consecutively from a single length of flat stock by flipping the material over for each cut.

Fig. 7-53. You also can use the miter-gauge miter box for making simple miters. The work is held flat on the jig's base instead of being secured in a tilted position.

Fig. 7-54. The miter gauge clamp demonstrated earlier in this chapter also can be used to hold work in a tilted position so a pass made with the blade at 90 degrees will result in a compound-angle cut.

Fig. 7-55. Taper cuts are made like rips but with a special jig that holds the work at an oblique angle. Commercial jigs, like the one being used, are available.

Fig. 7-56. Hold workpieces against the jig's adjustable leg and brace them against a stop block. Then move the jig and work smoothly past the blade. You can make this jig in your own shop.

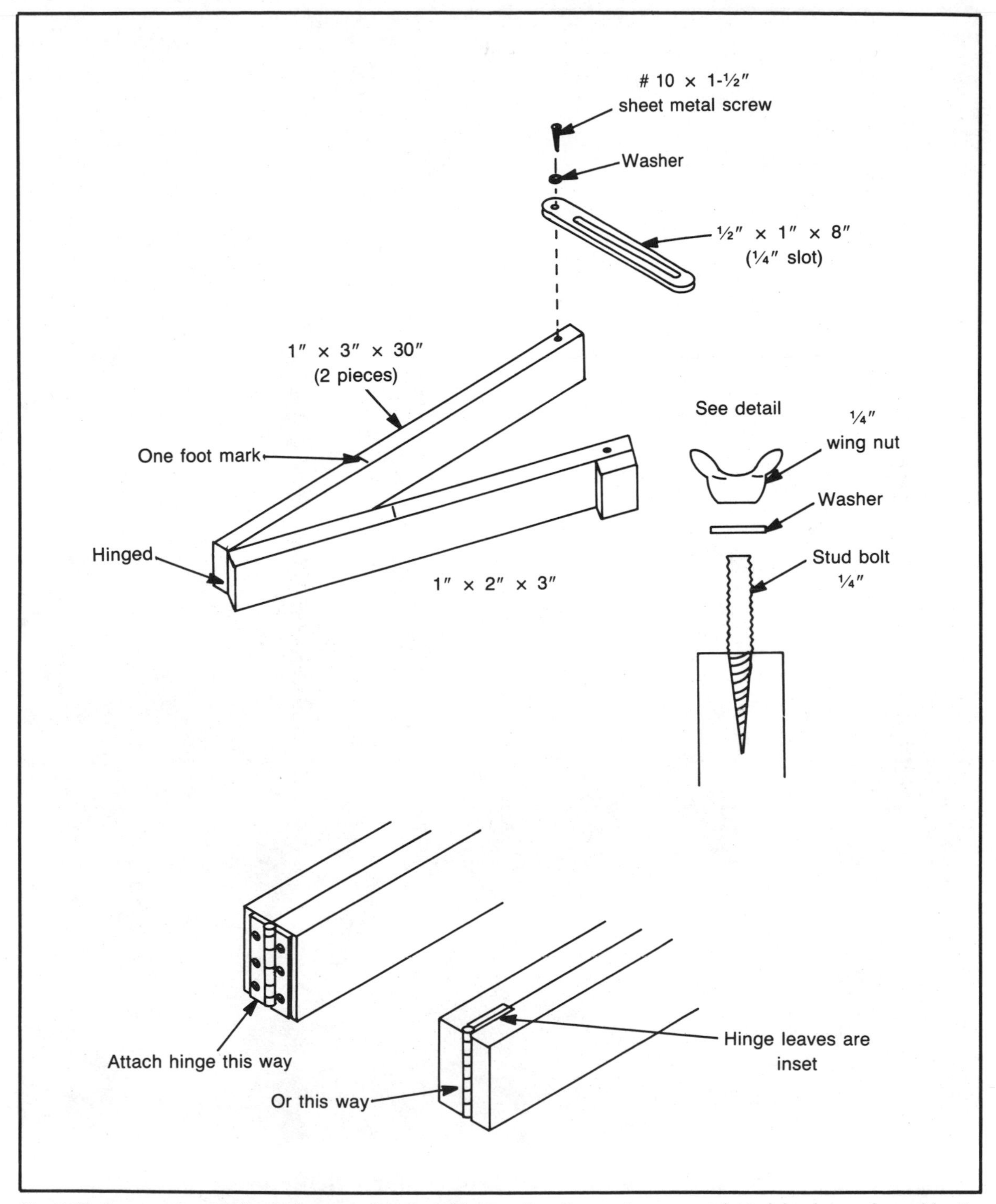

Fig. 7-57. Construction details of a homemade adjustable taper jig. Be sure to use sound, straight pieces of wood for the legs. The stud bolt is available in hardware or lumber stores.

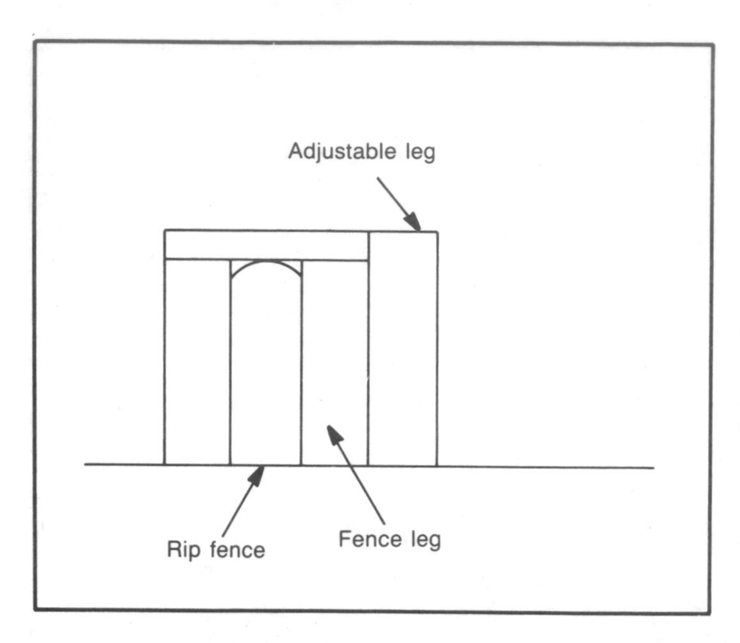

Fig. 7-58. You can modify the taper jig by adding components that allow it to straddle the rip fence. In this way, you can be sure the jig will not move away from the fence while you are sawing.

Fig. 7-59. The stop block is just a block of wood that is permanently attached to the back end of the jig's adjustable leg. The slotted crosspiece that is used to secure settings can be metal or hardwood.

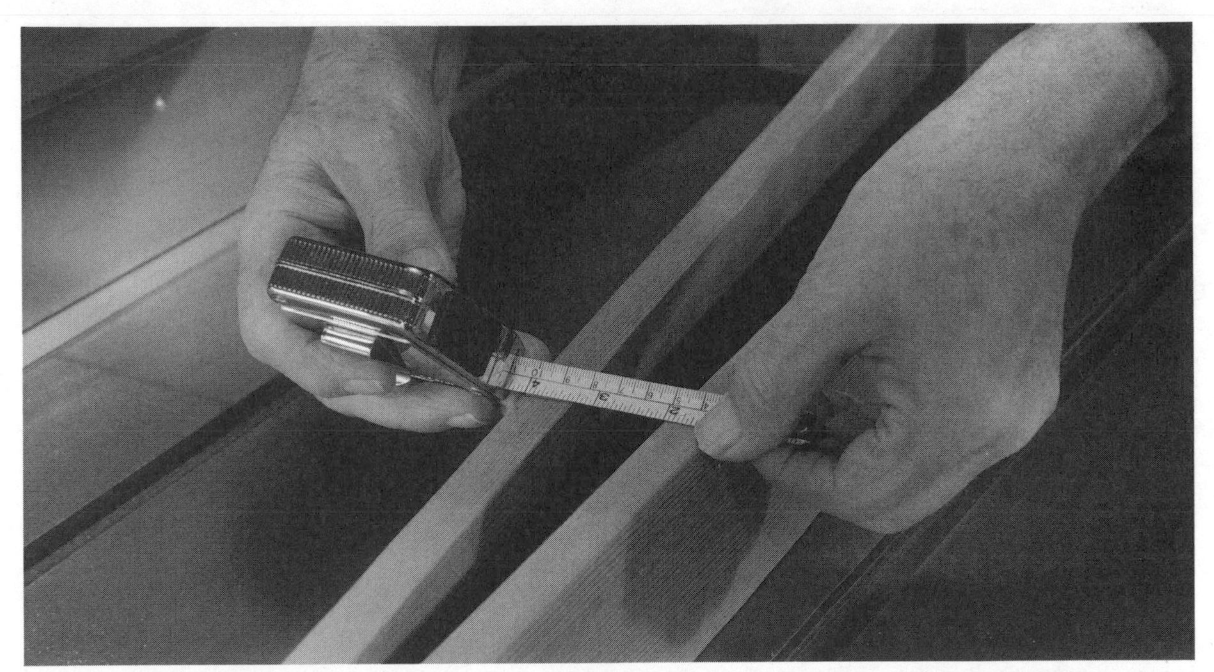

Fig. 7-60. Measuring between marks that have been made on the legs 12 inches away from the hinged end of the jig lets you adjust the jig for the amount of taper per foot you require.

Fig. 7-61. You also can use a taper jig with a tilted blade to saw compound angles, such as when the size of the workpiece makes it inconvenient to use other methods.

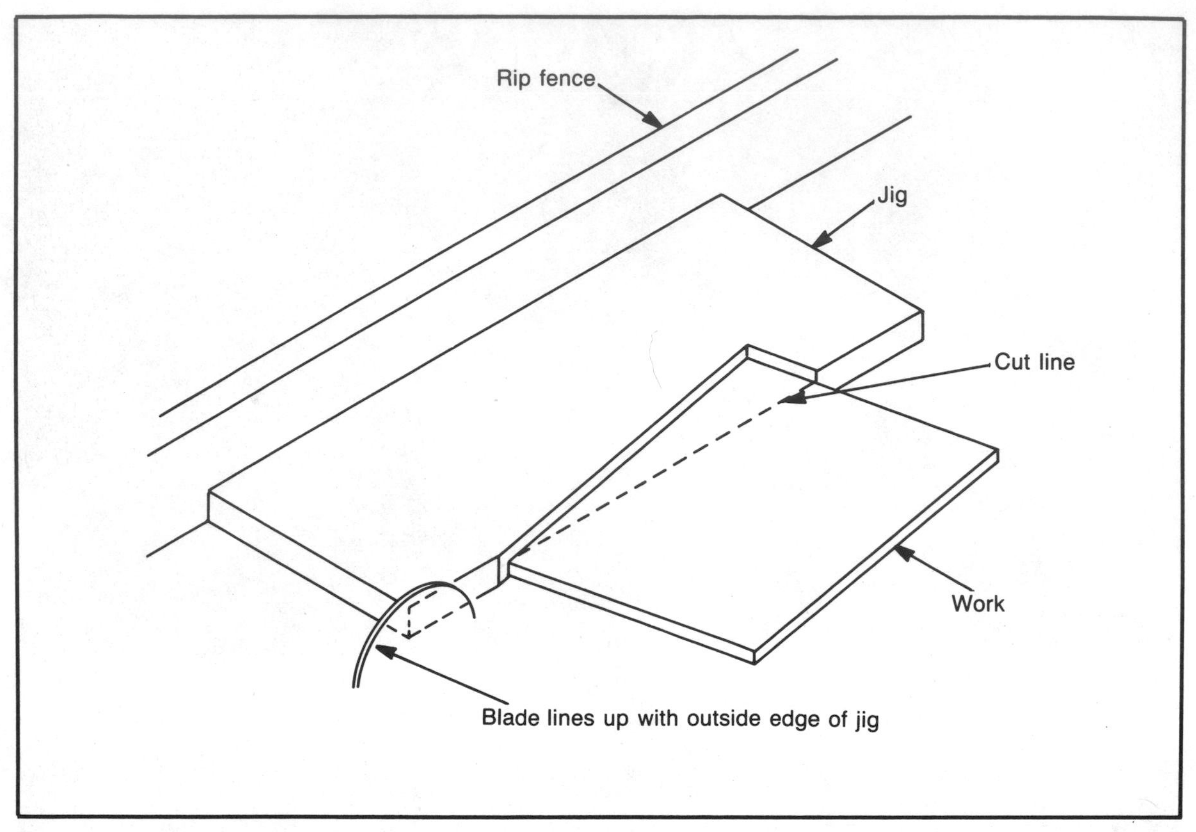

Fig. 7-62. A notching jig can be used to cut accurate tapers. Shape the bearing edge of the jig to suit the part you need or the part you wish to remove. Add a handle to the jig so it will be easier to move.

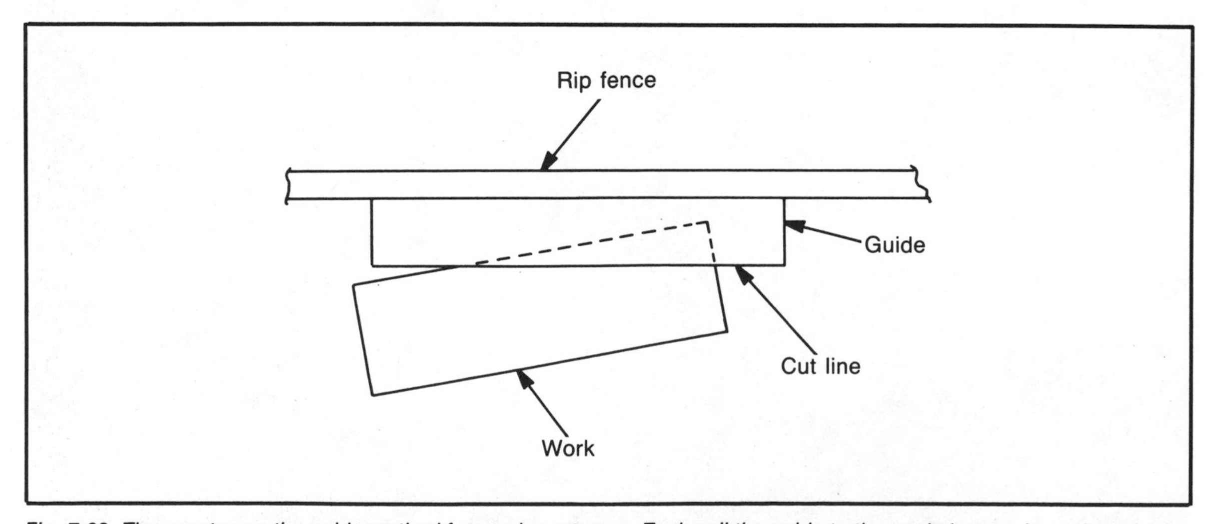

Fig. 7-63. The way to use the guide method for sawing a taper. Tack-nail the guide to the workpiece so its outboard edge is on the cut line. Be sure to use a push stick instead of your hand to move the work if the guide is a narrow piece. The distance from the fence to the inside of the blade equals the width of the guide.

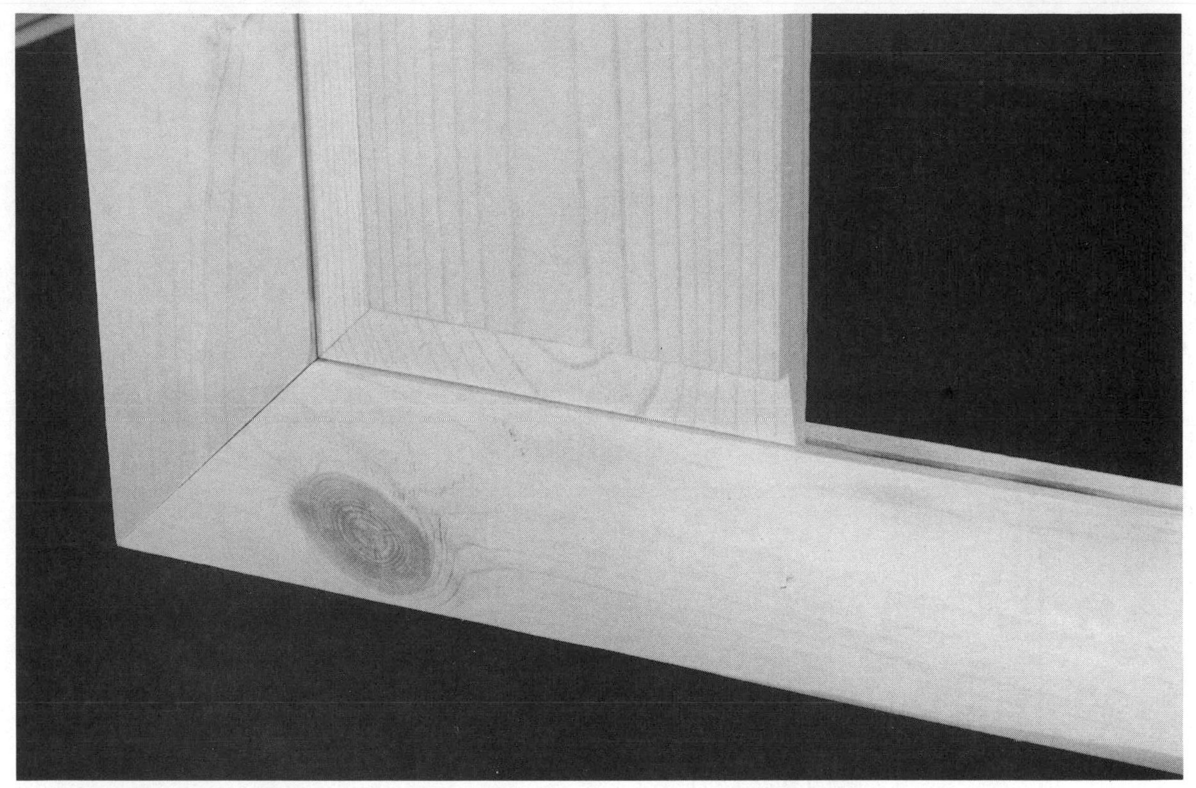

Fig. 7-64. A panel is "raised" to provide a decorative feature and an integral tongue that fits in the groove cut into the edges of the frame pieces.

Fig. 7-65. Panel raising is a common application for a shaping machine or a portable router. Because of the cutter that is used, the shaped edge has attractive contours.

Fig. 7-66. Panel raising can be done on the table saw with a conventional saw blade, but all planes of the design will be flat.

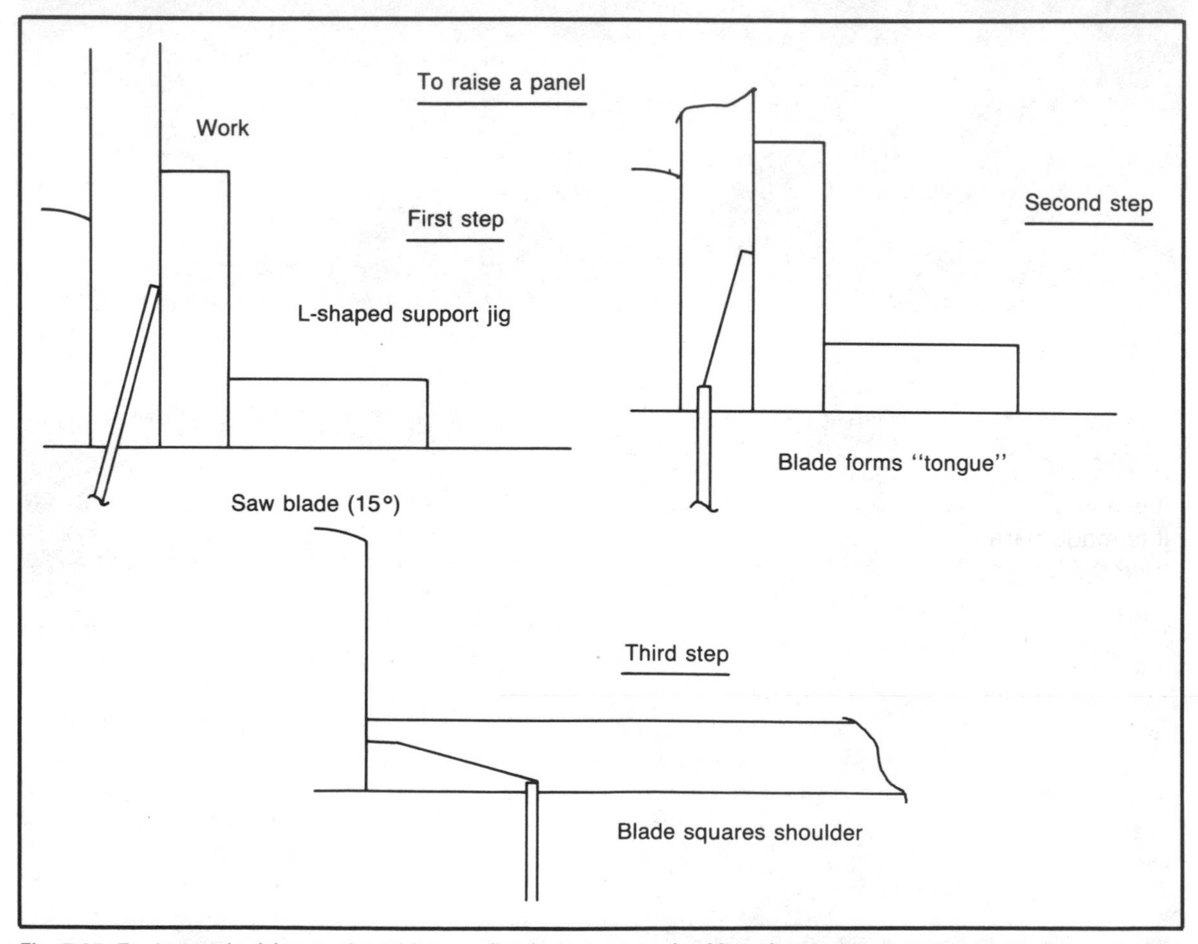

Fig. 7-67. To do panel raising on the table saw, first be sure to work with a sharp, smooth-cutting blade. The L-shaped support piece, which is clamped to the table, will keep the work in a vertical position for accurate cutting and provide a safety factor. Cross-grain cuts will have slight imperfections at end areas, so make them first. The cuts made with the grain will remove the flaws.

Chapter 8

How To Use a Dadoing Tool

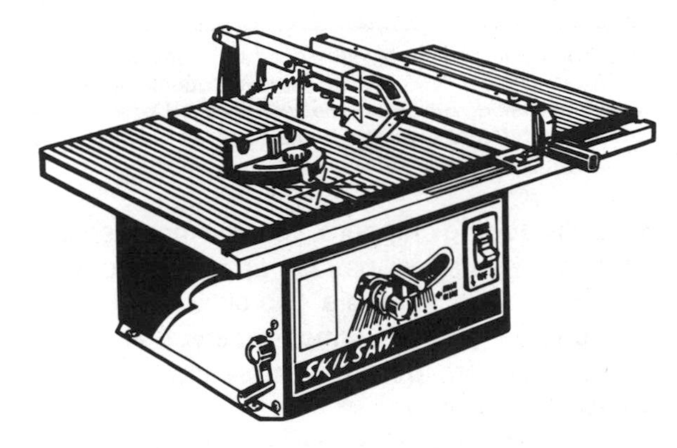

A DADOING TOOL IS ONE OF THE MAJOR ACCESSORIES for a table saw. When you organize a conventional saw blade so its elevation above the table is less than the thickness of the stock and then make repeat, overlapping passes, you form a U-shaped cut that is called a *dado* when it is made across the grain, and a *groove* when it is made parallel to the grain. You also might hear the term *ploughing* used to identify the cut when it is made with the grain. (These definitions might be questioned since many materials like hardboard and particleboard are grainless, but the point is made.) Regardless of the terminology and the direction of the cut, the form that is created is always U-shaped, and usually is sized to suit the thickness of the part to be inserted into it.

The setting of horizontal shelves in a bookcase presents a picture of when it makes sense to use the cut. If the ends of the shelves are inset into the side members through the use of a dado, they will have more strength than if they were simply abutted to the sides.

As I demonstrated in Chapters 5 and 6, you can form a dado or a groove by using the repeat-pass technique. This idea would be suitable for occasional use, but since such cuts are required quite often and often must be repeated with similar dimensions, it makes more sense to work with a special cutting tool designed for the purpose. Once set, the dadoing tool will cut any number of U shapes of equal width, and will accomplish each of them in a single pass. The dadoing tool also can be used for a wider range of jobs than basic dadoes and grooves.

TYPES OF DADOING TOOLS

A common dadoing accessory, called a *dado assembly*, consists of two outside cutters, which are actually heavy, combination saw blades each of which cuts a ⅛-inch kerf, and an assortment of *chippers*, which are always used between the blades (Fig. 8-1). Mount the units on the tool's arbor and secure them as you would a single saw blade. If you mounted just the outside blades, you would cut a U shape ¼ inch

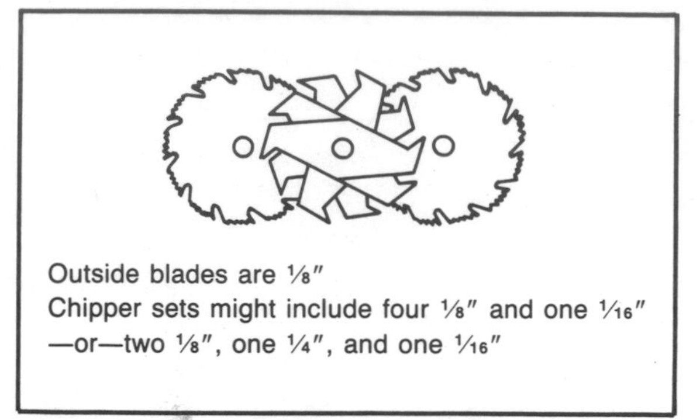

Fig. 8-1. A dado assembly includes two, matched outside blades and a set of chippers. Generally, each blade cuts a ⅛-inch-wide kerf. The two blades can be used for a ¼-inch cut. Wider cuts are available because of the chippers that can be placed between the blades.

wide. By adding chippers, you can increase the cut width in stages, usually up to a maximum of 13/16 inch. For example, two blades plus a ¼-inch chipper form a ½-inch-wide cut. Add two ⅛-inch chippers to the assembly and the cut increases to ¾ inch.

Chippers have *swaged* edges; that is, the edge that does the cutting is wider than the gauge of the chipper. Always mount chippers so the swaged cutting edges on the units that abut the blades are situated in the blade's gullets (Fig. 8-2). In action, the outside blades form the shoulders of the cut while the chippers, in overlapping fashion, cut away the waste between the outline cuts. Also mount chippers so their cutting edges won't be in line. Staggering the chippers will contribute to better balance of the assembly as a whole.

Not all dado assemblies are alike. In addition to differences in diameter—6- and 8-inch sizes are common—and in the number and sizes of the chippers that are supplied, the outside blades might be flat-ground with set teeth, taper-ground, or hollow-ground (Fig. 8-3), or the blades and chippers might have tungsten-carbide cutting edges (Fig. 8-4). A unit with flat-ground blades will cost less than the others, but will not be as efficient or cut as smoothly. Those with carbide-tipped teeth are usually the most expensive, but perform with optimum results and, like carbide-tipped saw blades, will stay sharp for much longer than all-steel cutters.

One problem with dado assemblies is that each of the units cuts a kerf of specific width, so the concept doesn't allow adjustment for odd-sized cuts. What do you do if the component for which you're cutting a dado for is a fraction thicker than ¾ inch? The solution, provided by many manufacturers, is to use paper washers (Fig. 8-5). The fact that the washers are paper is not significant. They can be cardboard or metal shim stock, whatever. The point is that you

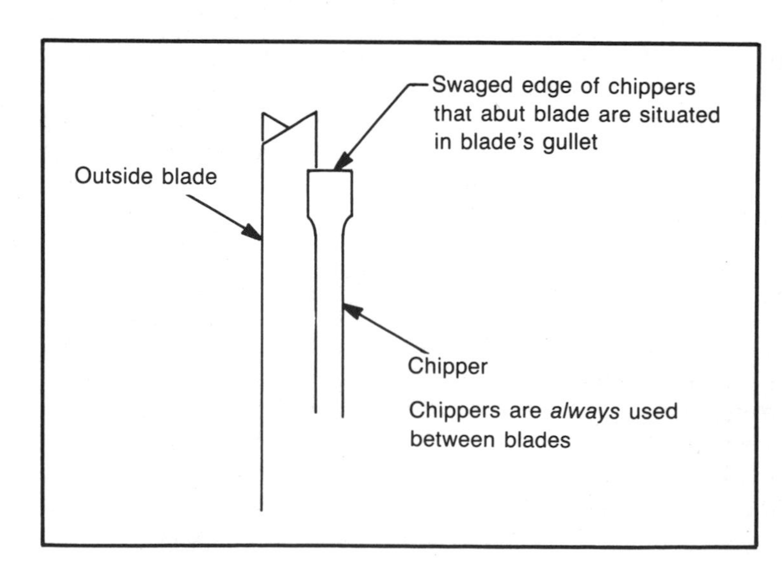

Fig. 8-2. Place chippers so their swaged cutting edges fall between the gullets on the blades. The chippers make overlapping cuts because their cutting edges are wider than the body gauge, resulting in total waste clearance.

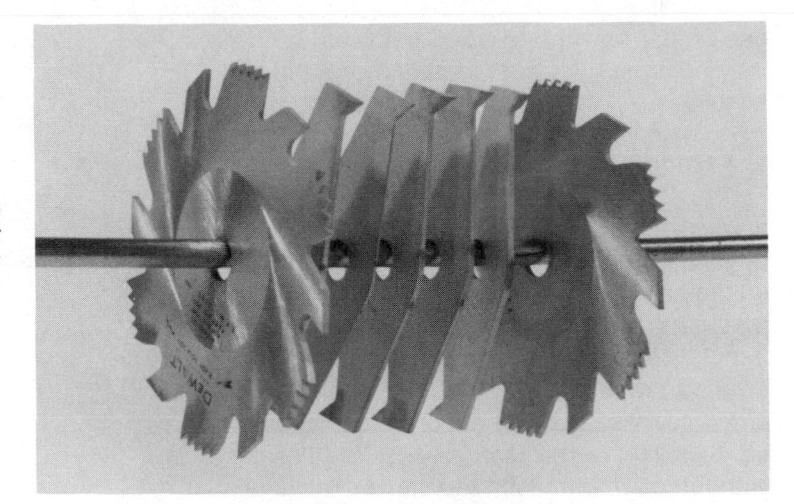

Fig. 8-3. The outside blades might be flat ground and might even have set teeth. The number of chippers provided with assemblies can vary.

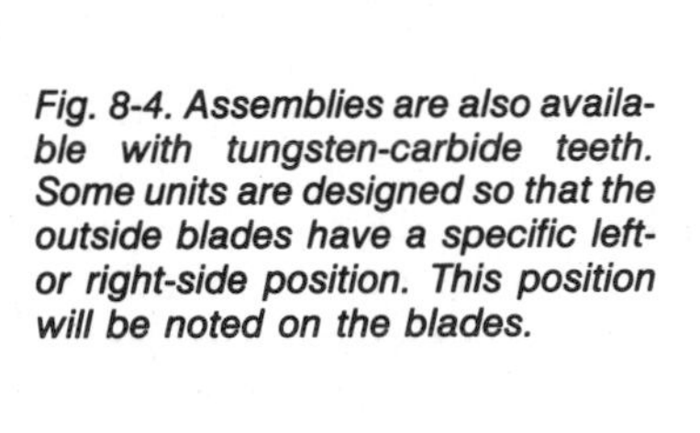

Fig. 8-4. Assemblies are also available with tungsten-carbide teeth. Some units are designed so that the outside blades have a specific left- or right-side position. This position will be noted on the blades.

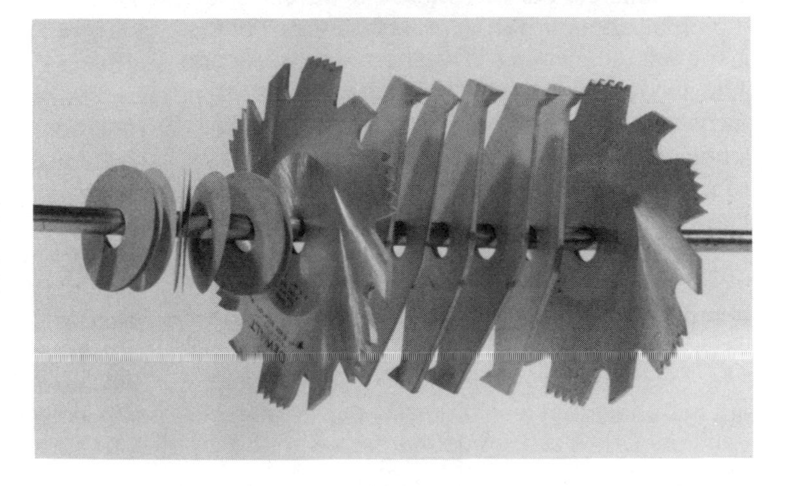

Fig. 8-5. Paper washers are often supplied with the cutters. They are used as part of the assembly so the cut made by a particular assortment of blades and chippers can be widened just a bit.

can adjust for an odd-sized cut width by making one or more of the washers part of the dado assembly. It doesn't matter if you place the washers between chippers or between chippers and blades.

A wise step is to mark the arbor holes of dado assembly components so they can consistently be mounted in the same relative position on the saw's arbor. When you send the tool out for sharpening, which is a job you should not attempt, point out the marks to the person who will do the work. *Jointing,* which is part of the sharpening procedure and ensures that all cutting edges will be on the same plane, should be done with the components in the same relative positions you use when mounting the parts for use. When the jointing procedure is followed and sharpening is done professionally, you will be able to cut dadoes and grooves with nice, flat bottoms.

Fig. 8-6. Most wobble dadoes are designed with a single blade that is adjusted for cut width by turning a central hub. This 8-inch, carbide-tipped version can be set for cuts from 1/8 to 15/16 inch. The calibrations on the hub are easy to read.

WOBBLE TYPES OF DADOING TOOLS

If you mount a saw blade so its cutting edges will move from side to side as it rotates, the blade will make cuts that are wider than the normal kerf. In essence, that is the concept of wobble-type dadoing tools. A major advantage of the concept is that settings between minimum and maximum are infinitely adjustable. Any cut width is possible.

The most elementary design consists of a pair or pairs of heavy washers that have a tapering cross section. When an ordinary saw blade is mounted between the washers, it will be tilted. The blade will wobble from side to side as it turns, making cuts that are wider than it does when used conventionally. The washers have registration marks so they can be set to tilt the blade for cuts of particular widths.

The Freud product shown in Fig. 8-6 is to wobble washers what a super automobile is to a scooter with skate wheels. For one thing, it works with a heavy-gauge blade containing a host of carbide-tipped teeth, so it can efficiently cut materials that would quickly dull an all-steel version. Cut-width adjustments are made by rotating the central hub, which is calibrated for settings from 1/8 to 7/8 inch. Like other products of its type, adjustments can be made without removing the tool from the arbor, although, of course, the arbor's locknut must be loosened before the cutter's adjustment hub can be turned. The tool has an 8-inch-diameter blade and, since it has a removable central bushing, can be mounted on a 5/8- or 3/4-inch saw arbor.

Two other types of wobble-type, carbide-tipped dadoing tools are shown in Figs. 8-7 and 8-8. Both are designed for mounting on a 5/8-inch arbor and, since the diameter of the blade is 6 inches—or 6 3/8 inches in the case of the Rockwell (now Delta) product—recommended maximum depth of cut is 3/4 inch. This is not a great liability since, obviously, it is more than enough when working with 3/4-inch material, and it is half the thickness of 2-inch stock which, when it arrives at the lumberyard, nets out at 1 1/2 inches thick.

An interesting version of a dadoing tool is shown in Fig. 8-9. This tool removes material with individual cutters, instead of saw-type teeth. This concept can't truly be placed in the wob-

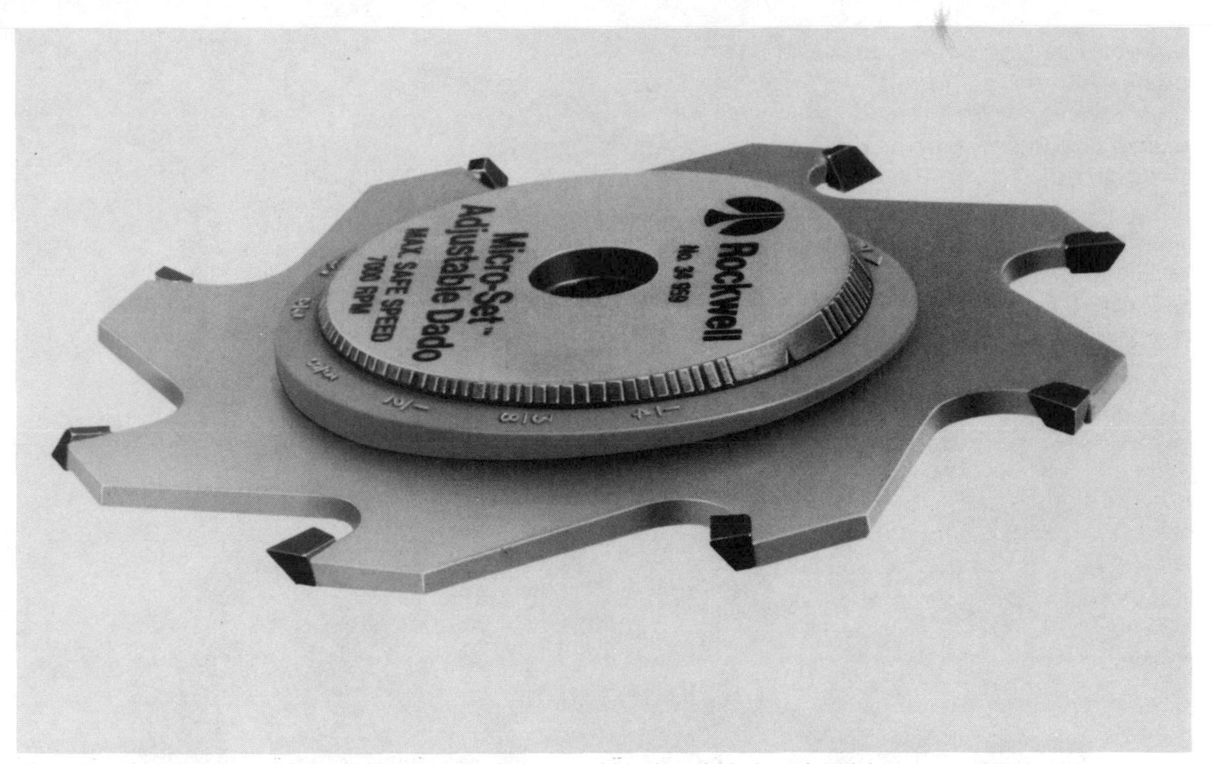

Fig. 8-7. This $6\frac{3}{8}$-inch dadoing tool has an extra heavy, carbide-tipped blade so its minimum cut width is $\frac{1}{4}$ inch and its maximum cut is $\frac{13}{16}$ inch. Tool diameter affects the maximum cut depth, in this case, $\frac{3}{4}$ inch.

ble category since the unit, as a whole, rotates without any side-to-side action. Each cutter travels its own plane, but the cuts overlap so the cut width can be determined by the adjustment hubs. These hubs, when turned, bring the cutters on line for a minimum cut, or spread them for wider cuts. This type of tool is available with all-steel cutters, or ones that have carbide-tipped cutting edges.

A new offering is the Craftsman (Sears Roebuck) product displayed in Fig. 8-10, which has the exotic name of *Excalibur*. Here we have twin, 8-inch, carbide-tipped blades that are infinitely adjustable for cut widths from $\frac{1}{4}$ to $\frac{13}{16}$ inch. Like other concepts, settings are determined by a central control hub that is calibrated for particular cut widths. Typically, settings can be made between calibrations for odd-sized cuts.

No matter what type of dadoing tool you use, it is always best to make test cuts in scrap stock before cutting good material. It is not likely that the calibrations on a control hub can be trusted as implicitly as a vernier scale. Unless you discover that it isn't necessary, use the hub only for approximate settings. Make final adjustments by actually measuring the cut, or after checking to see if the part you wish to insert in the cut will fit as it should.

Dadoing tools, like saw blades, are designed to operate safely at particular speeds. Always check to be sure that the allowable maximum rpm of the cutter is not less than the rpm of the machine on which you plan to use it. It is also important to determine the maximum tool diameter and the maximum cut width that is allowable on your table saw. You can discover these facts by reading the owner's manual.

SETTING UP FOR DADOING WORK

Dadoing tools, whether they are assemblies or wobble types, are mounted and secured on the arbor like saw blades, but there are particular

Fig. 8-8. Like other cutters, dadoing tools must not be used at speeds greater than that for which they are designed. The information should be right on the tool or available in the owner's manual.

points to keep in mind. Cutting teeth, of course, must always point toward the front of the machine. When you are mounting a dado assembly, place the right-hand blade against the arbor flange (or washer), then add the number of chippers you need, the left-hand blade, the second arbor washer, and finally, the locknut. When necessary, it's okay to work without the second arbor washer but don't work without the right-hand one. Never make a setup that doesn't leave enough threads on the arbor for secure tightening of the locknut.

The same system applies if you are working with a wobble-type tool. In this case, the locknut must be tightened after the tool has been adjusted for cut width. Don't go through any

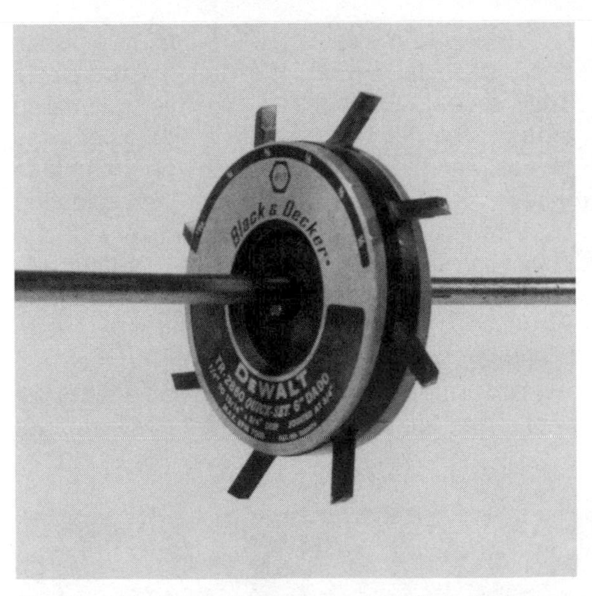

Fig. 8-9. *This dado-tool concept works with individual cutters instead of a saw blade. Like wobble types, it is infinitely adjustable between minimum and maximum cuts. Width of cut is controlled by turning outside plates.*

Fig. 8-10. *Craftsman's Excalibur works with twin, 8-inch, carbide-tipped blades for any cut width between ¼ and 13/16 inch. Markings on the blade identify cut-width settings. An extra scale is used to set the cutter's projection for depth of cut.*

cutter-mounting procedure without first unplugging the machine from its power source. It might seem that just being sure the tool's switch is in the "off" position is enough, but it isn't. Why take chances?

Dadoing tools make cuts that are much wider than a saw kerf, so the regular table insert must be replaced with a special dado insert, which is always available as an accessory (Fig. 8-11). Before plugging in electrical power and flicking the tool's switch to "on," rotate the dadoing tool by hand at least one full revolution to be sure that it turns as it should and does not make contact with any part of the insert, or of the guard when it is mounted. Be especially careful when you are elevating the cutter to be sure that its cutting arc is within the limits of the insert's slot.

On some dadoing jobs, especially when you are making narrow cuts, the opening provided by the dado insert might be greater than is needed. In such cases it is often wise, for accuracy and safety, to make a special insert that just allows the cutting portion of the cutter to poke through.

To make the insert, use the regular insert as a pattern to make a blank of hardwood or plywood. Then, with the cutter set below the table, secure the new insert with screws, if that is how the regular insert is attached, or use a system like the one in Fig. 8-12. Locate the fence and the springstick so they can't be damaged, and then turn on the machine and very slowly raise the cutter so it will cut its own slot. If the insert is not attached with screws, form the improvised one so it will fit very tightly in the table's opening. You don't want it to vibrate and chatter when you are working.

The depth of a dado or groove is determined by the height of the cutter above the table. You can establish this depth by using the height gauge that was demonstrated in Chapter 4, or by marking the setting on the edge of the stock (Figs. 8-13 and 8-14). It's always a good idea to determine the accuracy of cut width and cut depth by making a cut in some scrap material before working on good stock. Don't use a scale to check the dimensions of the cut. Instead, use the component that will be inserted in the cut to see if it will fit as it should.

CUTTING PROCEDURES

Dado or groove cuts remove considerably more material than a normal saw cut, so be sure to move the work across the cutter more slowly than usual. Like saw blades, the teeth on dadoing tools are designed to remove just so much wood at a given rate of feed. The ideal is a machine, not uncommon in industry, that automatically moves the work at a precise feed speed. You are not a machine, however, so you must be guided by the "feel" of the cut. Forcing the saw to cut faster is bad practice and can only harm the cutter and the work.

Trying to make very deep cuts in a single pass, especially when the dado or groove is quite wide, is another bad practice. It is much better to set the cutter's projection for a first pass to less than the depth you need and then make a second, or even a third, pass after adjusting the cutter's elevation for each one. You'll discover soon enough the depth you can cut efficiently and with optimum cut results in a single pass.

Important factors, aside from the fact that the cutter must be in prime condition are the cut width, the density of the material, and the

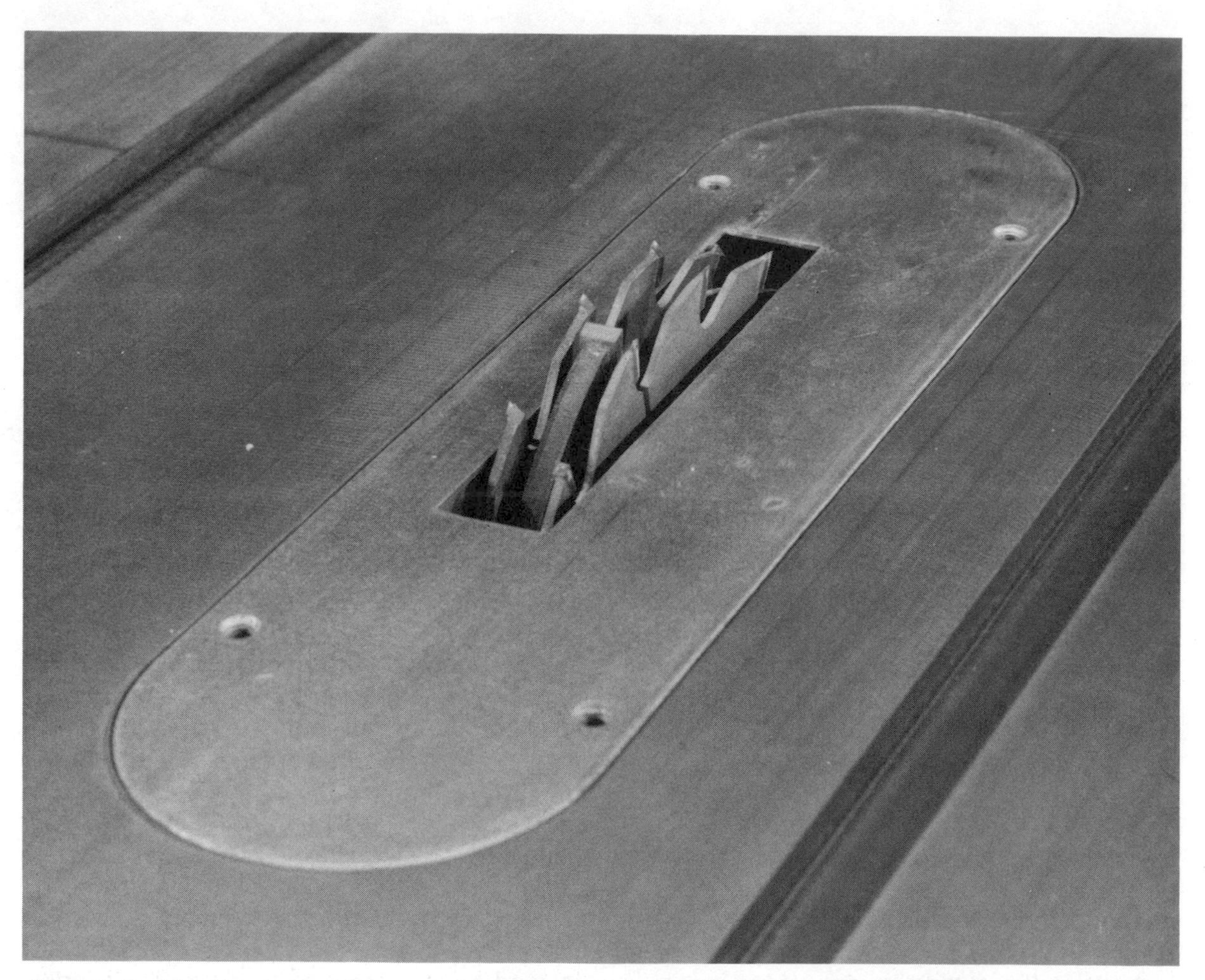

Fig. 8-11. Dadoing tools make wider cuts than saw blades, so a special insert must be used. Always rotate the cutter by hand at least one full revolution to be sure it won't hit the insert or any other part of the machine.

machine's horsepower. Work that chatters during the cut, the need to force the cut, and a cutter that slows up excessively are warning signals.

Dadoes and Grooves

Most times, you will cut dadoes by working as you would when making a simple crosscut. Hold the workpiece firmly against the head of the miter gauge and flat on the table as you make the pass (Fig. 8-15). Do not make a return. When the work and gauge are past the saw blade, stop the machine and wait for the cutter to stop turning before going further.

Tear-out is the splintering that can occur when the cutter emerges at the end of the cut. The flaw will be more apparent on plywood than lumber. To minimize it, if not eliminate it, end the cut by moving the work very slowly. You can solve the problem completely by attaching an extension to the miter gauge to back up the cut. Any tear-out then will occur on the extension, not the work. Another system calls for cutting the dado in stock that is a bit wider than necessary. Then you can remove the flawed edge by making a rip cut.

To form dadoes that are wider than the cutter can accomplish at its widest setting in a single pass, make repeat passes (Fig. 8-16). Overlapping the passes will help produce cuts with flat, even bottoms. If the bottom has any

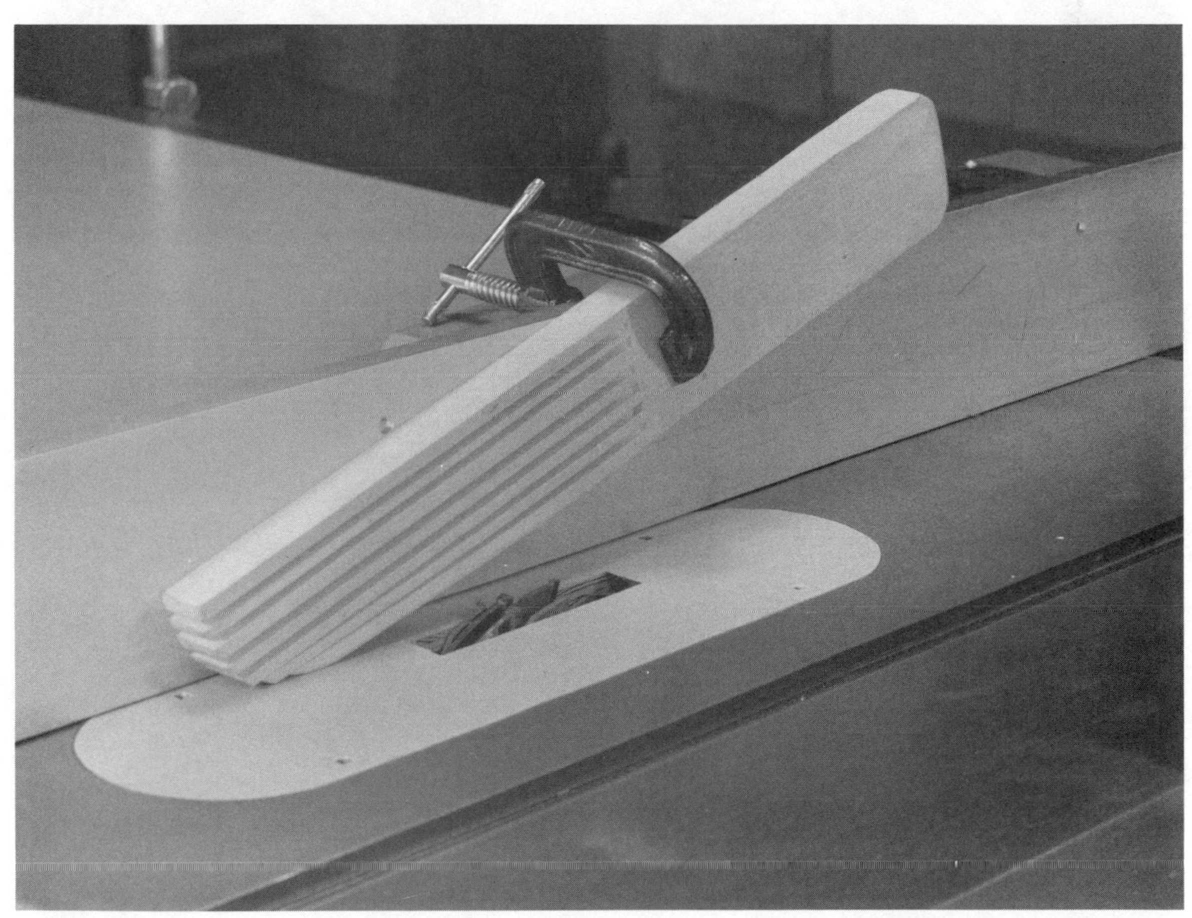

Fig. 8-12. A special insert, which you make, allows only the working part of the cutter to poke through. Be sure the insert is held securely before you start forming the slot.

Fig. 8-13. The projection of the cutter determines how deep a cut will be, which is as important as how wide the cut will be. You can set the projection accurately by using a homemade height gauge.

Fig. 8-14. Another way to set projection is to use a mark on the work as a reference point. If you work this way, look horizontally at the marked edge to be sure the reference line and cutter teeth are on the same plane.

Fig. 8-15. Most dadoes are cut by using the miter gauge as for a simple crosscut, but feed speed should be much slower since a lot of material must be cut away. Slow up at the end of the cut to minimize the tear-out that can occur when the cutter breaks through the rear edge of the workpiece.

Fig. 8-16. Use repeat passes to make cuts that are wider than the cutter's maximum cut width. You can cut to marks on the work or use the rip fence as a stop to gauge the cut locations. The latter method is wise when the same cut is required on several components.

kind of a taper and if the shoulders of the cut are slanted, check to see if the angle between the dadoing tool and the table is 90 degrees.

Projects often require parallel, equally spaced dadoes. An example is the side pieces of a bookcase or cabinet that will be fitted with equally spaced shelves. Making the cuts to layout lines on the work leaves much room for human error. It is better to set up a jig like the one in Fig. 8-17 to be sure of optimum results. The jig is a miter-gauge extension that is notched in two places by the dadoing tool. The distance between the notches equals the spacing that is required; the strip in the outboard notch is the control for positioning the work for the cuts. Place the first cut over the guide strip so the work is in correct position for the second cut. Repeat the procedure, placing the previous cut over the guide, for all cuts (Fig. 8-18).

To cut grooves, organize as you would for ripping operations. Make the pass while keeping the work flat on the table and snug against the fence. Figure 8-19 shows the setup, but doesn't demonstrate *how* to handle the procedure. To ensure accuracy and to provide a safety factor, it is always wise on operations of this nature to move the work with a combination pusher/hold-down and, when necessary, to use a springstick to keep the work against the fence. The distance from the fence to the *inboard* side of the dadoing tool determines the edge distance of the groove.

Forming extra-wide grooves is just a matter of making repeat, overlapping passes (Fig. 8-20). On grooving operations, there will be a tendency for the work to climb the cutter and for the cutter to lift the work. The actions will be more evident if the cutter is at all dull.

It doesn't hurt to use a device like a springstick to help hold the work correctly as the cut is made (Fig. 8-21). Be sure to place the springstick so its fingers won't slap down into the cutter when the pass is complete. If the device interferes with using a pusher/hold-down, get the work past the cutter by using a basic pusher stick.

Rabbet Cuts

Rabbets are L-shaped joinery forms that are called, logically, *end rabbets* when cut across ends of stock, and *edge rabbets* when formed on long edges. The depth and width of the rabbet is determined by the component that will be placed in it. For example, if you were sealing the back of a bookcase or a cabinet with a 1/4-inch-thick panel, and the project's case pieces were 3/4 inch thick, the rabbet would need to be 1/4 inch deep to match the thickness of the panel, and at least 3/8 inch wide so you would have room to use nails or screws as fasteners.

You can form end rabbets by working with the same setup as for dadoes. For extra-wide cuts, make repeat passes (Fig. 8-22). You can judge the width of the cut by working to lines marked on the work, or, since you are not cutting off material, you can use the rip fence as a stop. In either case, make the inside cut first. Then remove the remaining waste by making repeat passes that travel toward the open end of the workpiece.

To form edge rabbets, use a ripping setup. The safest way to work and one that will produce most accurate results is to attach a special wood facing to the fence by clamping or using screws that pass through the fence and into the auxiliary component (Fig. 8-23). Attaching the extra wood facing with screws is best since it will be easier to position it correctly whenever you need it.

You must cut a relief area for the dadoing tool into the auxiliary fence. To do so, set the dadoing tool so it is below the table's surface and lock the rip fence so about one-half of the wood facing's thickness will be over the cutter. Then, raise the cutter very slowly until it forms an arc about 3/4 inch high.

To form an edge rabbet, set the projection of the cutter so it will provide the depth of the cut, and lock the fence so the distance from the facing to the *outside* of the dadoing tool equals the width of the cut. Make the pass as shown in Fig. 8-24. Make sure the work is snug against the fence and flat on the table throughout the cut.

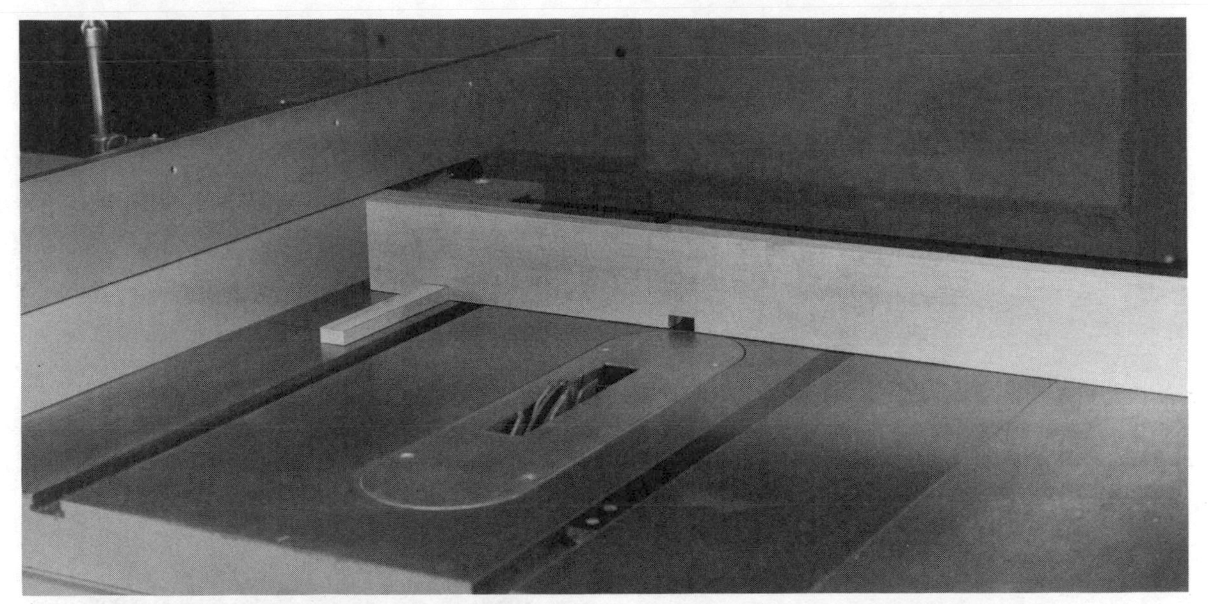

Fig. 8-17. The distance between equally spaced dadoes can be gauged automatically with this type of homemade miter-gauge extension. The distance between the guide strip and the open notch, which is in line with the cutter, equals the required spacing. The dimensions of the guide strip and the notch are the same as the dadoes that will be cut.

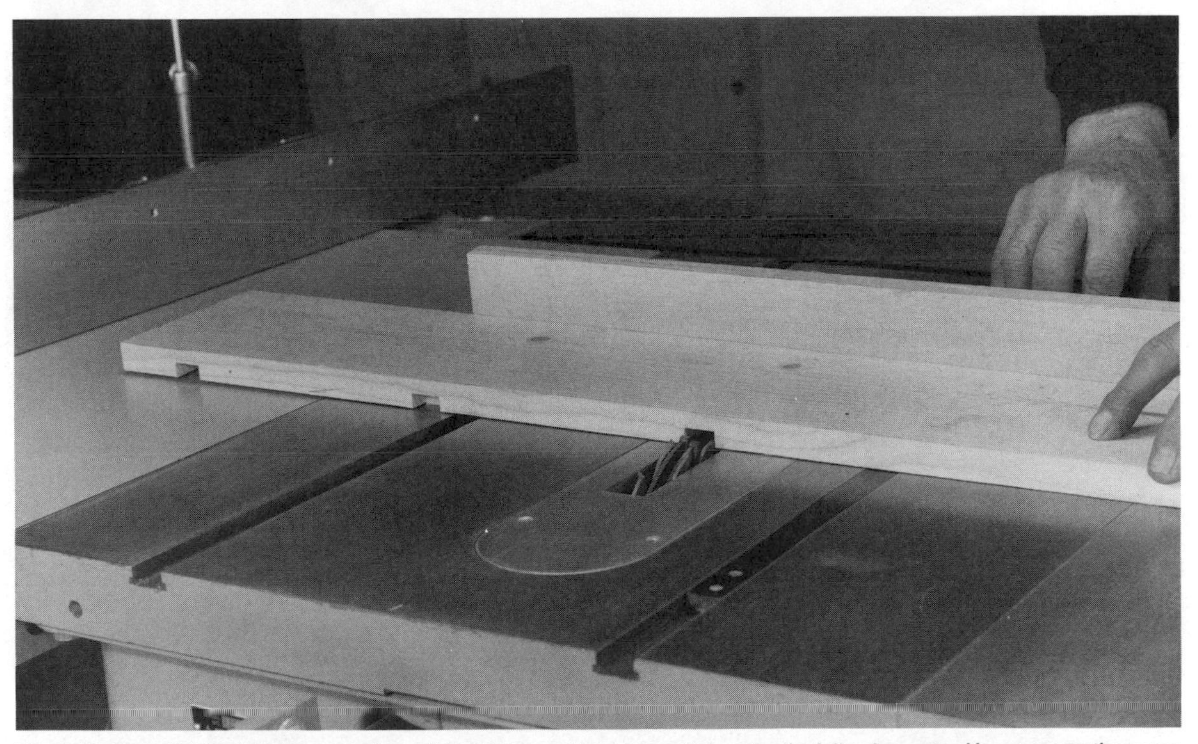

Fig. 8-18. Place finished dadoes over the guide strip to position the workpiece for the following cuts. You can use the same extension for jobs that require a different spacing so long as the dimensions of the dado remain the same. Reattach the extension after it has been moved for the new distance between cuts.

Fig. 8-19. Cut grooves by organizing the fence as you would for ripping. The distance from the fence to the inside of the cutter controls the edge distance of the groove.

Fig. 8-20. Form extra-wide grooves by making additional passes with the fence repositioned for each one. Overlapping the cuts will result in smoother groove bottoms. A pusher/hold-down provides a safety factor and helps you to cut accurately.

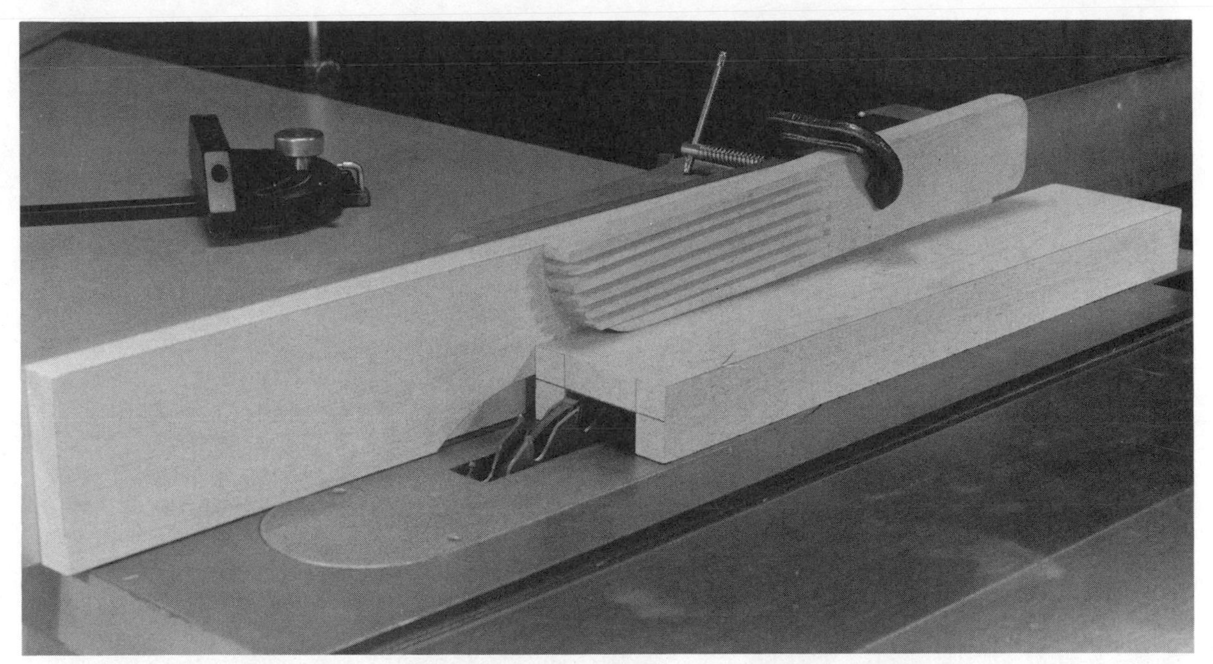

Fig. 8-21. Use one springstick to keep the work firm as you make the cut, and a second, clamped flat on the table, to keep the work from moving away from the fence. Use a push stick to move the work past the cutter.

Fig. 8-22. To form end rabbets, use the miter gauge to move the work past the cutter. Make repeat, overlapping passes for extra-wide cuts. For accuracy when the same cut is required on several pieces, use the rip fence as a stop.

You also can form edge rabbets by making the pass with the stock resting on its edge (Fig. 8-25). You might want to use this method when you can't make the required cut dimensions by making the pass with the workpiece in a horizontal position.

Forming Tenons

Tenons are projections formed on the end of components to fit square or rectangular cavities cut into mating pieces. A *stud*, or *open, tenon* is sized to fit a slot that is formed in the part to which it will be joined. A *true tenon* is shaped so it will be enclosed in a through, or stopped, cavity cut into its mate (Fig. 8-26).

The stud tenon can be viewed as end-rabbet cuts that are made in opposite surfaces of the workpiece (Fig. 8-27). Since the cut does not remove material that can be trapped and kicked back, you can use the fence as a stop to gauge the width of the cuts. The length of the tenon is established by the distance from the fence to the *outside* of the dadoing tool. Make the first cut—the shoulder cut—with the end of the workpiece abutted against the fence. Remove the remaining waste by making repeat, overlapping passes.

The true tenon is formed the same way, except that cuts are required on edges, as well as surfaces of the workpiece (Fig. 8-28).

Tenoning jigs, a homemade example of which is shown in Fig. 8-29, are super tools for accurate cutting of tenons. More information on this aspect of the work is available in Chapter 10.

Fig. 8-23. An auxiliary wood facing for the rip fence is needed for some dadoing operations. The relief cut in the facing is made by slowly raising the blade from beneath the table. Make the cut about ¾ inch high and into about half of the facing's thickness.

Stopped Cuts

A disadvantage of a dado joint is its unattractive appearance when it is seen from the front with the insert piece in place. Many times, the appearance of a project that has exposed dado cuts is improved by adding a frame or front facing strips that hide the U-shaped cuts. Another technique is to cut the dadoes so they don't run completely across the stock, and to shape the ends of the parts that will be inserted in the cuts to match the arc at the end of a stopped cut.

Stopping a cut is just a matter of using a stop block on the fence when cutting grooves (Fig. 8-30), or clamping a block to the saw's table to control the length of your cuts. The problem that remains is how to shape the end of the part that will be inserted so it will fit in the U-cut as it should. This problem is solved with a pattern made by using the idea demonstrated in Fig. 8-31. Select a long, slim piece of scrap and advance it over the cutter until the arc of the cut is established. Use this scrap cutout to mark the insert piece so it can be shaped to fit the arc of the stopped dado cut (Fig. 8-32).

You also can use a stop block to control the length of extra-wide cuts that do not go completely across the stock (Fig. 8-33). The smoothness and levelness of the bottom of the cut will indicate the quality of the cutter you are using and the care you use to control the operation (Fig. 8-34).

Slots

Dadoing tools are not designed to cut completely through materials; they are not used like

Fig. 8-24. It's always best to cut edge rabbets with the work edge riding the fence. Operating so the workpiece does not need to be moved between fence and cutter just about eliminates the possibility of kickback.

Fig. 8-25. You also can form edge rabbets making the pass with the stock on edge. If the same cut is required on opposite edges, invert the stock while keeping the same surface against the fence and make a second pass.

Fig. 8-26. You can use a dadoing tool to form tenons. The example on the right is a stud, or open tenon. The other is a true tenon.

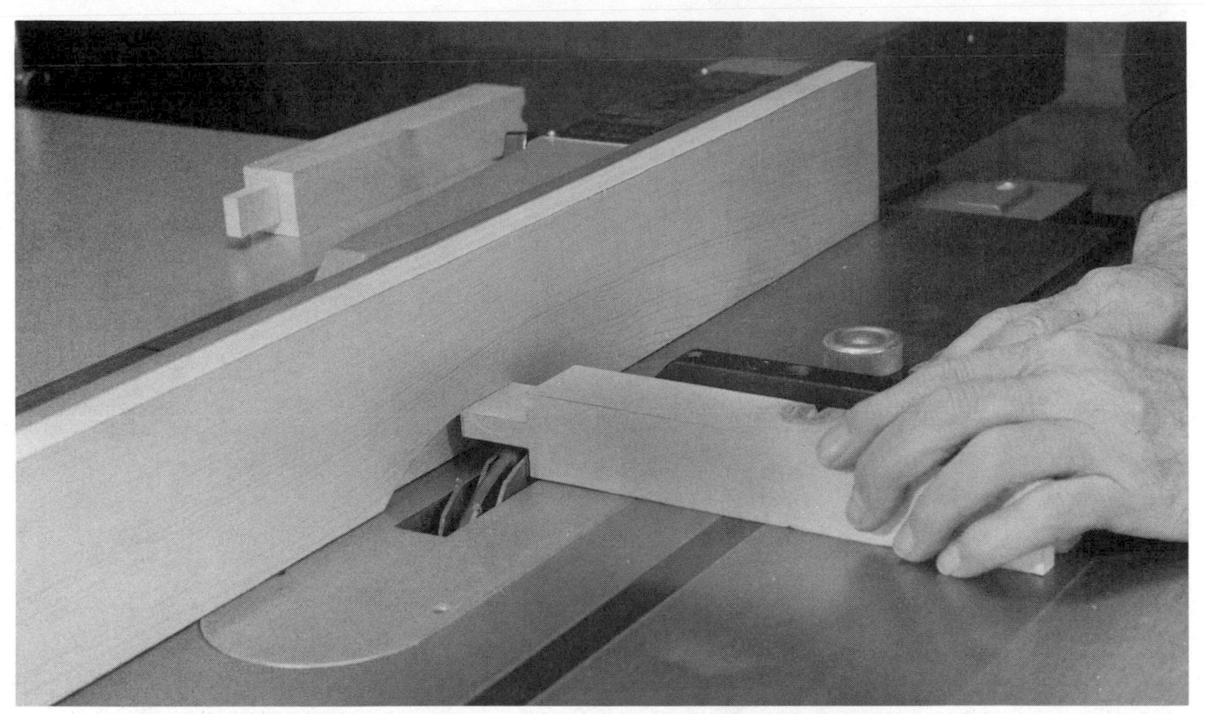

Fig. 8-27. To form the stud tenon, make back-to-back rabbet cuts. The distance from the fence to the outside of the cutter determines the length of the tenon. The depth of cut will establish the thickness of the tenon.

Fig. 8-28. The true tenon requires cuts on all four surfaces of the component. It fits a square or rectangular cavity that is formed in the mating piece. The stud tenon requires only a slot.

Fig. 8-29. There are other ways to form tenons. The plans for this homemade tenoning jig and information on its use are given in Chapter 10.

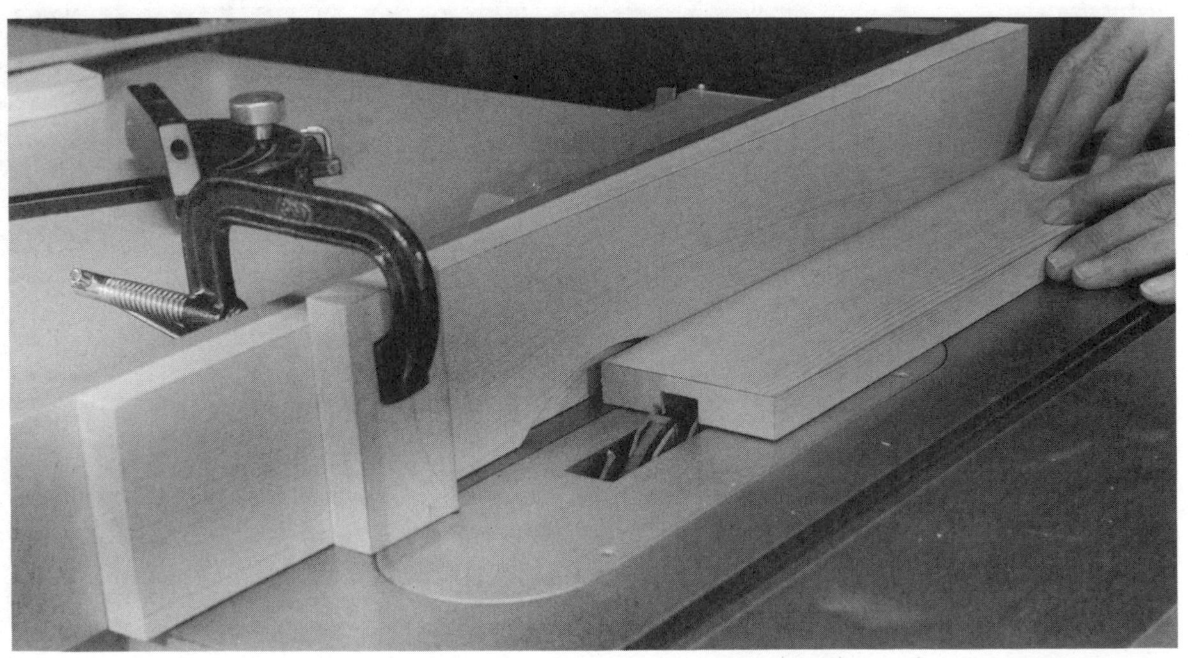

Fig. 8-30. A stop block clamped to the fence can be used to gauge how long a groove will be. A stop block clamped to the table can be the control mechanism for stopped dadoes.

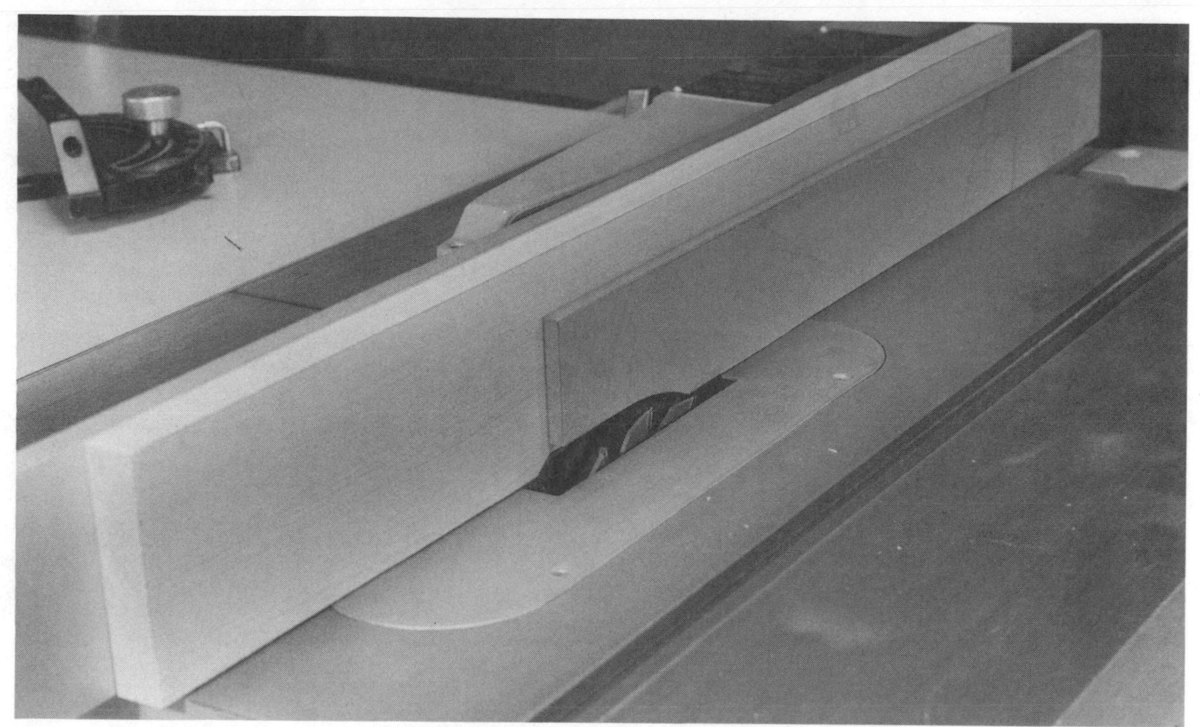

Fig. 8-31. The end of the component that will be placed in the groove or dado must be shaped to fit the arc which remains at the end of a stopped cut. Make a short cut into the edge of a slim piece of material to establish the shape of the arc.

Fig. 8-32. Then use the slim piece as a pattern to mark the shape required on the insert piece (arrow). Shape the arc by cutting with a jig, band, or coping saw.

Fig. 8-33. You also can use a stop block on the fence to control the length of extra-wide cuts. Adjust the position of the fence for each cut so that they will overlap.

Fig. 8-34. How smooth and even the bottom of a cut is will tell a lot about the quality of the cutter you are using and the care you take. An important factor for smoothness is keeping the work firmly down on the table as cuts are made.

Fig. 8-35. The stopped-cut technique is also used for slots needed between the extremities of a component. Start the cut by raising the cutter from beneath the table until it pokes through the workpiece.

Fig. 8-36. The slot is elongated as the workpiece is moved toward the second stop block. The block clamped at the center near the cutter serves as a hold-down.

Table 8-1. Common Dadoing Problems and Corrective Steps.

PROBLEM	POSSIBLE CAUSE	CORRECTION
Bottom of dado or groove uneven	Chippers uneven/Outside blades poorly matched	Chippers and blades should be jointed and sharpened as a set and kept in prime condition
Tool chatters or stalls	Inconsistent placement on arbor	Mark blades and chippers so they can always be mounted the same way
	Cutting too fast	Use slower feed/Allow tool to cut at its own pace
	Cutting too deep	Make repeat passes when necessary
	Dull cutters	Sharpen and keep sharp
Burn marks on cutters or wood	Forcing cut	Slow up feed/Allow cutter to work efficiently
	Cutting too deep	Make repeat passes when necessary
	Dirty cutters	Clean cutters regularly/Maintain in prime condition
	Dull cutters	Sharpen and keep sharp
Excessive feathering at end of cut	Breaking out of cut too fast	Finish cut by feeding very slowly, especially when cutting cross-grain
Cut-width inaccurate	Poor setup	Dado assemblies may require paper washers for exact settings/Set adjustable dadoes very carefully/Make test cuts
Dado or groove is slanted	Poor work handling	Keep work flat on table throughout pass

saw blades. If, however, you work carefully and use correct setups, you can use them to form wide slots. You can control the length of the cut with a single stop block on the rip fence when you start the cut at one end of the material and halt it before it reaches the opposite end. Just feed the work against the cutter until it reaches the stop block that is clamped somewhere at the rear end of the fence.

When the slot must be contained within the extremities of the workpiece, use two stop blocks. Clamp a stop block at the front and the rear of the cutter so the distance between them equals the length of the slot you need. With the cutter set below the table, place the workpiece in the position shown in Fig. 8-35. It should be snug against the fence and abutted against the front stop block. Use a block of wood or a springstick as a hold-down. Turn on the machine and raise the cutter until it pokes through the workpiece, and then very slowly move the work until it contacts the second stop block (Fig. 8-36). Remove the work after you have turned off the machine and lowered the cutter until it is below the saw's table.

Some of the problems that can occur when doing dadoing or grooving operations, and possible solutions for them, are listed in Table 8-1.

Chapter 9

How to Use a Molding Head

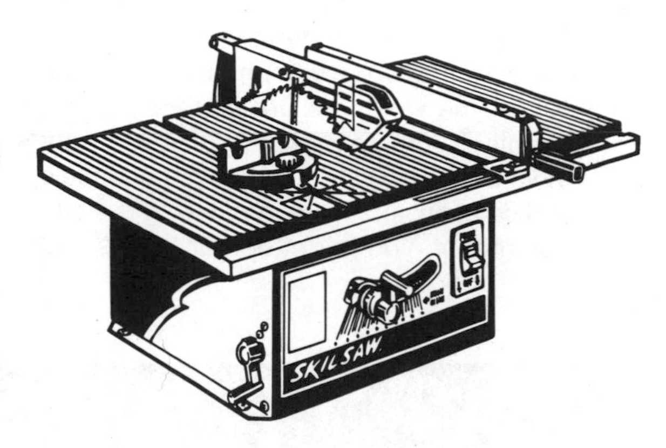

A MOLDING HEAD, SOMETIMES CALLED A *MOLDING cutter head* or a *shaper head*, is another major accessory for the table saw. With a molding head and a reasonable assortment of knives, or *cutter bits*, which must be used with it, you will be equipped to produce a host of classic and original molding designs and will be able to do the unique kind of surface "carving" shown in Figs. 9-1 and 9-2.

In addition, you can use the tool for practical applications like forming edge-to-edge joints, cabinet-door lips, and tongue-and-groove joints. You can perform many of these standard woodworking operations, and others that usually require the use of an individual shaping machine, by adding a molding head to your equipment.

TYPES OF MOLDING HEADS

Two of the several molding-head designs that are available are displayed in Figs. 9-3 and 9-4. They may be thick or thin, heavy or light, and work with two-knife or three-knife sets, but they all are used the same way on a table saw. The head mounts on the arbor like a saw blade, but, as you might imagine, there are special considerations, which will be discussed in a bit. Actually, the head itself is but a holder for the knives, which do the cutting. The knives are secured by some means in slots that are equally spaced about the perimeter of the tool.

Some table saws, especially very small ones, can't handle molding heads, or must be used with concepts of a particular size and shape. Generally, molding heads are interchangeable among various brands of table saws, but it is wise to determine the suitability of any particular one for your machine. For example, the length of the arbor has a bearing on how thick the accessory can be. These pertinent questions are answered in the owner's manual. Many operators feel more comfortable choosing major accessories that are offered by the manufacturer of their machine.

KNIVES

The assortment of molding knives from which

you can choose can be bewildering. The types whose profiles are shown in Fig. 9-5 are typical, but don't present the entire picture. There are facts to consider, however, that will help you toward wise selections, which is important if only because knives are expensive.

For one thing, it isn't necessary to have every type of knife. Start with a few basic designs, that all of us who are involved in table-saw work can use, and add others only when they solve a problem or contribute to furthering your expertise. The very first ones you choose should be practical for the work you are currently doing. For example, if you are making tables and want decorative edges on slabs, you might choose a bead-and-cove design. If cabinetmaking is a current assignment, you might opt for knives that will form a door lip.

An important factor is that molding knives are of two designs. They are either combination types, like those shown in Figs. 9-6 and 9-7, which result in different shapes depending on what portion of the profile you cut with, or they are designed primarily for full profile cuts, which produce a standard shape like the glue-joint knives shown in Fig. 9-8.

Another consideration is that some knives are sold in matching sets, each group of knives forming a cut that mates with the shape formed by the second set of knives. Examples are the knives shown in Fig. 9-9, which are designed specifically to form a drop-leaf table joint, and those in Fig. 9-10, which are used for joining boards edge to edge with a tongue-and-groove connection. The practice of making test cuts in scrap stock is especially important when you are

Fig. 9-1. The molding head, fitted with blank knives, can be used to create faceted surfaces. For results to be as clean as this, the knives must be in prime condition and the cuts must mate precisely.

working with matched sets of knives. The work must be done so that shapes formed by each set will be in perfect alignment. If not, the mating components won't come together as they should.

There is no rule that says you must use the entire profile of a knife designed for a particular purpose or a specific portion of a combination knife. You can use any part of any knife if the shape that results suits your purpose. For example, the three-bead design shown in Fig. 9-11 can be used for three beads, two beads, or one bead. You can cut into the surface of workpieces, making parallel cuts to striate the entire area, or you can work on edges. It all depends on how you organize the relationship between the cutting tool and the work.

SETTING UP

Be sure that the molding head is clean before and after you use it. Pay special attention to the slots that receive the knives and to the knives themselves. Any accumulation of sawdust or wood residue can interfere with correct placement of the cutters. The knives are held in place with screws, but methods of attachment will vary, so be sure to follow the manufacturer's instructions for the right way to go. Remove the knives when you are through working. Clean the components and then store the knives so cut-

Fig. 9-2. Knives that have shaped profiles can also be used to do surface carving for effects like this. Cut with the grain when making parallel cuts. Make the cross-grain cuts first when the pattern has intersecting lines.

Fig. 9-3. This solid-steel, heavy-duty molding head is drilled to distribute weight for better balance. The cutting knives are secured in three equally spaced slots. Knives and slots must be kept clean.

Fig. 9-4. Another typical molding head design. Knives are slotted to fit the back edge of the slot that receives them to ensure good alignment of the knives. Like other heads, this one may be purchased alone or in a set with an assortment of knives.

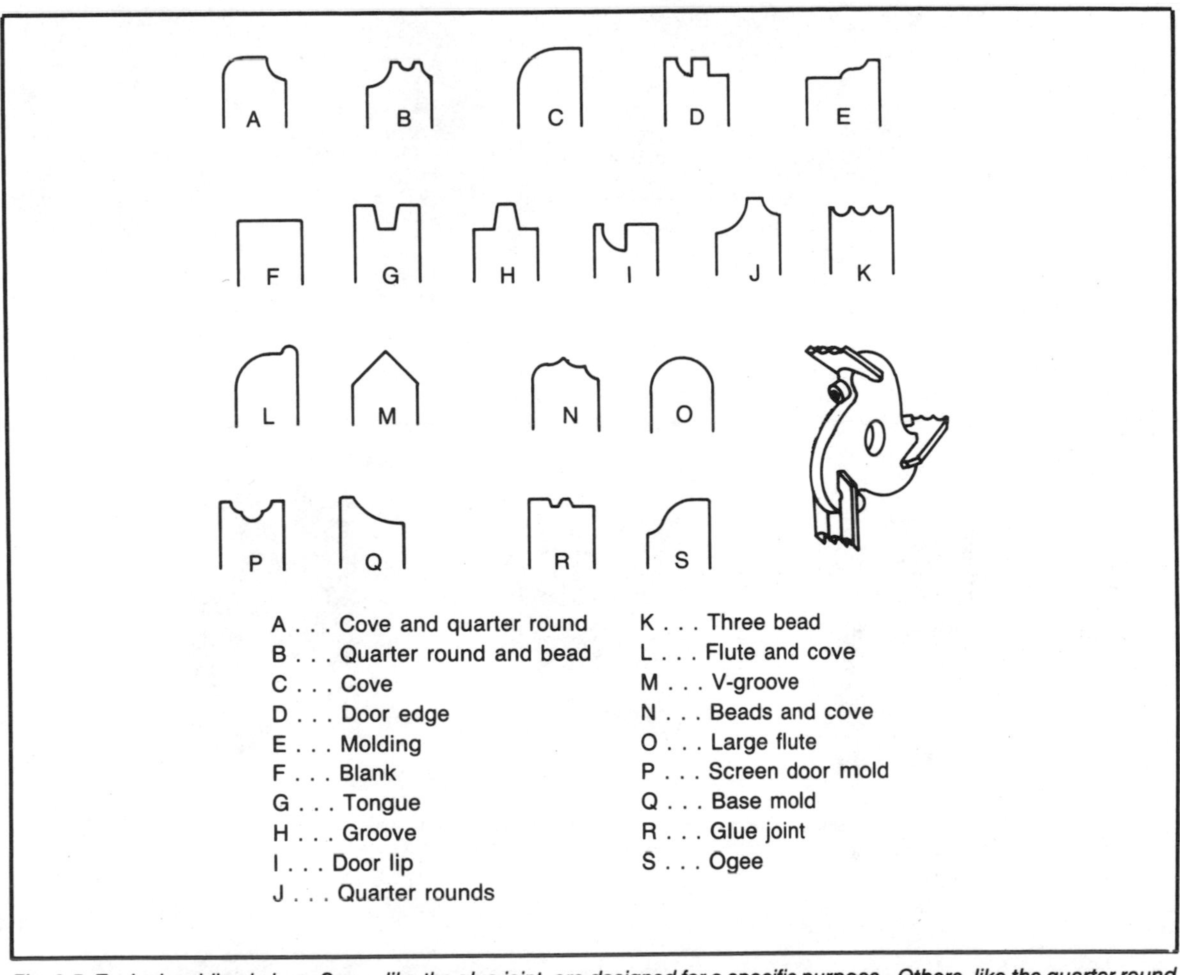

Fig. 9-5. Typical molding knives. Some, like the glue joint, are designed for a specific purpose. Others, like the quarter round and bead, are combination types. The section of the cutter's profile you use determines the shape you get.

Fig. 9-6. This set of combination knives can be used to shape various types and sizes of beads and coves. Remember that the shape the knives produce is always the opposite of the profile.

Fig. 9-7. This popular set of combination knives can be used for two sizes of quarter-round shapes plus a bead. The bead and large quarter-round portion of the profile are often used to shape edges on projects like tables and trays.

Fig. 9-8. The glue joint is used to shape mating edges of components that will be joined edge to edge. Careful setups are required so the shaped edges will come together as they should.

Fig. 9-9. Some knives are sold in sets that produce mating shapes. These are used to form a drop-leaf joint. One knife shapes the table's edge; the other produces the mating shape on the hinged leaf.

Fig. 9-10. This matching set of knives is used for shaping tongue-and-groove joints. Many operators will have two molding heads for convenience and accuracy when working with sets of matched knives.

ting edges can't be nicked. Coating them with a film of light oil will help keep them in prime condition.

All of the thoughts expressed in Chapter 8 for correct mounting of a dadoing tool apply to a molding head. Don't have the tool plugged in when you are organizing the cutter on the arbor. Whether or not you can use one or both of the arbor washers will depend on the thickness of the head and the length of the arbor, but you can garner this information from the owner's manual. In any event, it's critical that you leave enough thread on the arbor so you can tighten the locknut securely.

Like a dadoing tool, a molding head makes much wider cuts than a saw blade, so it must be used with a special insert (Fig. 9-12). Be sure that you mount each knife so cutting edges face the front of the machine. Before you plug in the saw, turn the head at least one full revolution by hand to be certain it will rotate freely in the slot without hitting any part of the saw. Follow this procedure each time you change the head's elevation.

It is also a good idea to stand aside after turning on the saw and letting the molding head run free for 30 seconds or more. Then unplug the tool and check to see if the knives are seated as they should be. Do this occasionally even while you are working. A nuisance? Maybe, but working safely isn't.

Don't neglect to make special inserts anytime you are doing work that can be done more safely and accurately by minimizing the open-

Fig. 9-11. This set of knives can be used to produce three-bead edges or to striate a surface. In either case, the projection of the cutter must allow only the bead profiles to perform.

Fig. 9-12. When you buy a molding head, be sure to acquire the special insert that must be used with it. Always be sure that the setup doesn't allow the cutter to hit the insert or any part of the machine.

ing around the cutting tool. As for other special inserts for saw blades or dadoing tools, be sure the homemade insert is held securely in the table opening and then very, very slowly raise the cutter from its starting position under the table (Fig. 9-13).

Many molding operations, especially those that utilize only part of a knife's profile, are performed by moving the work along the rip fence. This makes it necessary to equip the fence with an auxiliary wood facing like the one that was suggested for dadoing work. After attaching the facing, preferably with screws through the rip fence, position it so a clearance arch can be cut about halfway into the facing's thickness. The arch doesn't need to be more than ¾ to 1 inch high. Be careful when you are raising the cutter to form the arch since you don't want to cut into the insert (Fig. 9-14).

WORKING

Like a dadoing tool, a molding head removes a

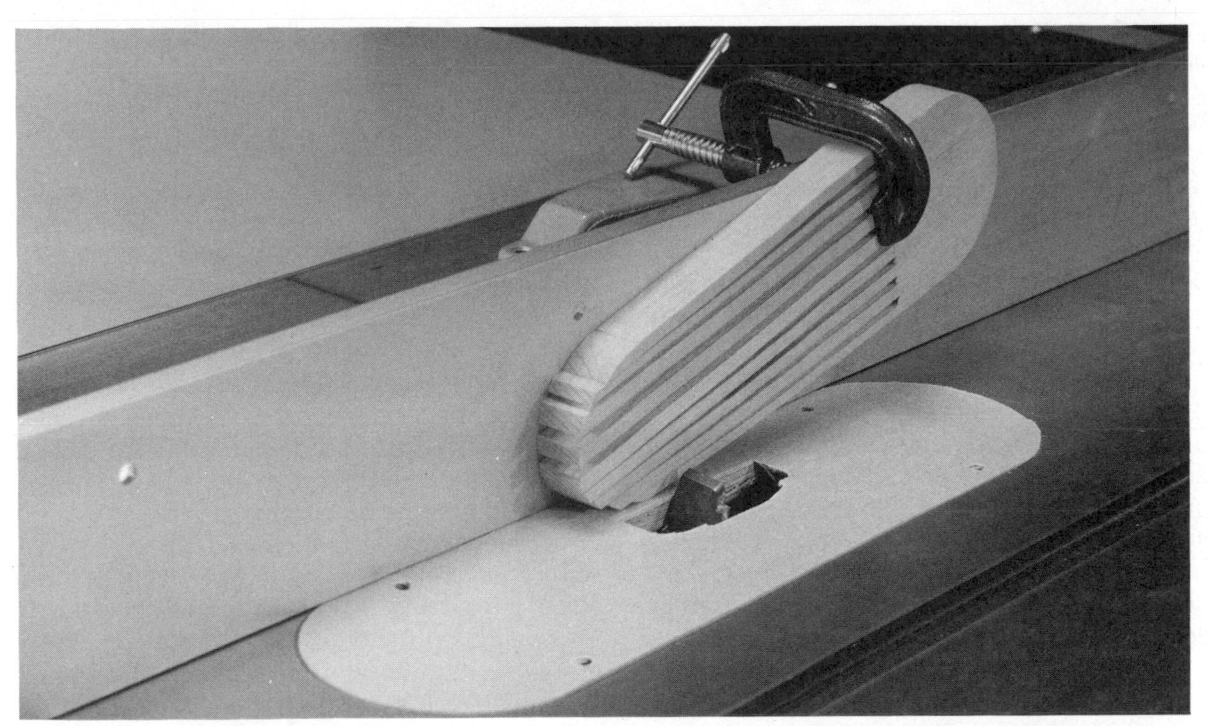

Fig. 9-13. Don't hesitate to make special inserts when the operation can be done more safely and more accurately by minimizing the open area around the cutter.

Fig. 9-14. The special wood facing required for shaping operations is simply a clean, straight piece of wood secured with screws that pass through the rip fence. The relief arch for the cutter should be one-half to three-quarters into the thickness of the facing.

lot more material than a saw blade, so feed speeds must be much slower to allow the knives to cut efficiently. Forcing is bad practice and can only result in rough cuts and, probably, burn marks on the wood and the cutters. Trying to make very deep cuts in a single pass also will result in poor work and can harm the motor and the cutter. It takes a little more time to get to the depth you need by making repeat passes with the cutter raised for each one, but it's the way to go if quality is a goal.

If the saw's motor slows up excessively or if the work vibrates or chatters so it is difficult to hold it steady, you can be pretty certain you are cutting too fast or too deeply. The right procedure results in a smooth operation, with the knives removing fine shavings instead of chunks of wood. When repeat passes are in order, you'll get optimum results by making the last pass as light as possible.

Whenever possible, do molding operations by using the fence for work guidance and support. Since the majority of shaping is done on the edges of stock, you'll find this suggestion suitable for most work. The depth of the cut is controlled by the cutter's elevation above the table. The width of the cut is determined by the position of the fence.

When you are making end cuts, which will be across the grain, use a miter gauge to advance the workpiece unless the work is too large for the use of a gauge to be practical. Employing a miter gauge is especially important for accuracy and safety whenever you must shape the end of narrow pieces (Fig. 9-15).

Alignment of components is another critical factor. The fences must be parallel to the table slots, and the angle between the fence and the miter gauge must be 90 degrees.

It is characteristic of cross-grain work that some splintering or feathering will occur at the end of the cut (Fig. 9-16). The degree of the flaw will depend on the density of the material, the makeup of its grain, and the ease with which you move the stock at the end of the pass, but it's difficult to avoid completely. If you require a

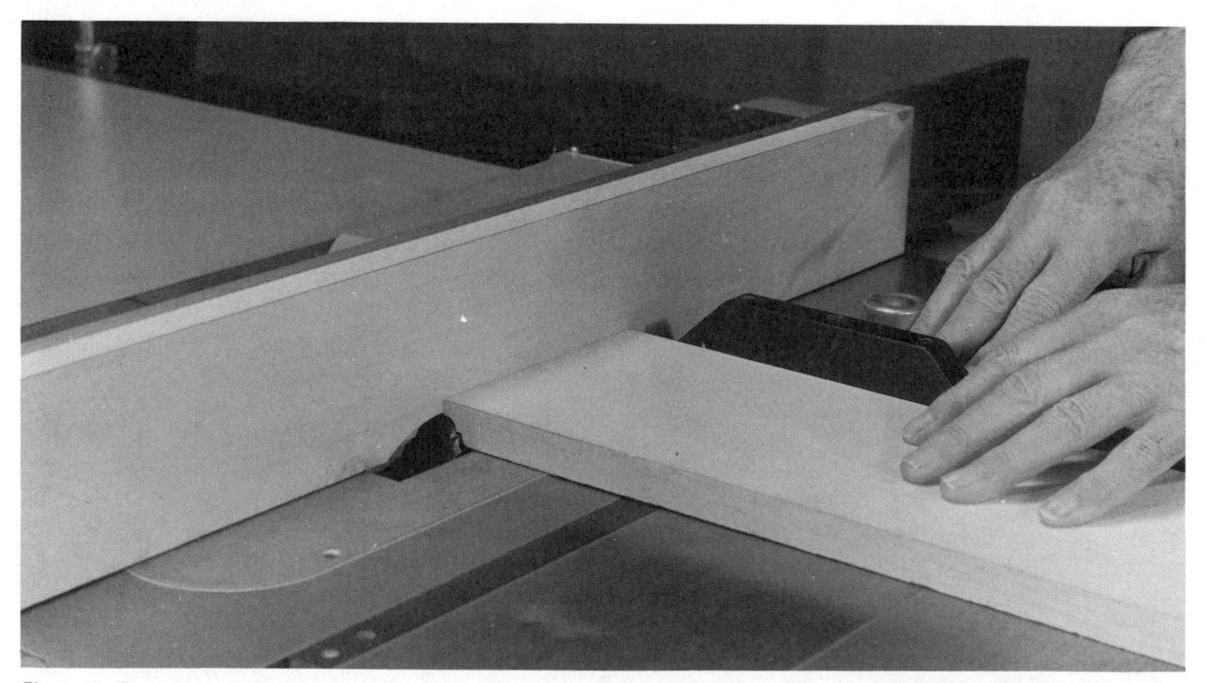

Fig. 9-15. To make end cuts, advance the work with a miter gauge. This step is very critical for safety when you are working on narrow pieces of stock. Make the pass slowly, especially when the cutter starts to break through at the end of the pass.

shape only on the end of the stock, you can solve the problem by shaping a piece of material that is a bit wider than you need and then removing the flaw by making a rip cut. When you need a shape on four edges or two adjacent edges, make the cross-grain cut first. The following cut, or cuts, made parallel to the grain will remove the imperfections (Fig. 9-17).

The relationship between fence and cutter doesn't change when you are doing edge shaping. The procedure changes because, during edge shaping, the stock is moved as if you were making a rip (Fig. 9-18). Use a springstick to maintain the stock's contact with the table. Don't use your hands to move the work past the cutter unless the work is wide enough for safety. When necessary, work with a pusher stick or with a combination pusher hold-down.

When you need the same shape on many pieces, you can make a setup that will ensure accuracy. Clamp one springstick so it bears down on the work in front of the cutter, and a second one at the back of the cutter. Clamp a third springstick to the table to hold the workpiece firmly against the fence. You also can organize commercial spring-type hold-downs for this procedure (Fig. 9-19). Whatever hold-downs you use, be sure to place them so they don't snap into the cutter after the pass is completed.

You also can do edge shaping by making the pass with the stock on edge instead of flat on the table (Fig. 9-20). This variation might re-

Fig. 9-16. There will always be some splintering or feathering at the end of cross-grain cuts. If the end shape is the only one required, make it on stock that is a bit wider than you need. Then you can rip to remove the flaw.

Fig. 9-17. If shaping is needed on four edges or two adjacent edges, make the cross-grain cuts first. The following cuts, made with the grain, will remove imperfections.

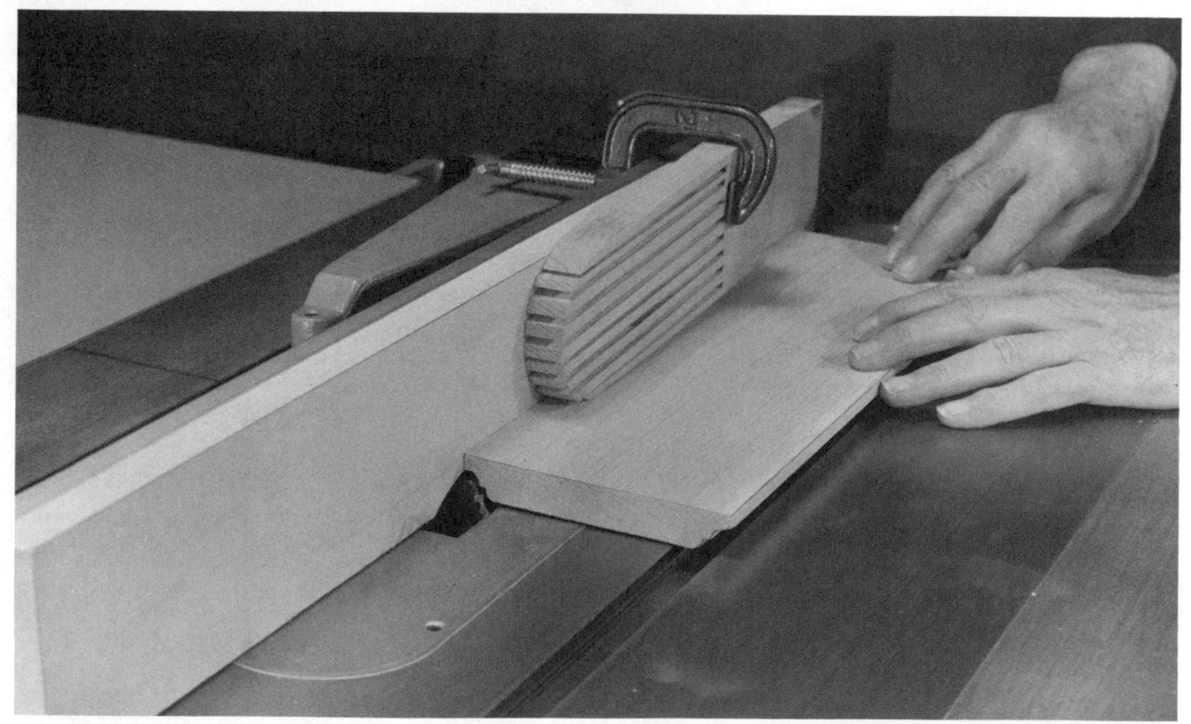

Fig. 9-18. To make edge cuts, move the work along the fence. A springstick helps to keep the work flat on the table. Cuts are always smoother when they are made with, instead of against, the grain. You can't always work this way, but working with the grain is a good basic rule.

Fig. 9-19. Commercial spring-type hold-downs do a good job of keeping the work flat on the table and snuggly against the fence. Using three springsticks will provide the same kind of assistance.

Fig. 9-20. Another way to shape edges is to make the pass with the surface of the stock against the fence.

sult in a "different" shape from that which results when the stock is flat on the table, even though the same knives are used.

Shaping does not *always* need to be done on individual pieces. For some projects, you can add the final touches to components that are already assembled (Fig. 9-21). Picture frames, frames for paneled doors, and preassembled facings for cabinets are examples.

Making Slim Moldings

You can shape slim strips with the molding head, but you must follow special techniques in order to work safely. One system calls for forming the shape you need on a piece of stock you can safely handle and then sawing off the edge you need (Fig. 9-22). Continue the procedure, after smoothing the sawed edge if necessary, until the stock becomes too narrow for hand feeding. This method involves alternating between shaping and sawing, and possibly sanding, which can be time-consuming and a nuisance when you need many similar pieces.

You can set up for production work by making one of the jigs shown in Figs. 9-23 and 9-24. Saw the strips for the molding to the right width first. The jig has an opening that will accommodate the thickness and the width of the prepared strips. The fit of the strips in the opening should be reasonably snug so they won't chatter as the cut is made. Feed the strips through at the front of the jig, using succeeding strips as pushers. Finish the cut on the last strip by pulling it out from the back of the jig.

Shaping Circular Work

You can do shaping operations on the perimeter of circular workpieces by organizing the jig displayed in Fig. 9-25. This jig consists of two matching blocks, each cut at a 45-degree angle on one edge. A backing piece is attached to each block so the components can be clamped to the rip fence to form a *V* that will cradle the work so it can be rotated for the cut. The blocks are separated only enough to accommodate the size of the work and are placed so the centerline of the *V* is on line with the center of the cutter.

Hold the work securely, flat on the table, and very slowly move it forward until it contacts the bearing edges of the jig. Contact with the cutter will have been made at this point so you can rotate the work, slowly, in a counterclockwise direction to complete the cut (Fig. 9-26). Be sure that contact between the work and both legs of the *V* is firmly maintained throughout the pass. It is also important to be sure that the cut leaves enough solid edge on the work for good contact between the work and the guide edges of the jig. If you wish, you can tack-nail a strip across the *V* to serve as a hold-down. This strip will help you keep the work flat on the table as you do the shaping.

Rabbets and Tongues

Rabbets and tongues, forms required in basic joinery, can be cut very efficiently with a molding head fitted with blank knives. Using a table saw for joinery operations will be covered in Chapter 10, but it seems apropos to make the point that the molding head can be used for more than decorative work.

You can shape rabbet cuts along edges by working as shown in Fig. 9-27. The depth of the rabbet depends on the projection of the cutter; its width is controlled by the distance from the fence to the outside of the knife. Blank knives are made with a clearance slope on outer edges so you can be sure that the shoulder of the cut will be just as smooth as the bottom and that the corner between them will be clean.

To form tongues, make back-to-back rabbet cuts. To do so, make passes with the work flat and invert it for opposite cuts, or by working with the stock on edge (Fig. 9-28). Here, the *width* of the cut depends on cutter projection, while its *depth* is controlled by the relationship between fence and knife.

You might find that making cuts of this nature with a molding head and blank knives will result in smoother work than that achieved with a dadoing tool.

Fig. 9-21. Picture frames and similar assemblies can be shaped on edges after the components have been assembled. Be careful if you have used fasteners in the joints. You don't want the knives to cut into metal.

Fig. 9-22. You must never shape slim pieces by trying to control the operation with your hands. A safer way is to shape the edge of a wide piece of stock and then saw off the part you need.

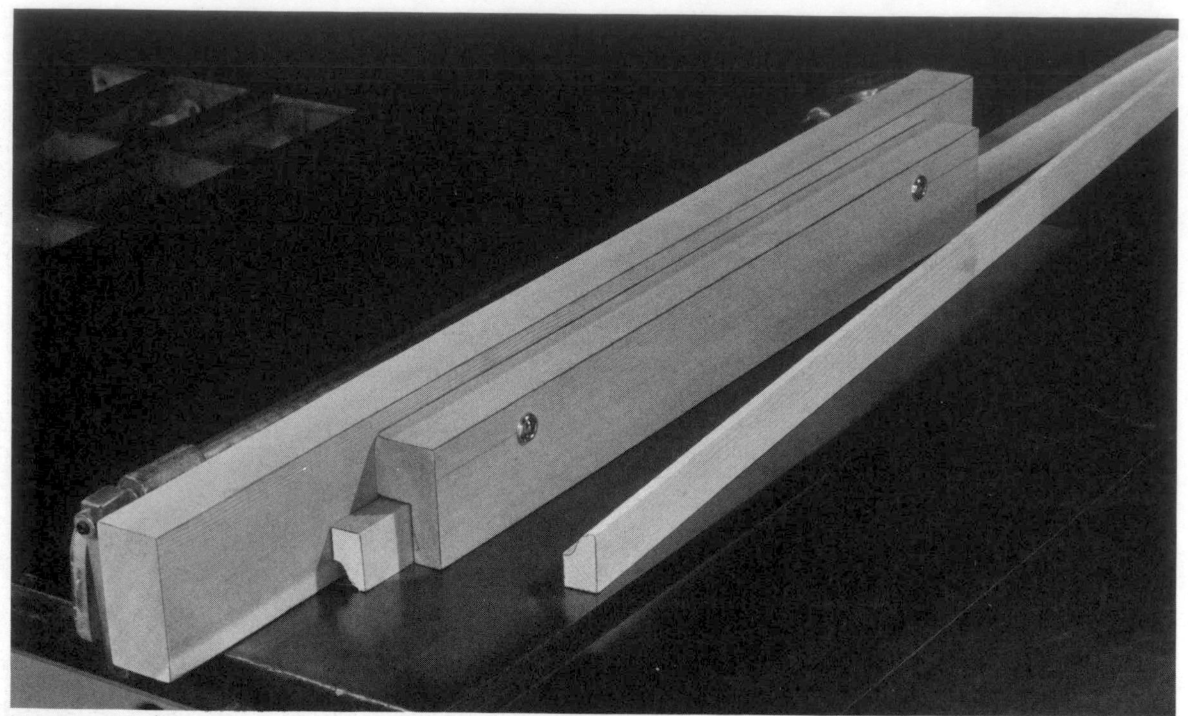

Fig. 9-23. Making a jig of this type is a more professional way to go when you need many similar pieces of slim molding. The technique allows precutting of workpieces, which saves a lot of time and energy.

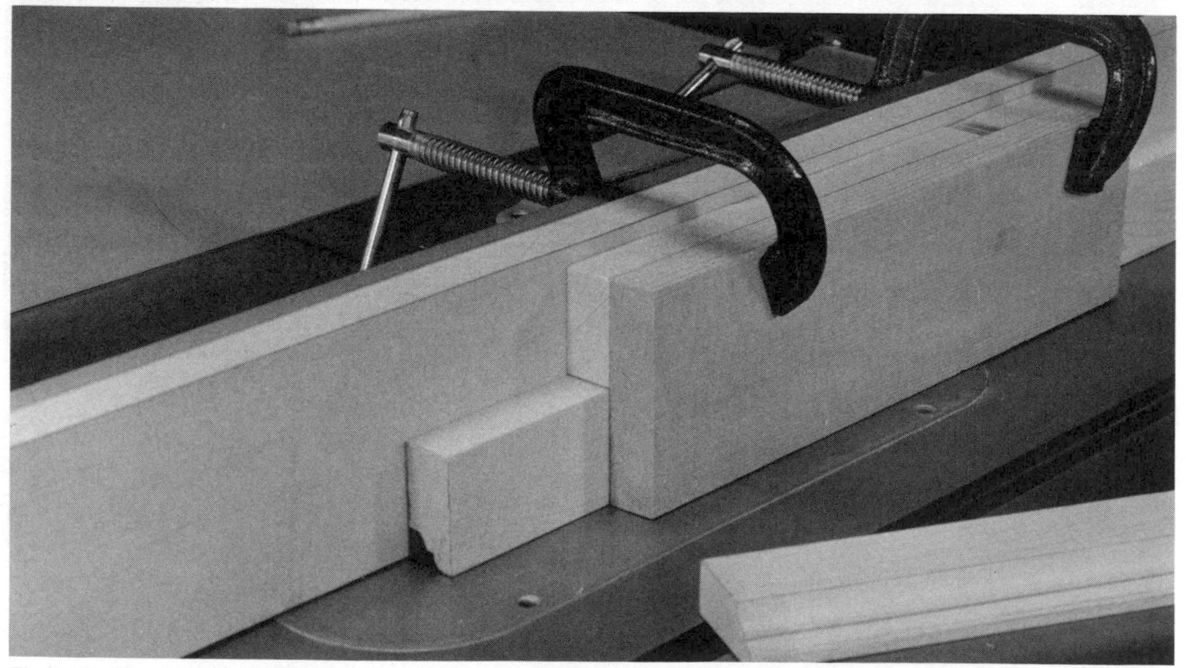

Fig. 9-24. When a particular slim-molding operation is a one-time chore, you can improvise the control jig this way. Be sure the L-shaped opening that is created by the positions of the clamped pieces will provide a reasonably snug fit for the workpieces.

Fig. 9-25. You can shape the perimeter of circular pieces if you provide a V-block jig. Place the two parts of the jig so the V and the cutterhead are on the same centerline.

Scalloping

Scallops are semicircular, equally spaced, decorative cuts formed in the edges of workpieces (Fig. 9-29). You can form them with a molding head fitted with knives of your choice, but the technique calls for very careful work handling and accurate placement of the control jig shown in Fig. 9-30. The jig requires two matching guides that have L-shaped cutouts at one end. The guides are attached at the L-shaped end with a *spacer*, which is a piece of stock whose thickness equals that of the workpiece. This assembly is then attached, on the right-hand side, so the total jig can be secured to the rip fence with clamps.

Position the fence so the cutter will be in the space between the guides. Make cuts by bracing the work against the spacer and very slowly lowering it over the turning cutter until it is flat on the table. At this point, turn off the machine and move the jig forward to establish the spacing for the next cut.

One way to establish spacing is to take measurements from the front of the table to the forward edge of the guides. For example, if you wish the cuts to be 2 inches apart, then move the guide forward 2 inches for each cut. Be sure to hold the work very firmly as you position it and lower it for each cut.

Fig. 9-26. Start the cut by advancing the work very slowly until it contacts the knives and seats firmly against the control edges of the jig. Then, just as slowly, rotate the work in a counterclockwise direction (arrow). Keep the work flat and held firmly throughout the pass.

Surface Carving

You can think of other names for this technique—striating, faceting, panel raising—but whatever the name, the idea is to make cuts of limited depth into the surface of workpieces, instead of along edges. The effects you get have everything to do with the knives you use. The cuts can be parallel, like those in Figs. 9-31 and 9-32, or they can intersect like the examples that were shown in Fig. 9-1.

The important factors to consider follow. The workpiece must be flat to begin with. Smoothest cuts result when you make passes so the knives cut *with* the grain of the wood. When you are making passes that intersect, make the cross-grain cuts first. When the design calls for working with a tilted cutter, be sure the knives have enough clearance in the insert's slot (Fig. 9-33). Provide for hold-downs to help you keep the work flat on the table. Whenever possible, use a combination pusher/hold-down to move the workpiece. Don't make cuts so deep that the stock will be thinned excessively. Stock that is too thin will result in a component that is weak and that becomes too flexible for safe control. You must be very precise when you are setting up for cuts that must match or that require a particular spacing.

Check Table 9-1 when you're not happy with the results of a molding operation. It will lead you to the cause of the problem, and will help you correct it.

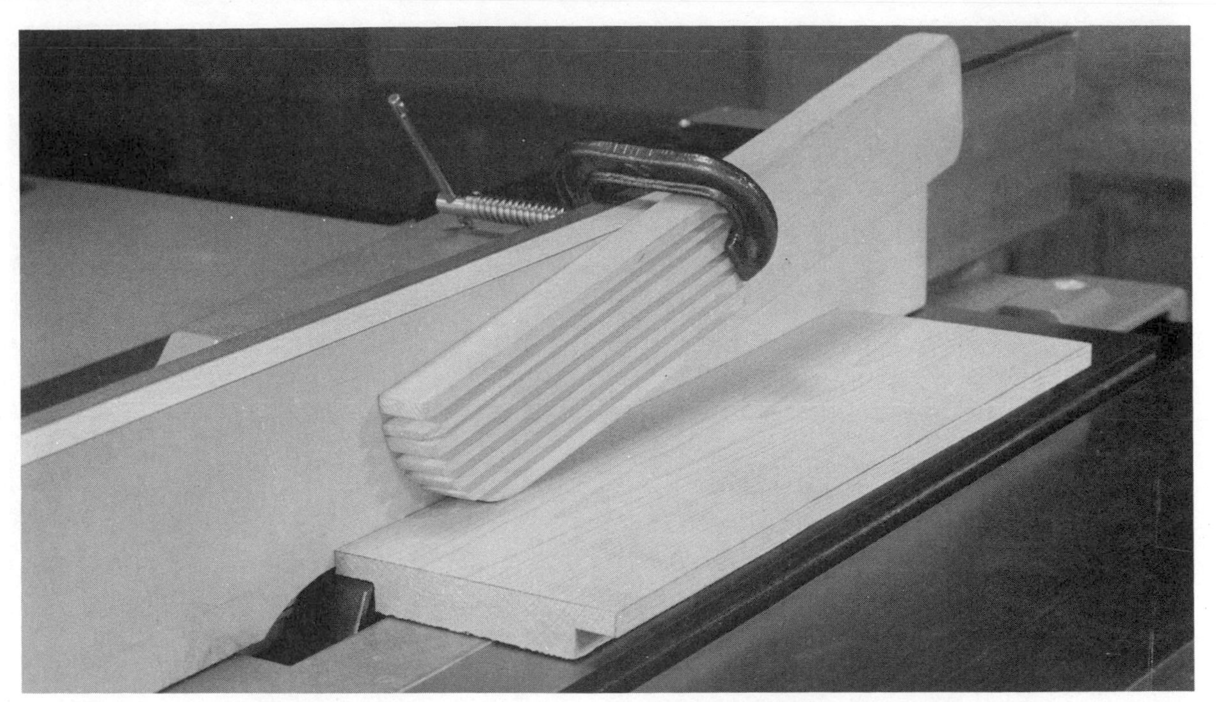

Fig. 9-27. Blank knives are very useful for joinery forms such as rabbets. It's not unusual that, when used so, they will do a better job than a dadoing tool.

Fig. 9-28. Back-to-back rabbet cuts form a tongue. When working this way, you want to be sure that the thickness of the tongue will be enough to provide safe bearing surface on the table. Don't allow the work to tilt at the end of the pass.

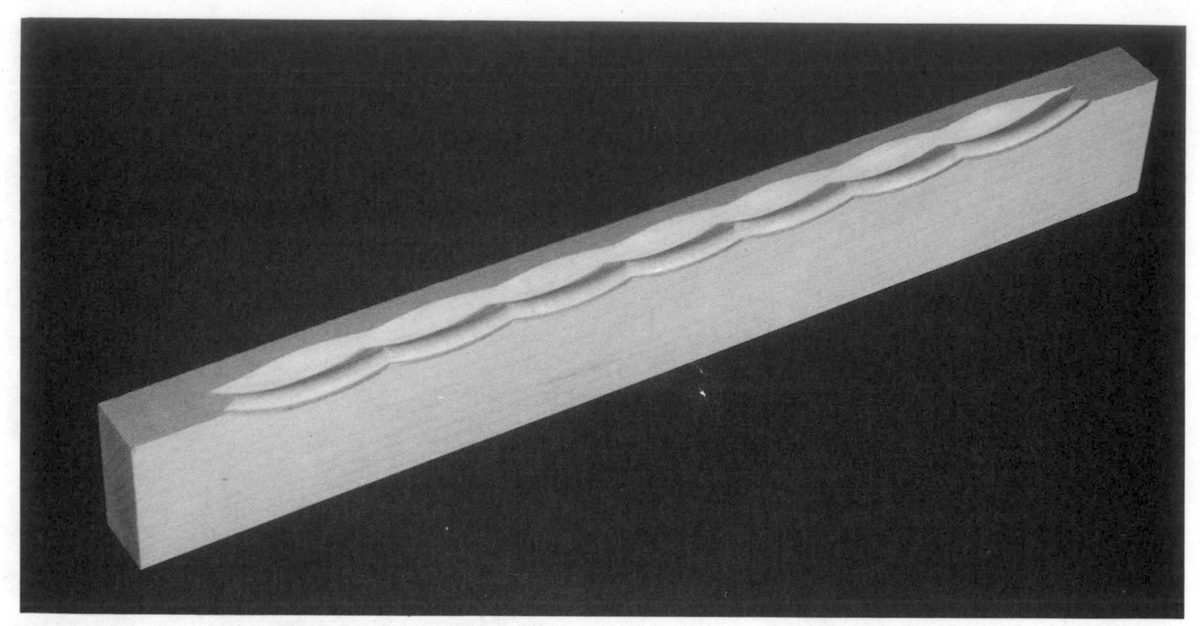

Fig. 9-29. You can use a molding head to do decorative edge-scalloping.

Fig. 9-30. Scalloping calls for a very special setup and careful attention to the procedure described in the text. Each cut requires that the control jig be moved a specific distance.

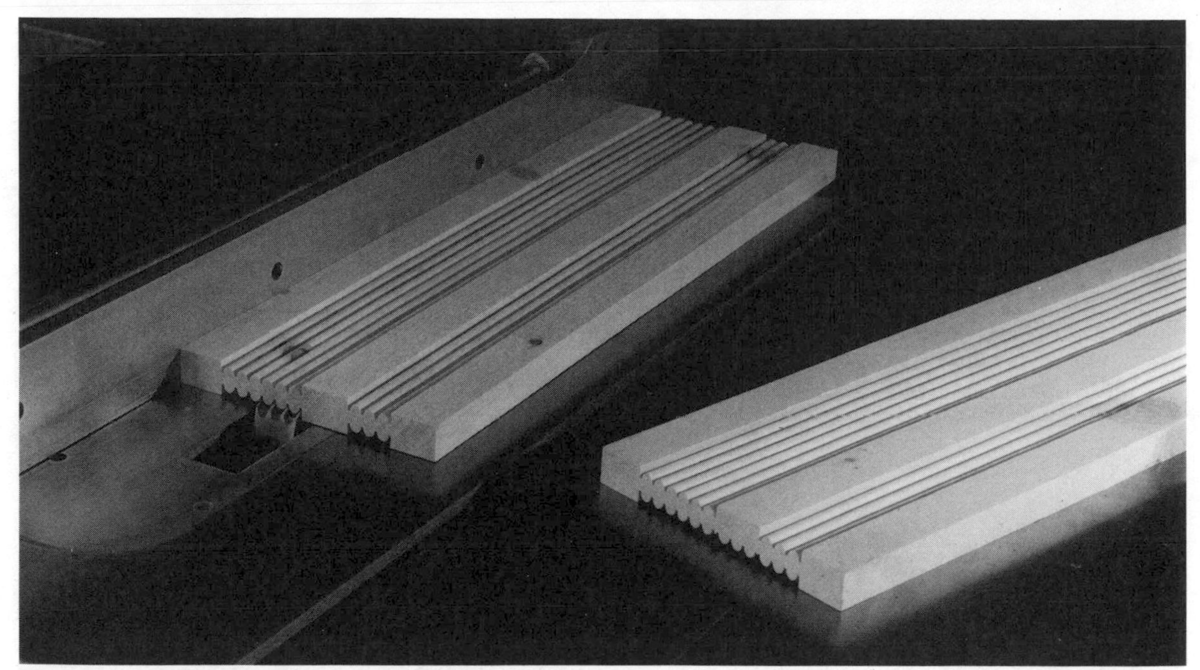

Fig. 9-31. These projects illustrate the type of surface carving that can be done with three-bead knives. On work of this nature, there is no problem to make all the passes so that cutting is done parallel with the grain of the wood.

Fig. 9-32. Keeping the work flat on the table is essential for quality results. Here, a mortising hold down is used, which is possible on a Shopsmith rip fence.

Fig. 9-33. After tilting a cutterhead, be certain that the knives will have freedom in the insert's slot. Working with a pusher/hold-down is a good idea. Being certain that the workpiece is placed accurately for each cut is essential for optimum results.

Table 9-1. Troubleshooting Chart for Molding-Head Operations.

PROBLEM	POSSIBLE CAUSE	CORRECTION
Molded edges not uniform	Poor work-handling	Keep work firmly down on table and against fence throughout pass
	Poor work edge	Edge to be molded must be straight and smooth/Be sure work has uniform thickness
Work chatters/Tool stalls	Cutting too fast	Use slower feed/Allow cutter to work at its own pace
	Cutting too deep	Make repeat passes when necessary
	Dull tools	Keep molding knives in sharp condition
	Dirty tools	Keep knives free of wood residue
Inaccurate cuts	Poor setups	Make test cuts before cutting good stock/Be especially careful with matched cutters like tongue and groove and glue joint cutters
	Knives not set correctly Poor work-handling	Be sure knives are firmly seated and correctly aligned Keep work down on table and firmly against fence throughout pass
Excessive feathering at end of cut	Breaking out of cut too fast	Slow up feed at end of cut especially when cutting cross-grain
Burn marks on knives or wood	Cutting too fast	Slow up feed
	Cutting too deep	Make repeat passes when necessary
	Dirty tools	Keep knives clean, free of gum and pitch
	Dull knives	Keep knives in sharp condition/Have knives sharpened in sets

Chapter 10

Practical Woodworking Joints

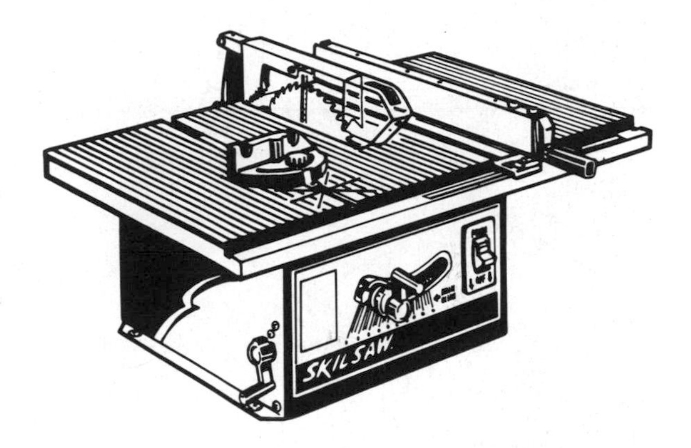

THERE ARE TWO IMPORTANT ASPECTS TO ANY woodworking project: how it looks and how it is put together. The first, its appearance, concerns aesthetics, something that is highly personal and difficult to define. The other, possibly more important, concerns engineering—how the components are connected—which determines whether the project will become an heirloom or be disposed of in the fireplace after a period of time.

I am not suggesting that every project should involve intricate joinery. In fact, the reverse is true. There is little point in using anything but a butt joint if the sides and ends of a temporary box are going to be secured with nails. If the box form is the basis for a hope chest or even a small jewelry box and the material used is a fancy, expensive hardwood, then you might think about a lock corner, splined miter, or finger joint, which would contribute to a long-lived project and indicate your dedication and respect for the wood.

Regardless of design, all the cuts required for a joint must be made precisely. Even the prosaic butt joint won't hold as it should unless the mating surfaces are smooth, flat, and square.

Not all woodworking joints can be made entirely by using a table saw. The rectangular or square cavity for a mortise-tenon joint requires the use of a drill press or portable router, or must be made by hand by drilling a series of overlapping holes and then cleaning out the waste with a chisel. On the other hand, the open mortise-tenon, in which the tenon fits a slot, is a table-saw function.

Dovetail joints can be accomplished on a table saw but the chore is tedious, leaves much room for human error, and ultimately requires some attention with hand tools. I believe that anyone interested in dovetail joinery should be equipped with the tools that automatically supply accuracy with minimum fuss, namely, a portable router with dovetail bits and a readymade dovetail jig. The cost of the equipment can be less than the price of a super, carbide-tipped saw blade.

All of the joints shown in Fig. 10-1 and a few others that will be demonstrated in this chapter can be accomplished on the table saw. It is likely that you will find viable substitutes for joints which might require additional woodworking equipment. A good rule is to choose the least complicated joint that is suitable for the job.

CONCEALING RABBETS AND DADOES

A point was made in Chapter 8 about the dado joint being unattractive when viewed from the front. The solution given was to cut a stopped dado and then shape the insert piece to fit the arc that results where the cut ends. A second solution is shown in Fig. 10-2. You can shape

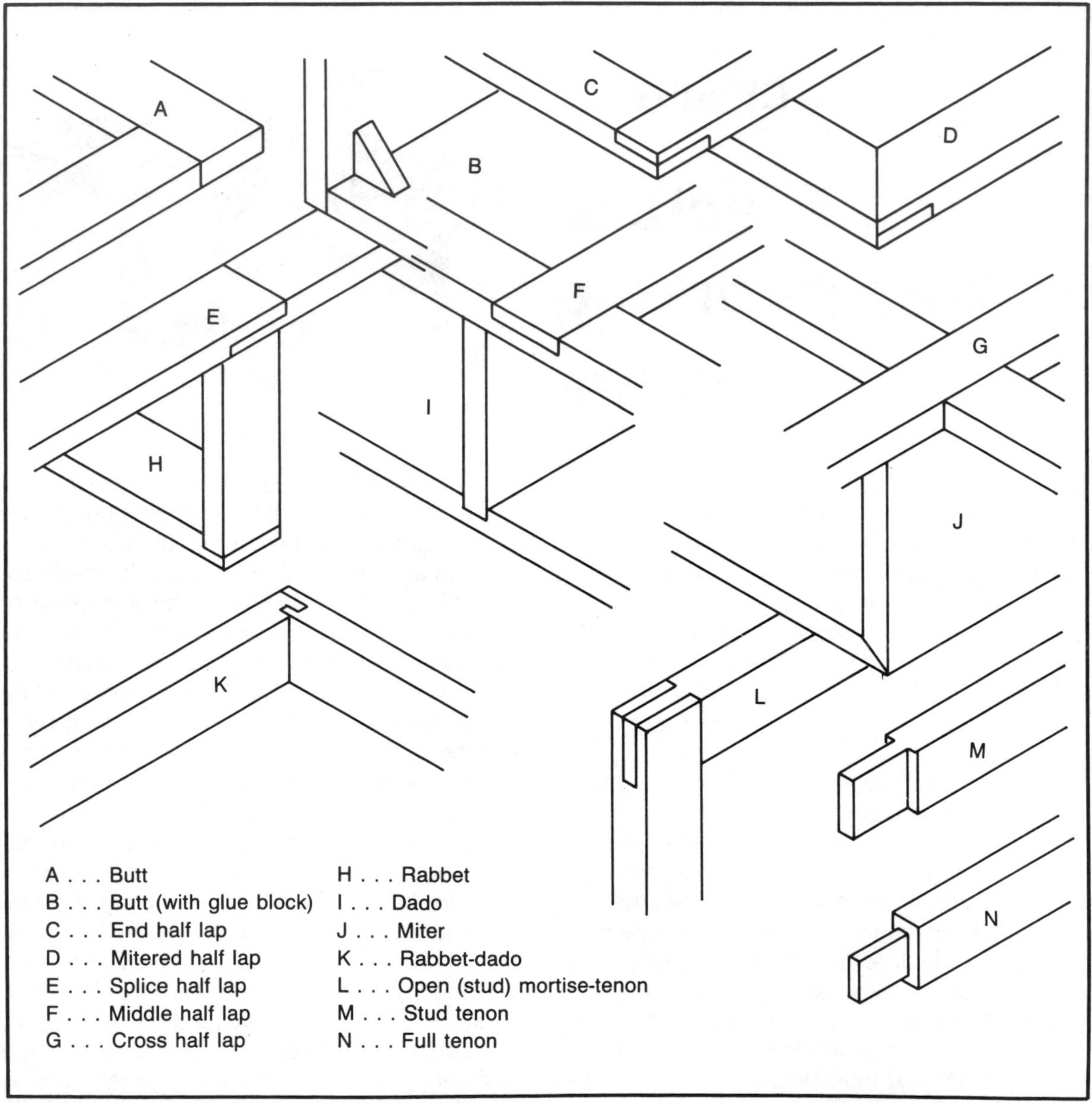

Fig. 10-1. The only tool you need to make the cuts that are required for these joints, and some others that will be shown in this chapter, is the table saw. Cutting can be done with a saw blade or with a dadoing tool.

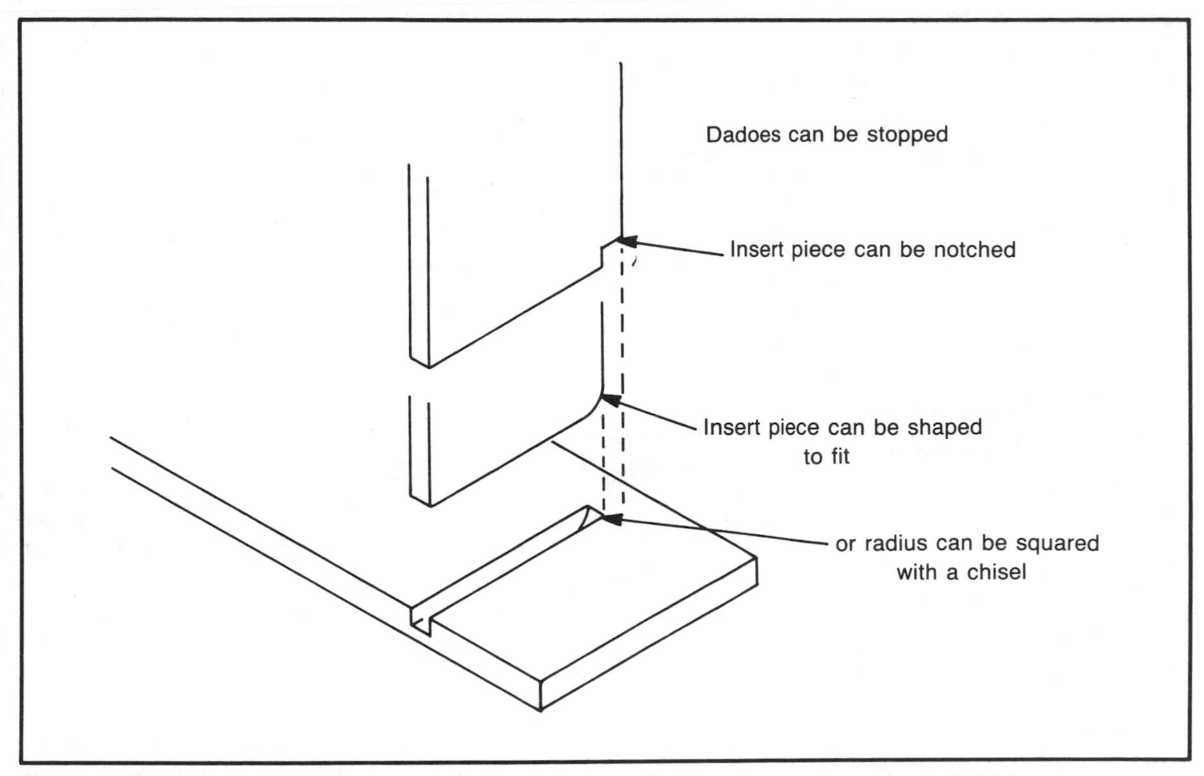

Fig. 10-2. These are methods you can use to conceal the unattractive end appearance of a dado joint.

the insert piece with a notch that is long enough to span across the arc, or you can square the cut at its end by working with a chisel so that the insert piece would require no attention at all. When using this method, the width of the insert—say a shelf—would be less than the width of the vertical component.

The same thoughts apply pretty much to rabbet cuts. You can stop the rabbet and shape the insert part to fit, or you can square the arc end with a chisel, or you can use a filler to simulate a stopped cut (Fig. 10-3).

A *rabbet joint* is often used as the connection between the front and side of a drawer whether the drawer front will be flush or lipped (Fig. 10-4). Actually, this is not the strongest or the most attractive joint for the purpose—a rabbet-dado or a lock-corner joint is better but it will do for utility drawers that won't get much abuse. Reinforce the joint by driving nails or screws or by drilling holes and inserting dowels through the side member into the shoulder of the rabbet. Of course, you should also use glue.

Rabbeted pieces are often used to reinforce or hide a corner joint or to add a decorative detail to a project (Fig. 10-5). Often, the corner pieces are made long enough to serve as legs, say for a planter box, a small bench, or a table.

CONCEALING PANEL EDGES

Man-made panels like those displayed in Fig. 10-6, are super materials for woodworkers, but edges, if left exposed, don't contribute much to the appearance of a project. A solution, especially when working with plywood, is to cut a rabbet in one part that is deep enough to leave only the surface veneer (Fig. 10-7). This does a good job of concealing core plies, but establishes a weakness at the base of the rabbet's shoulder. If you try the concept, be sure that other project components will supply necessary rigidity and strength, or add glue blocks to reinforce the joint.

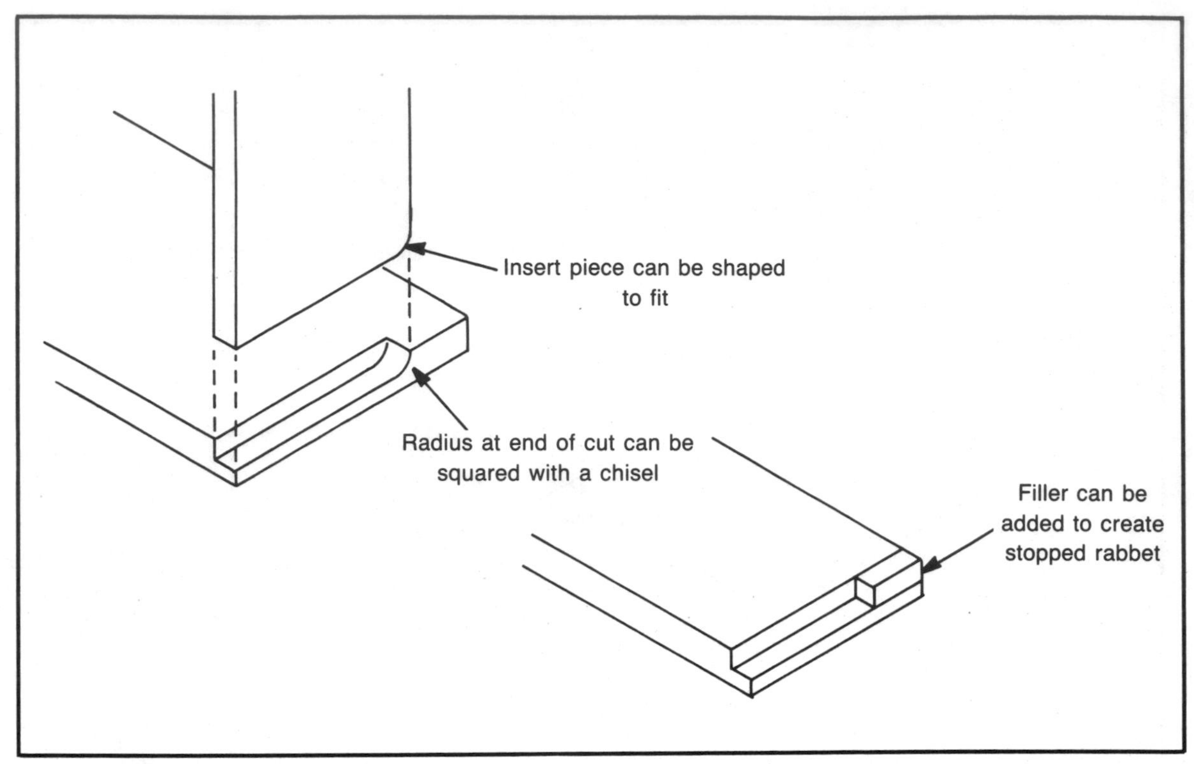

Fig. 10-3. Like a dado, a rabbet joint is not attractive when viewed from the front. These are typical methods that are used to conceal the joint.

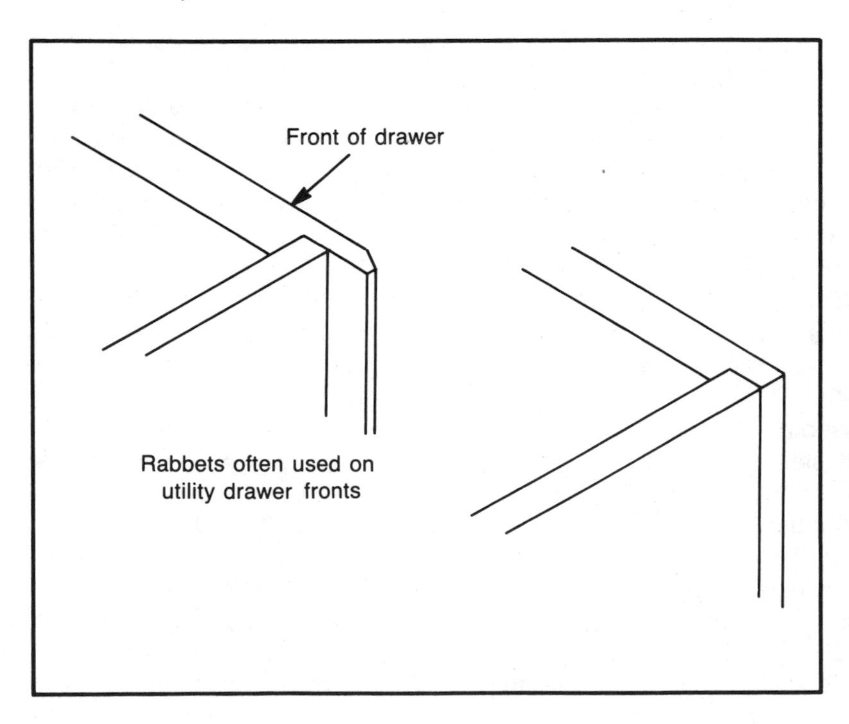

Fig. 10-4. Drawer fronts are often rabbeted to receive the side of the drawer. For the joint to be as strong as possible, nails, screws, or dowels are driven through the side of the drawer into the shoulder of the rabbet.

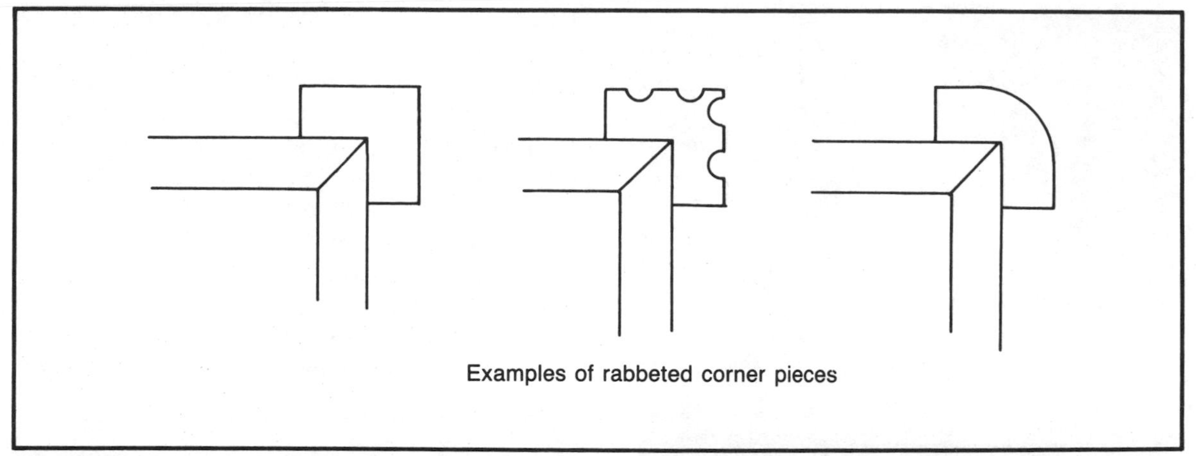

Fig. 10-5. Rabbeted pieces can be used to hide and reinforce a corner joint or to provide a decorative detail. Many types of molding with rabbeted corners are available and can be used as is for this type of application.

An interesting way to hide panel edges is illustrated in Fig. 10-8. The first step is to make a 45-degree miter cut along the edge of the panel. The cut must not reduce the thickness of the cutoff. Then take the triangular piece that is cut off, turn it end over end, and rotate it a quarter turn on its axis before you glue it back from where it came. The result is that the veneered undersurface of the triangular piece becomes the panel's new exposed edge.

An excellent technique for a panel joint that hides edges and results in minimum disruption of the grain's direction and pattern is something I call the *waterfall* joint (Fig. 10-9). Two cuts are required with the saw blade tilted to 45 degrees. The first cut removes the piece that will be reattached to the parent stock. The second cut, made on the cutoff, removes a triangular piece, which is discarded. Essentially, since a wedge of material has been removed, the cutoff can be "folded" down so the grain pattern seems to flow over the edge of the joint. The process is easy to understand if you visualize a piece of wood so soft that you can remove a V-shaped section with a knife while leaving a thin veneer at the base of the *V*. Then see the wood being bent to close the *V*.

Tapes, which are actually strips of flexible veneer, are commonly used to cover panel edges. Readymade types are available in almost any wood species. Some are attached by using a contact cement; others, which are more convenient to work with, are self-adhesive and can be secured with the heat from a household electric iron (Fig. 10-10).

A problem with ready-made edge tapes is that the grain runs longitudinally, so the grain direction of the panel can be matched on only two edges. On the remaining edges, the grain direction of the tape would be at right angles to panels' surface veneer. Persnickety woodworkers will often custom-make tape by making saw cuts that slice off strips of surface veneer from the parent panel. The technique is also useful when ready-made tapes are not available for the panel material on which you are working, for example, hardboard-surfaced plywood (Fig. 10-11). Slicing off surface veneers calls for very careful work and the use of a sharp planer-type saw blade.

Homemade edge tapes can be attached with conventional woodworking glues, but this method calls for a type of clamping that can be a nuisance. It is better to work with a contact cement since it eliminates the need to use clamps. Be sure to read and to follow the instructions on the products of this nature.

Concealing the edges of a panel and at the same time making it appear to be thicker than it is, which is a possible consideration for the

Fig. 10-6. Panel materials like plywood, particleboard, flakeboard, and others, are a boon for woodworkers, but edges need special attention for them to be visually appealing.

tops of tables and chests, calls for the use of special perimeter pieces that can be as simple as those shown in Fig. 10-12. When making and attaching the border components, allow the top edges to project just a fraction above the surface of the panel. Doing a bit of sanding after you have assembled the parts will ensure that all top surfaces are flush.

Figure 10-13 suggests other ideas you can use to frame panels. Those that project above the panel's surface are suitable for serving trays and similar projects where a perimeter lip is practical.

HALF-LAP JOINTS

There is a difference between a *lap joint* and a *half-lap* joint. The lap joint is a wood connection where an area of one component actually laps over part of a mating piece. The connection is actually surface to surface, essentially a butt joint. For the sake of clarity, it should be called a *surface lap.* The strength of the joint lies in the area of glue contact and whatever reinforcements, like nails or screws, are used.

The half-lap joint generally is a design, where the thickness of each component at the contact area is reduced by one-half so that, when the parts are put together, the exterior surfaces will be on the same plane and the pieces will interlock. When the necessary cuts are made accurately, the joint will resist lateral and twisting stresses.

When the parts to be joined with a half lap are equal in thickness and width, the width of the cut required in each piece equals the width of the parts. The depth of the cut is one-half of the stock's thickness.

The cuts that are needed to form *end half laps,* often called *frame laps* since they are frequently used as the corner connection on square or rectangular frames, are no more than wide rabbets (Fig. 10-14). The fence can be used as a stop since nothing is being cut off. The distance from the fence to the outside of the cutter, which can be a dadoing tool or a molding head fitted with blank knives, equals the width

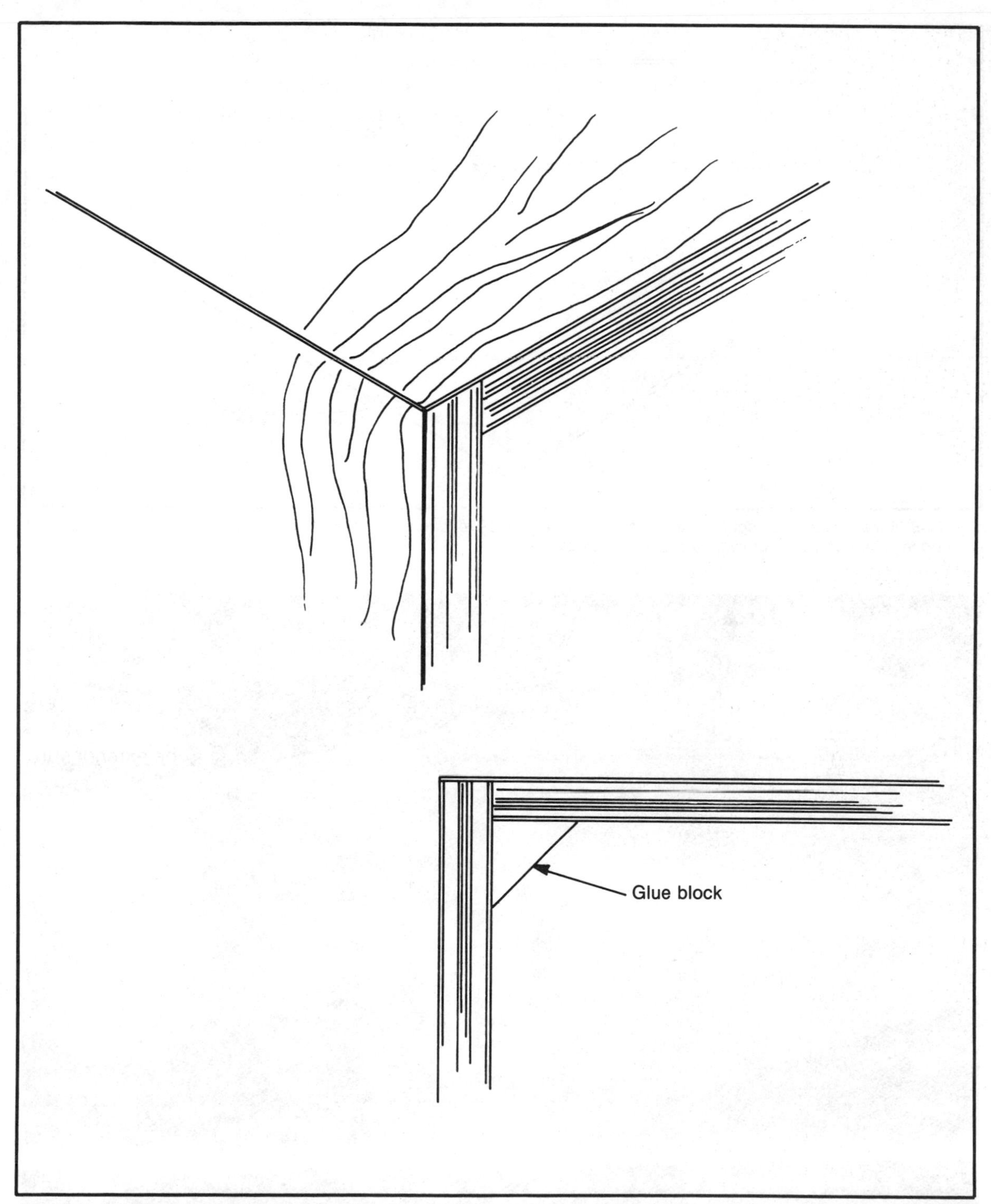

Fig. 10-7. A deep rabbet will conceal unattractive plies, but the joint is weak and should be reinforced with glue blocks, unless other project components supply necessary strength and rigidity.

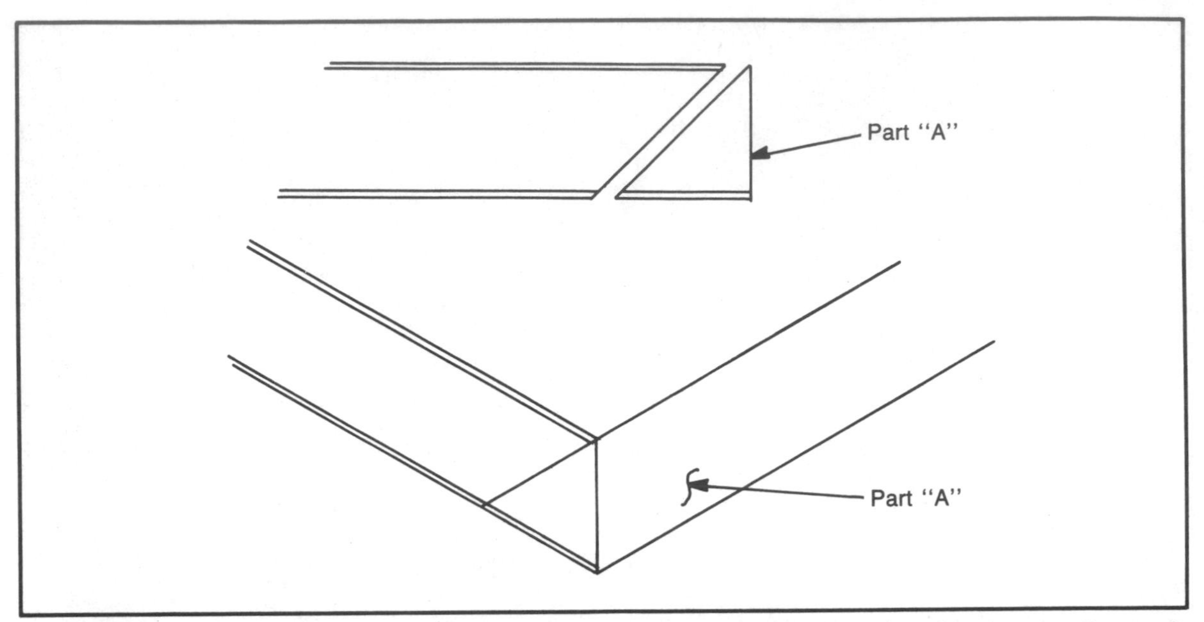

Fig. 10-8. A triangular piece cut from the edge of a panel can be reattached so that its undersurface becomes the panel's new edge. The cut must not reduce the thickness of the cutoff.

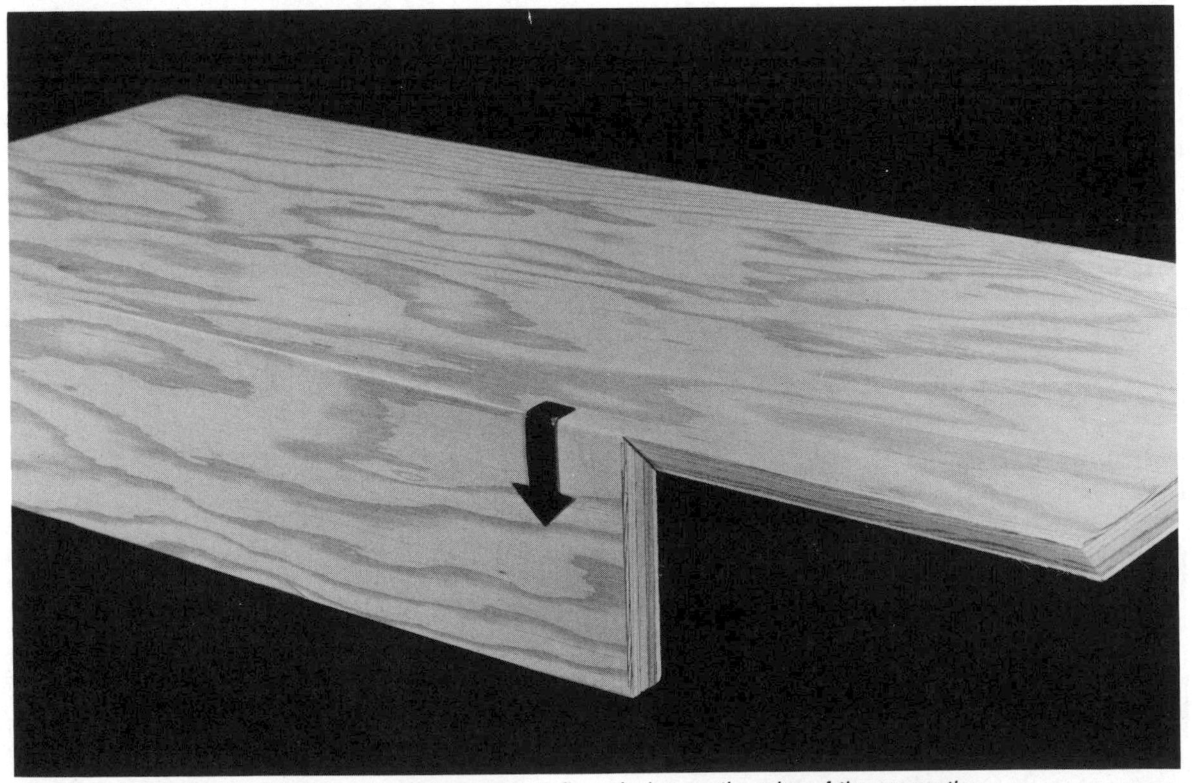

Fig. 10-9. The waterfall joint allows the grain pattern to flow nicely over the edge of the connection.

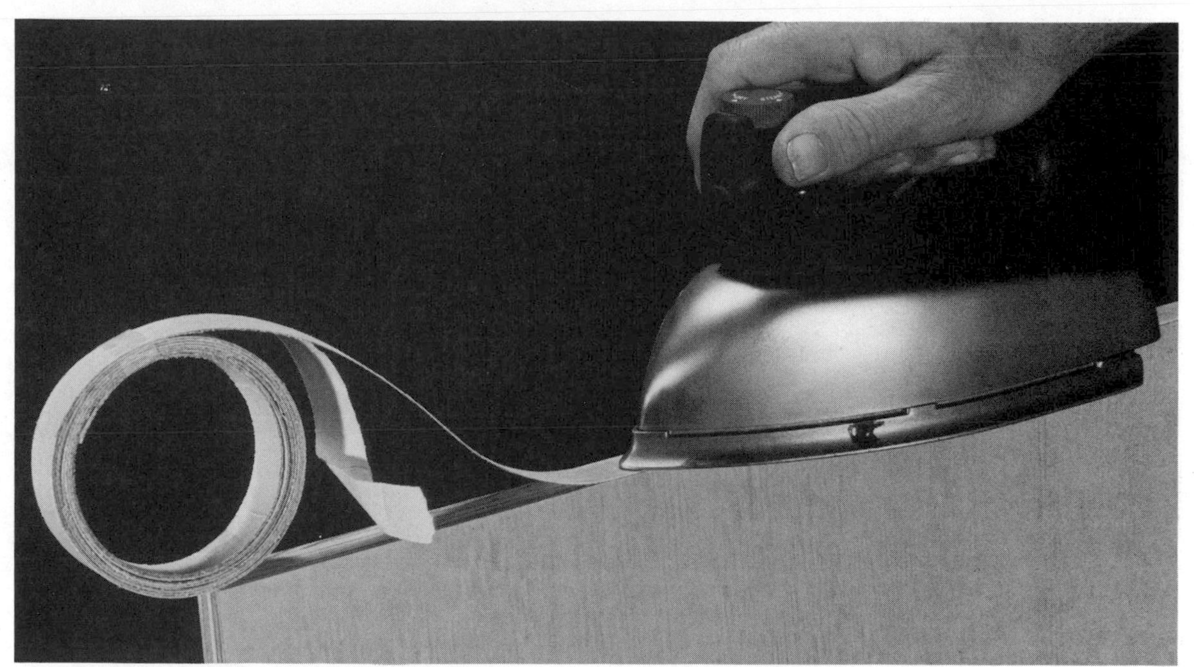

Fig. 10-10. Banding tapes, actually flexible veneers, are available in many wood species. Some are attached with glue or contact cement, others are self-adhesive and require only heat, which can be supplied by a household iron.

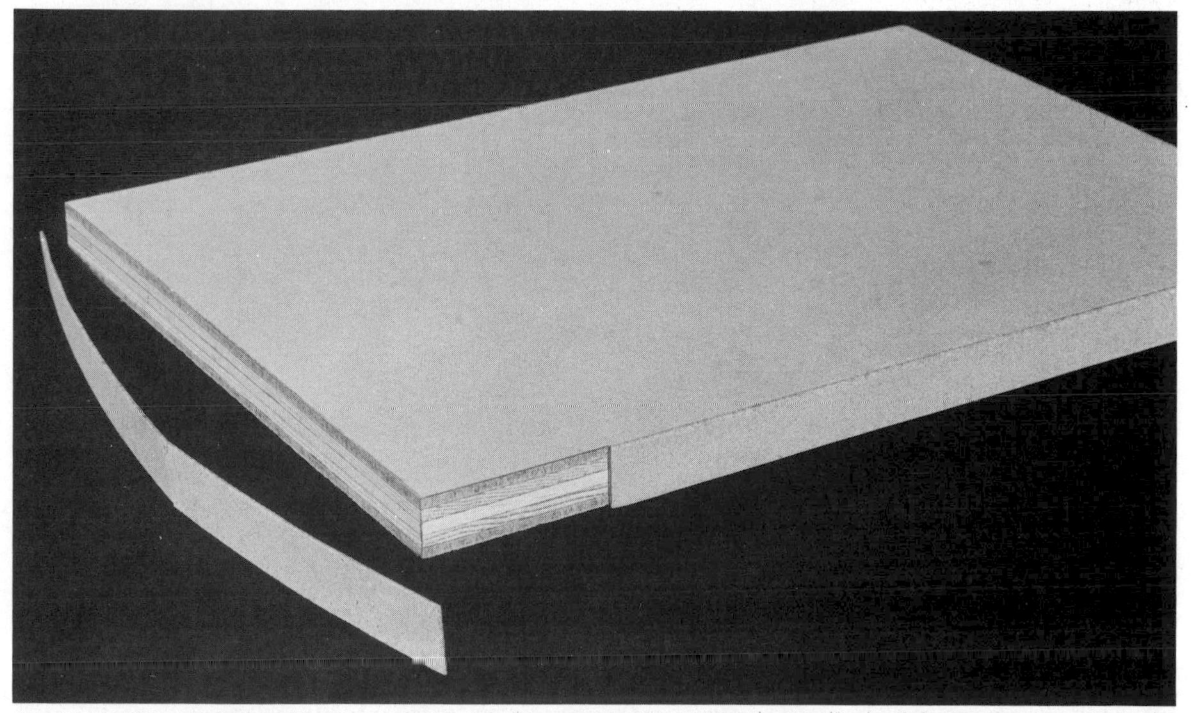

Fig. 10-11. Many workers make their own edge bandings, especially when a readymade product that matches the parent panel is not available. Careful cutting with a planer-type saw blade is necessary.

Fig. 10-12. You are not limited to flexible tapes for concealing panel edges. These examples show how you can hide unattractive plies while adding bulk to the panel's edge.

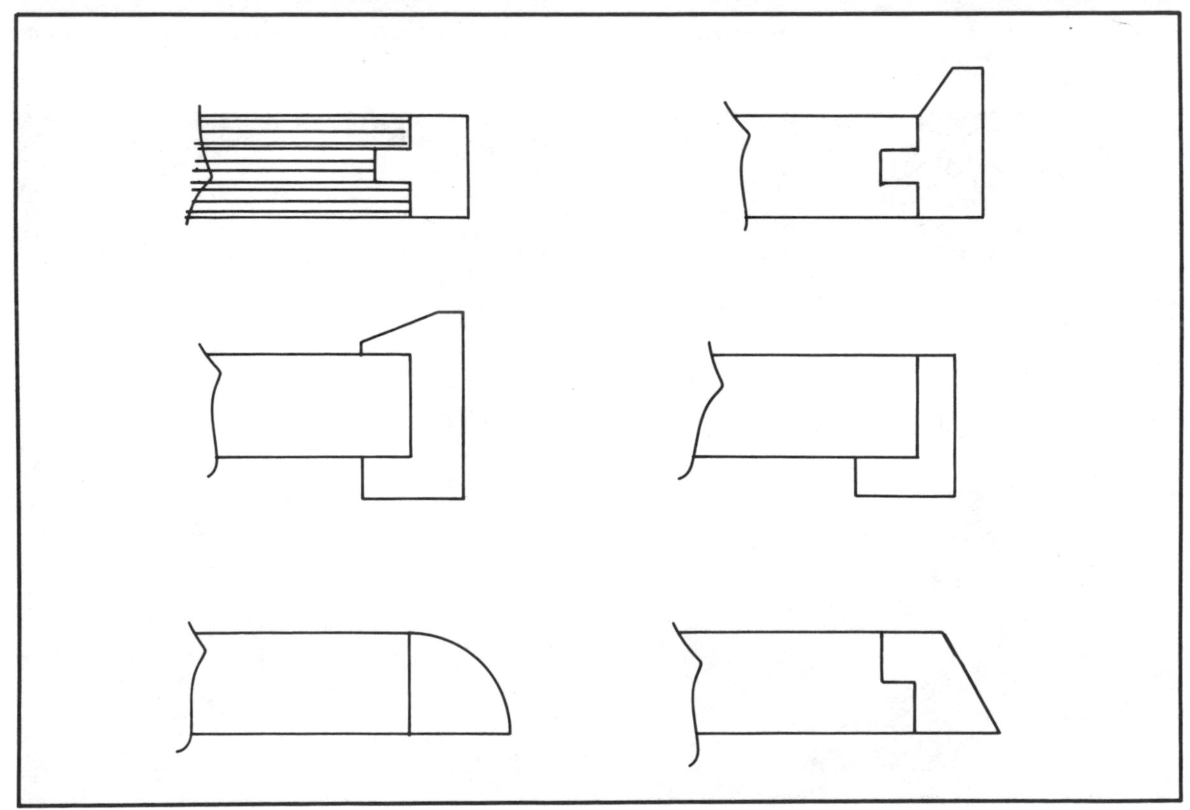

Fig. 10-13. Other design ideas that can be used when a panel is framed. Those with lips that project above the panel's surface should be considered for projects like trays and small tables.

Fig. 10-14. The cuts required for end half laps, or frame laps, are just wide rabbets. To form them, make repeat passes with a dadoing tool. It's okay to use the fence as a stop since a cutoff is not involved.

of the cut. Cut depth is controlled by the cutter's projection. The waste is removed by making repeat, overlapping passes. Careful cutting will result in flush surfaces and edges when the parts are assembled (Fig. 10-15).

The *splice half lap* (Fig. 10-16) is accomplished in the same way. There can be a difference in the width of the cuts since the distance the parts will overlap is optional.

The *cross half lap* (Fig. 10-17) is accomplished in much the same way, except that the cuts are made like wide dadoes. You can cut to layout lines on the work (Fig. 10-18), or use the fence as a control for the shoulder cuts and then clean between them by making repeat passes. The latter method is the one to use when the cut is required on many similar pieces.

A *notch joint* (Fig. 10-19), is something like a cross half lap. The difference is that the cuts are made into the edges of the workpieces, rather than across surfaces. The *depth* of the cut is one-half the stock's width; the *width* of the cut equals the stock's thickness. The cuts required for this type of work can be made very accurately by clamping the parts together and cutting through both at the same time (Fig. 10-20).

One step contributing to clean half-lap assemblies is the use of a cutting tool that will form flat, smooth bottoms (Fig. 10-21). Unless you have discovered that your dadoing tool will accomplish this, it might be best to do the work with a molding head fitted with blank knives.

CORNER LOCK JOINT

A major advantage of a *corner lock joint* is that the parts interlock, which means the components will stay together even if the glue fails. The design is often selected for drawer and box corners and even for case goods. It is not as difficult to make as its appearance might indicate, but it is essential to be precise when cutting. A

Fig. 10-15. Similar cuts made in components that are equal in width and thickness result in a joint that is uniform on all surfaces and edges.

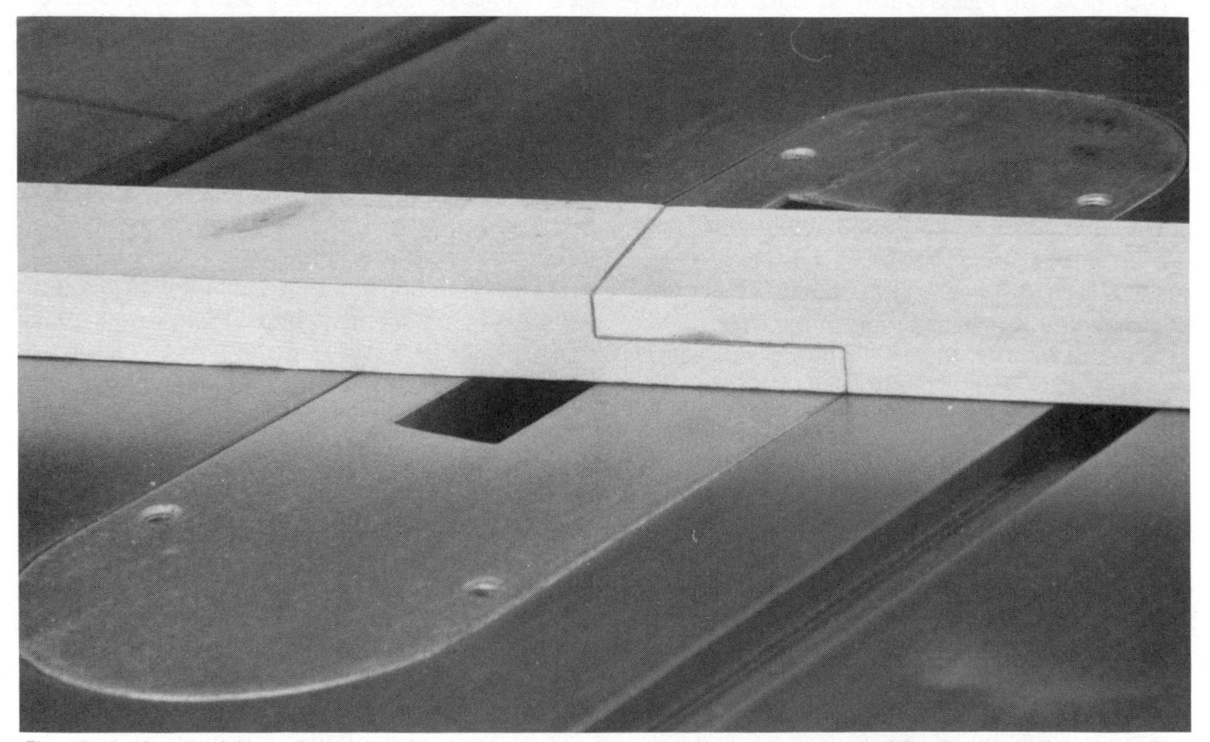

Fig. 10-16. You can join workpieces end to end by making the same kind of cuts required for the end half-lap joint. The length of the cut depends on how much you want the pieces to overlap.

Fig. 10-17. A half-lap joint is often used as a connection for components that cross each other. The cuts required are simply dadoes that are as wide as the parts and that have a depth equal to one-half the part's thickness.

Fig. 10-18. You can arrive at the shape needed for a cross half lap by making repeat passes with a dadoing tool. You can also use a molding head fitted with blank knives for work of this nature.

Fig. 10-19. A notch joint is the same as a cross half lap, except that cuts are made into the edges of the workpieces, rather than across their surfaces.

Fig. 10-20. Cuts for the notch joint will be very accurate if you hold the parts together so they can be cut simultaneously. Cut depth is one-half the stock's width. Cut width equals the stock's thickness.

Fig. 10-21. One of the secrets of obtaining precise half-lap joints is to work with a cutting tool that will produce smooth, flat bottoms in the cuts. Also, be sure the angle between the side of the cutter and the table is 90 degrees.

typical procedure done on 1-inch-thick stock is shown in Fig. 10-22.

Do all work with a dadoing tool set to cut a ¼-inch-wide groove. Make the first cut on one component (Part A), with the stock on edge. If the workpiece is narrow, feed it across the cutter by using a tenoning jig, which is an accessory that you can make (see Chapter 11). To make the second cut on the same part, use a miter gauge and position the stock flat on the table.

For all the cuts on the mating piece (Part B), move the work with a miter gauge. Notice that only the projection of the dadoing tool is changed; the width of the cuts remains constant. Careful attention to positioning the parts for each of the cuts is critical for the pieces to mate as they should. The parts should fit without wobbling, but without needing to be forced.

RABBET DADO JOINT

The *rabbet dado joint* is similar to the corner lock joint, but does not have the same degree of interlock (Fig. 10-23). It is a strong joint with many applications, a common one being the connection between the sides and front of a drawer.

The depth of the cut that is made into the end of one part should equal the thickness of the second part. Its width should be one-third the stock's thickness. Then cut the same part as shown in Fig. 10-24. The length of the bottom projection that remains should equal one-half the thickness of the mating piece. Make the final cut, which is a simple dado, on the second part (Fig. 10-25.) The depth, width, and edge distance of the cut is determined by the dimensions of the form that has been completed in the part to which it will be attached.

THE FINGER JOINT

The *finger joint*, sometimes called a *box joint*, is frequently found on treasured antiques and is a very popular wood connection among modern woodworkers because it has both visual and structural appeal (Fig. 10-26). The joint does not have the interlocking feature provided by dovetails, but it has great holding power because of the unusual amount of gluing surface it provides.

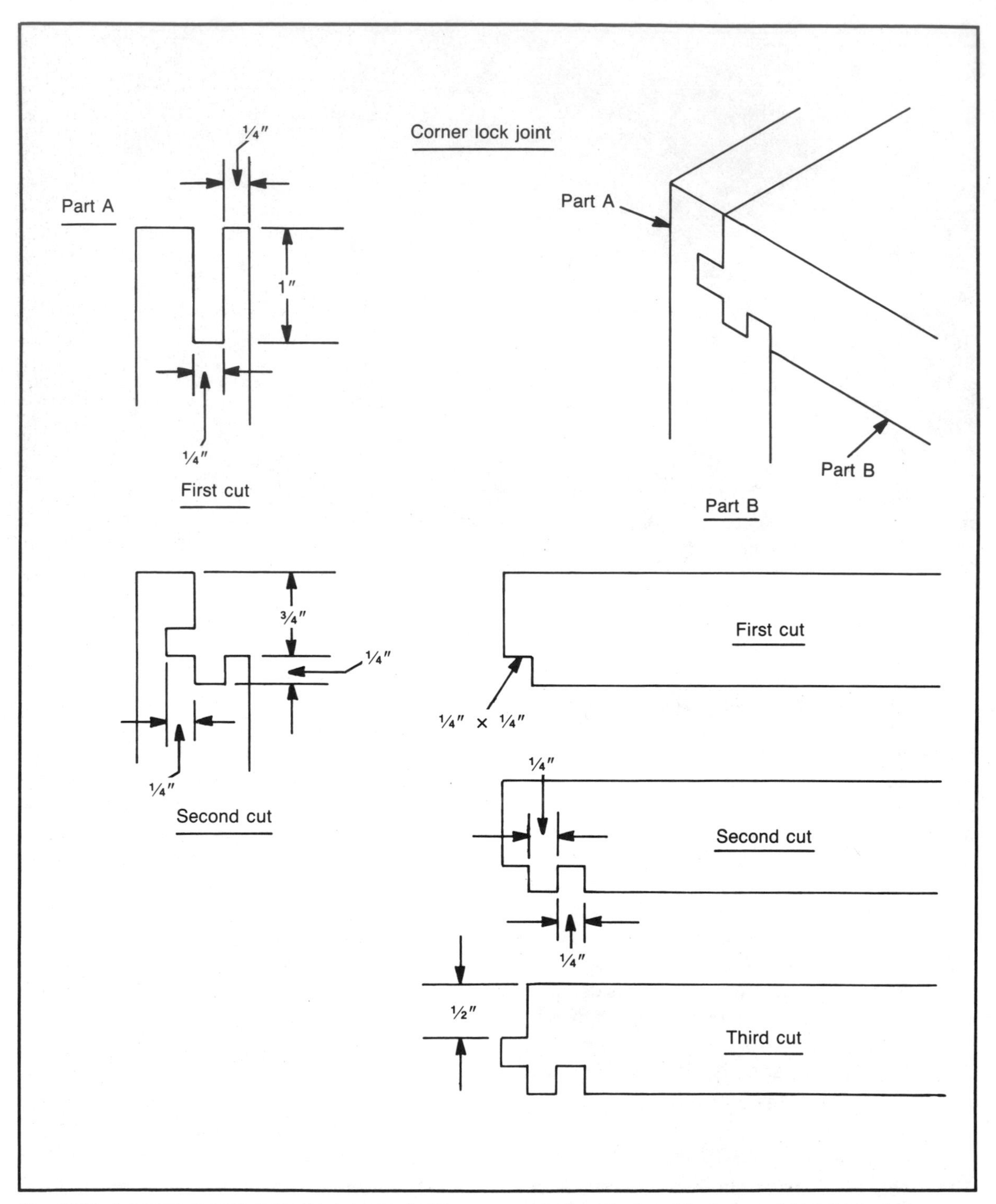

Fig. 10-22. A typical procedure for the cuts required for a corner lock joint when it is done on 1-inch-thick stock.

Fig. 10-23. The rabbet dado joint is often used as the connection between the front and sides of a drawer. It is a good joint for this purpose, even though it does not have the degree of interlocking supplied by a corner lock joint.

It is often used on exposed areas of projects as a sign of dedicated craftsmanship, but it is just as frequently used in internal, hidden areas simply because of its strength. Typical applications include drawers, box and case corners, tool chests, small jewelry cases, and music boxes. The list goes on.

A common suggestion for the design of the joint is that the fingers and the notches should equal the thickness of the stock (Fig. 10-27). There is nothing wrong with the idea structurally, but I feel that being more flexible can result in more visual appeal and can even add to the strength of the connection. One idea you can use when the parts to be connected are equal in thickness is to make the thickness of the fingers one-half the thickness of the stock (Fig. 10-28). This method makes the joint look less bulky and adds to the gluing surface. When the parts are not equal, the thickness of the fingers can match the thickness of the thinner component (Fig. 10-29).

Finger Jigs To Make

Making finger joints by hand is a tedious chore, and making the cuts with a power saw by working to layout lines on the work leaves considerable room for human error. Establishing a control that makes the job go faster and ensures accuracy is the way to go. That's where a finger jig (Figs. 10-30 and 10-31) enters the picture.

To make the jig, cut a straight piece of wood so it can be attached to the miter gauge like any extension. Mount a dado assembly on the arbor. Set the dadoing tool for the cut width that is needed; set its projection so the depth of the cut it will make equals the thickness of the stock on which you will work. Attach the extension to the miter gauge and make a cut somewhere near the extension's center. Next, duplicate the cut, spacing it a groove width away from the first one. Glue a guide block, with dimensions that equal the height and width of the cut in the extension,

Fig. 10-24. The first cut is a groove across the ends of the drawer front, which should be made by using a tenoning jig. This second cut leaves a bottom projection whose length is equal to one-half the thickness of the mating part.

Fig. 10-25. The final cut, made on the sides of the drawer, is a simple dado. Its depth should equal the length of the projection that was formed in Fig. 10-24.

Fig. 10-26. The finger joint is attractive enough to be left exposed. It has great strength because of the unusual amount of gluing surface.

into the second cut. The length of the guide block is not critical, but it should project from the extension a distance that is at least double the combined thickness of the workpieces. Before you construct the jig, be certain that the angle between the miter gauge and the side of the dadoing tool is 90 degrees.

How To Use the Finger Jig

Have all parts for the project on hand and marked for correct relationship. This step is important since you will cut mating pieces simultaneously.

Place one piece of a pair against the extension and make a cut that will form an L-shaped notch into the edge of the work (Fig. 10-32). The width of the cut from the edge of the work must match exactly the cut width of the dadoing tool. For better control over this first step, you can make a spacer strip with dimensions that match

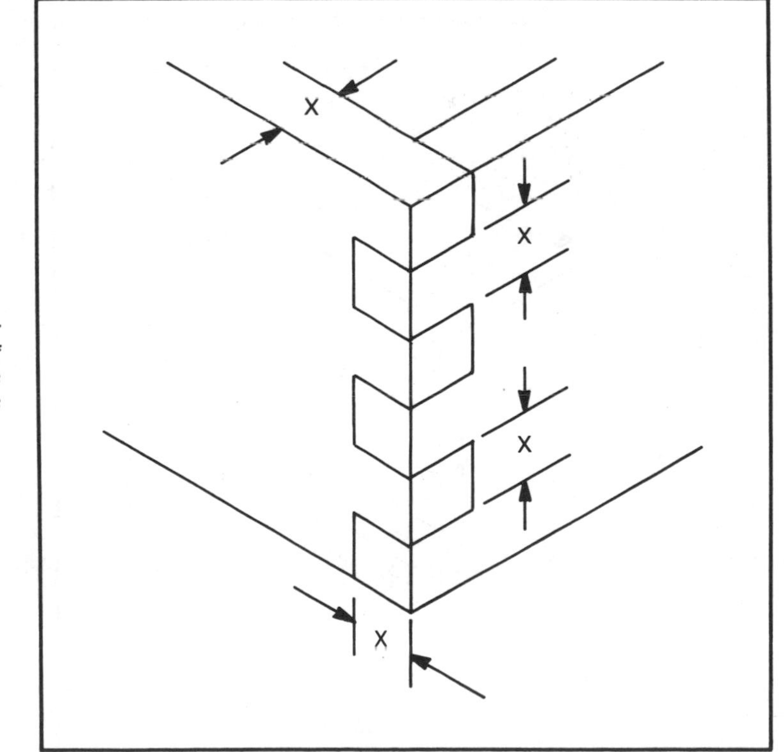

Fig. 10-27. A general rule for finger joints is that the depth and width of the cuts should match the thickness of the stock. This rule doesn't always lead to the most attractive results.

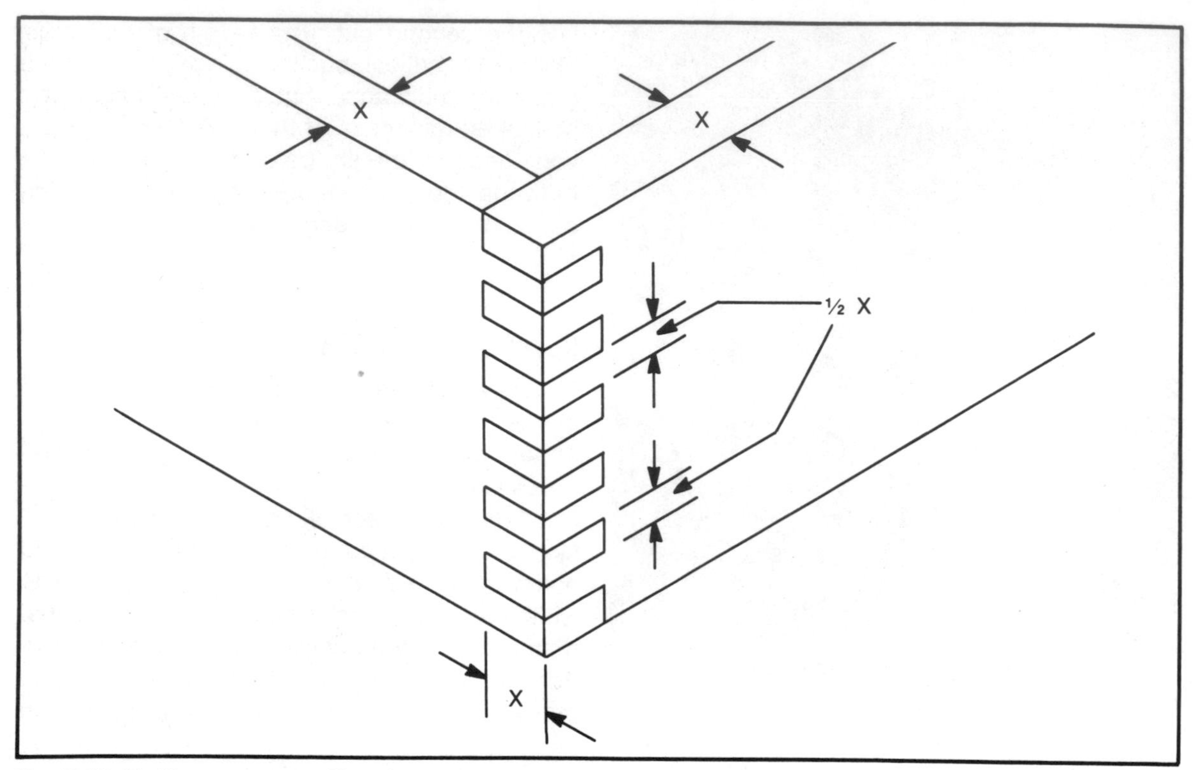

Fig. 10-28. Working as suggested here leads to a more visually pleasing connection, and even supplies additional strength.

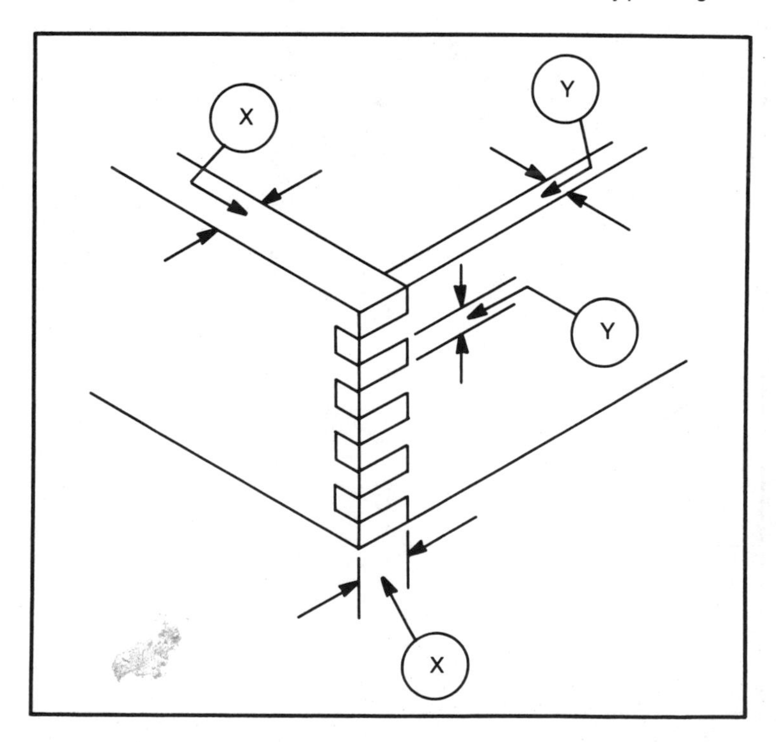

Fig. 10-29. This is the way to work when the joint components do not have the same thickness.

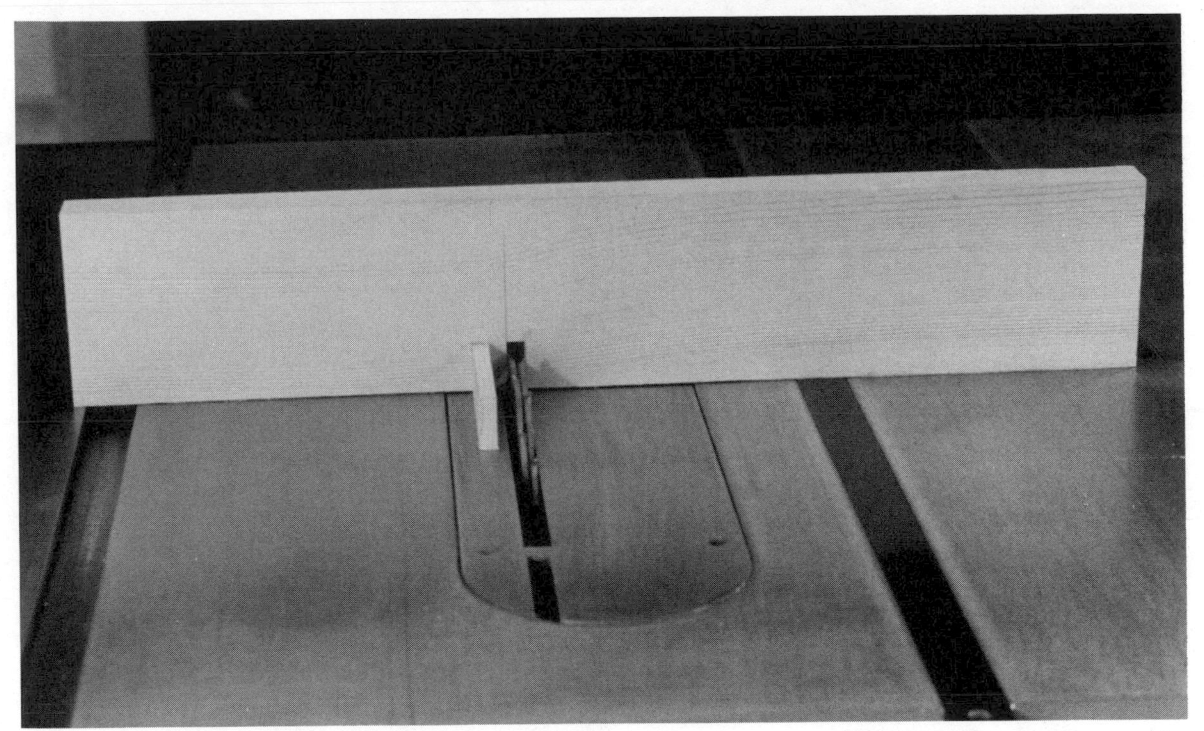

Fig. 10-30. The basic finger jig is an extension that is attached to the miter gauge. Be sure that the angle between the face of the extension and the side of the cutter is 90 degrees.

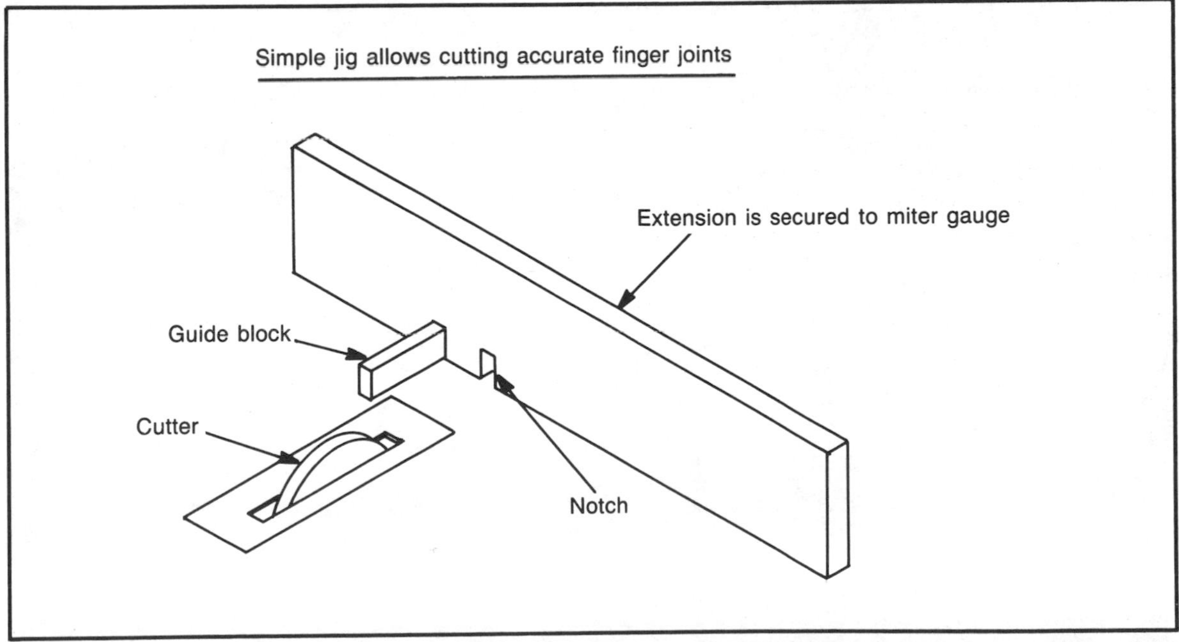

Fig. 10-31. The text explains how the cuts in the extension should be made. The distance between facing surfaces of the guide block and the cutter equals the spacing required between cuts.

Fig. 10-32. Make the first cut on one component with the edge of the workpiece in line with the outside surface of the cutter. The text explains how you can use a spacer for better control over this first step.

Fig. 10-33. The second step is to move the workpiece so the first cut is over the guide block.

Fig. 10-34. Make the second cut after you abut the mating component against the guide block. Cut the parts simultaneously. Be sure to hold them firmly; use a clamp if you wish.

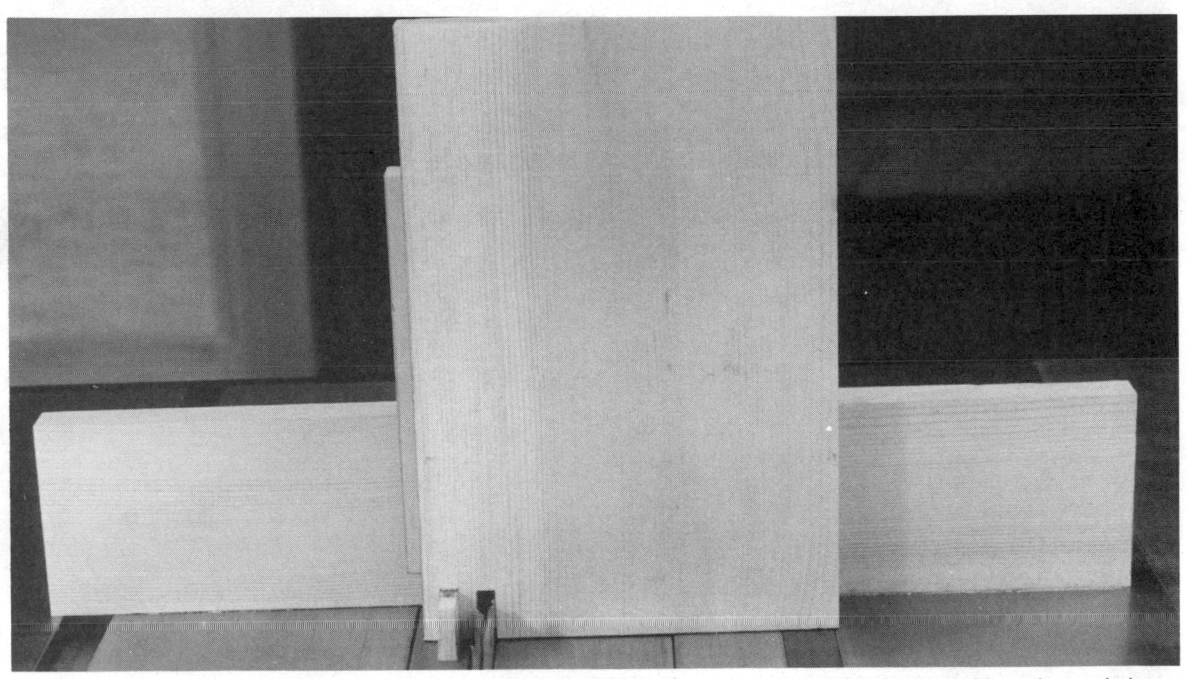

Fig. 10-35. Making the third cut for the finger joint. The second cut, placed over the guide block, positions the workpieces correctly.

the groove width, and place it between the guide block and the edge of the work.

Next, move the workpiece so the L-shaped cut will be *over* the guide block, and add the mating piece so it abuts against the guide block. At this point, you can use a clamp to hold the two pieces together, if you wish. Next, move the parts together for the second cut (Figs. 10-33 and 10-34). Continue the cutting process, placing the groove that was just formed over the guide block so the workpieces will be correctly positioned for the next cut (Figs. 10-35 and 10-36).

One of the problems you might encounter when you are following this procedure is that, since the first cut is a full-groove width, the last cut might be an odd size. You can equalize partial cuts at each end of the assembly by working as follows. Draw a vertical line on the extension midway between facing sides of the guide block and the cutting tool. Draw a second vertical line on the extension to indicate the centerline of the groove. Position the workpieces so that one part is in line with the first mark and the mating piece is in line with the second mark. Make the first cut while holding the two pieces firmly together. Make the second cut after moving the parts so that each is abutted against the guide block. Position the workpieces for all remaining cuts by placing the last cut that was made over the guide block.

An Alternate Finger Jig

The jig shown in Fig. 10-37 is an independent unit because it is mounted on, and guided by, twin bars that ride in the table slots. An advantage of the design is that it will supply work support on both sides of the cutter. The cutting

Fig. 10-36. Make all the remaining cuts by following the same procedure.

procedure is the same as described for the jig used with a miter gauge. If you make it, be sure that the angle between the fence and the side of the cutting tool is 90 degrees. This concept might be especially appealing to workers who have very small table saws. Construction details are given in Fig. 10-38.

Modifications of the Finger Joint

The finger joint doesn't need to end in a sharp corner. There are many times when shaping the joint after assembly, along the lines suggested in Fig. 10-39, can add to the appearance of the project. This kind of work should be done after the glue used for assembling the joint is thoroughly dry.

You can reinforce the joint, or hold it together should the glue fail, by drilling a hole through the assembled joint and inserting a dowel (Fig. 10-40). You should form the hole on a drill press, but if you work carefully, you can accomplish the job in good style with a portable drill. Use an extension bit if necessary. If the hole is very long, you can work more accurately by drilling from both ends of the joint.

You also can modify the finger joint so it becomes a swivel joint (Fig. 10-41). First hold the parts together correctly as you drill the hole for the dowel. Then "dress" the ends of the fingers on each component to provide clearance for the turning action.

The same jigs used for finger joints can be

Fig. 10-37. You can make an independent finger jig by mounting a fence on twin bars that will ride in the table slots. The procedure for making the cuts does not change.

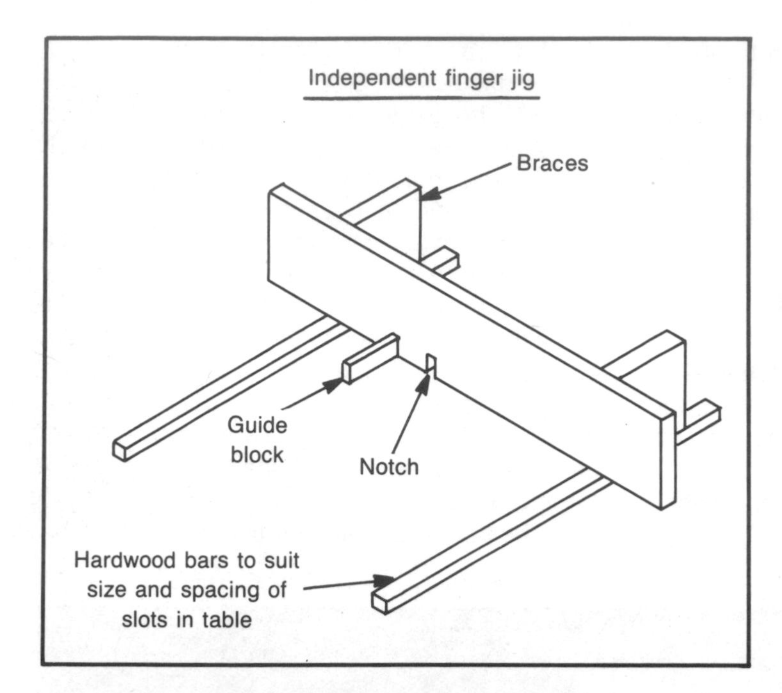

Fig. 10-38. Method for making an independent finger jig. When you add the braces, be sure that the angle between the fence and table is 90 degrees. The angle between the fence and the side of the cutter must also be 90 degrees.

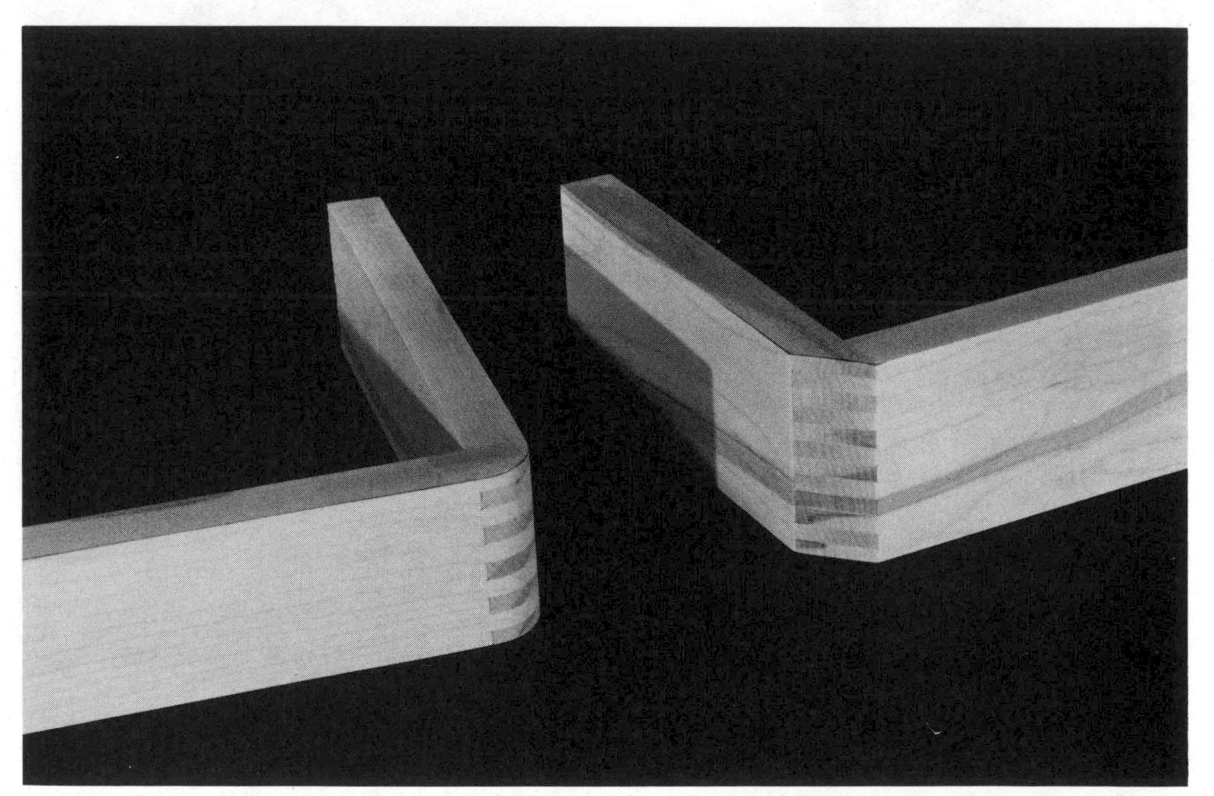

Fig. 10-39. The finger joint can be shaped like these examples. The work is done after the glue used in the joint is thoroughly dry.

Fig. 10-40. You can reinforce the finger joint with a dowel. Drilling from both ends of the assembly will help you work accurately if you use a portable tool.

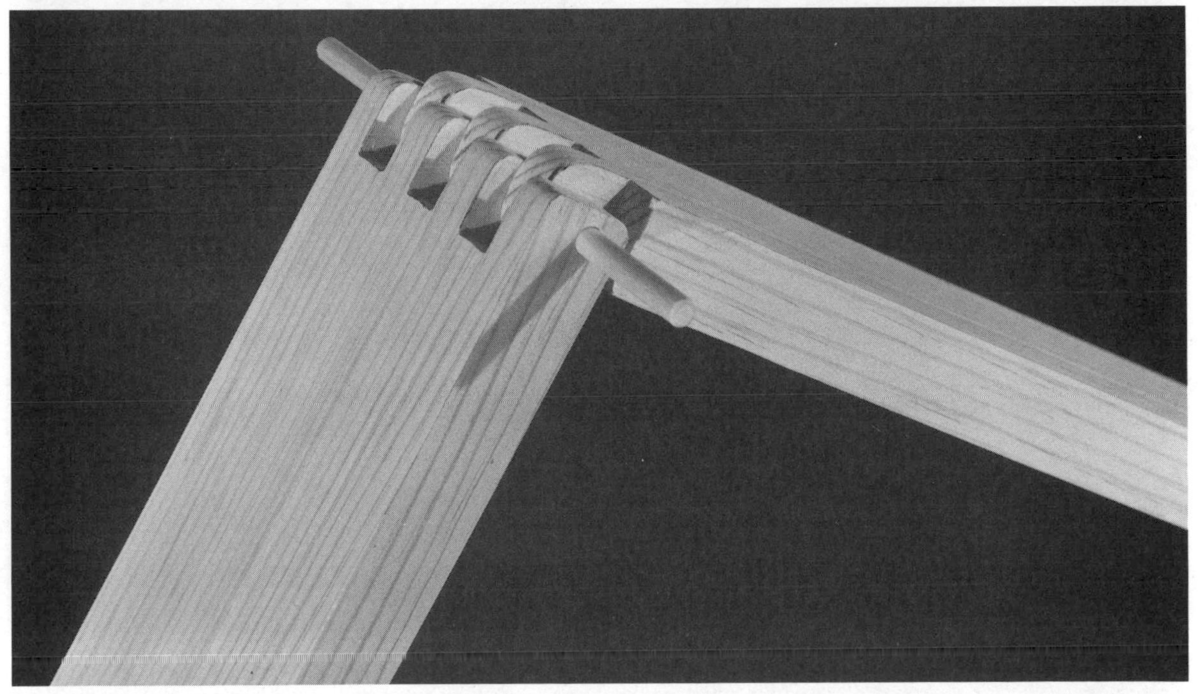

Fig. 10-41. Modifying the finger joint so it can swivel is a question of dressing the fingers for clearance. Do this step after drilling the hole for the dowel.

used to form the notches required for grid-type assemblies, like the example in Fig. 10-42. The opposite pieces are held together with clamps or by tack-nailing so that mating cuts can be made simultaneously. Working this way ensures accuracy in the size of the cuts and in their alignment.

EGG-CRATE ASSEMBLIES

The way to do this kind of work quickly and accurately is to make a miter extension jig that is brother to the jigs used for the finger joint (Fig. 10-43). The only difference between the concepts is in how you arrive at the dimensions for the cut in the extension and the guide block. The height of the guide block and the depth of the cut are equal and must be one-half the *width* of the partitions. The thickness of the guide block and the width of the cut are also equal, but must match the *thickness* of the partitions. The distance between the facing sides of the guide block and the cutter determines the spacing of the partitions.

Make the first cut with parts abutted against the stop block. Make the other cuts by situating the last cut *over* the guide block. This method ensures that parts will mesh together accurately on assembly and that all sections will have the length and width you decided on (Fig. 10-44).

HOW TO USE TENONING JIGS

It is often necessary to make cuts like true and stud tenons, slots for the open mortise-tenon joint, and others like them on the ends of narrow pieces of work. You can make some of the cuts by using the miter gauge and positioning

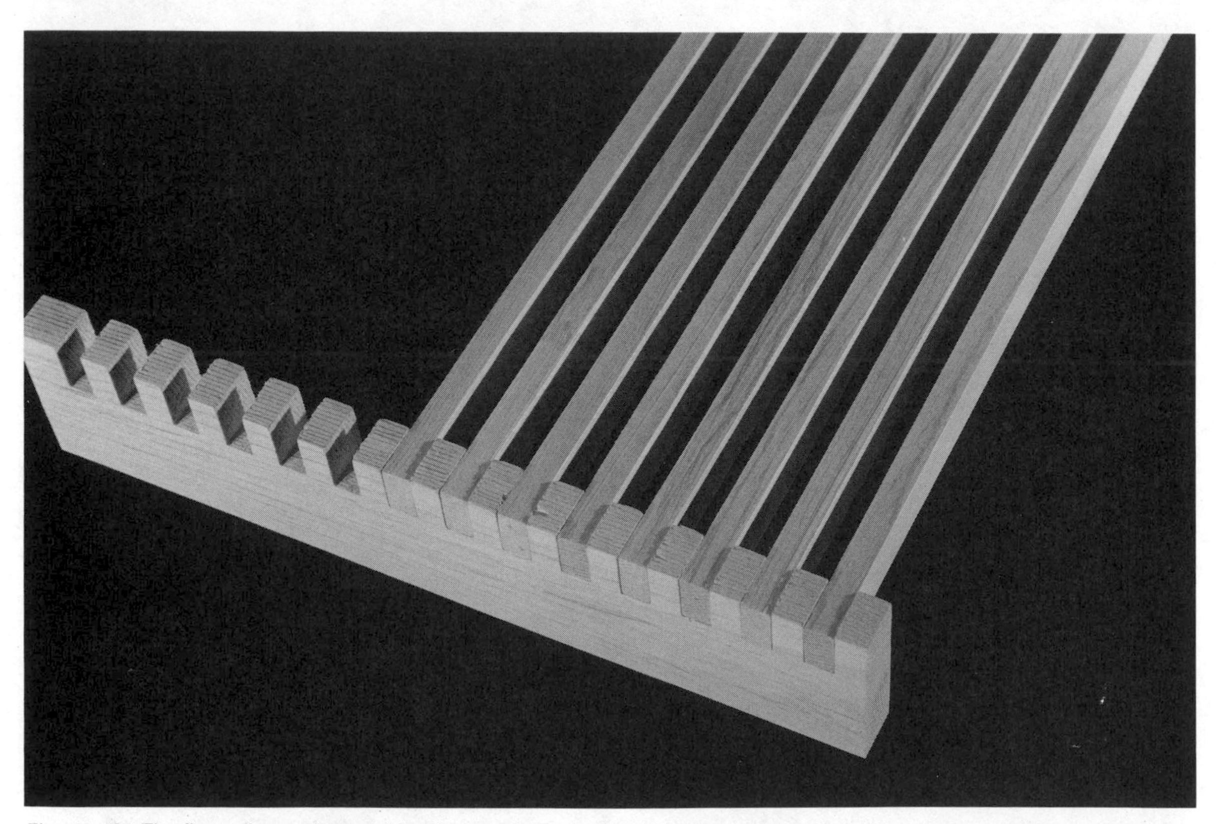

Fig. 10-42. The finger jig can be used for the cuts required for grid-type projects. Opposite parts are held together so the cuts in each can be made simultaneously.

Fig. 10-43. The cuts required for egg-crate projects can be made accurately using the same jig design as for finger joints.

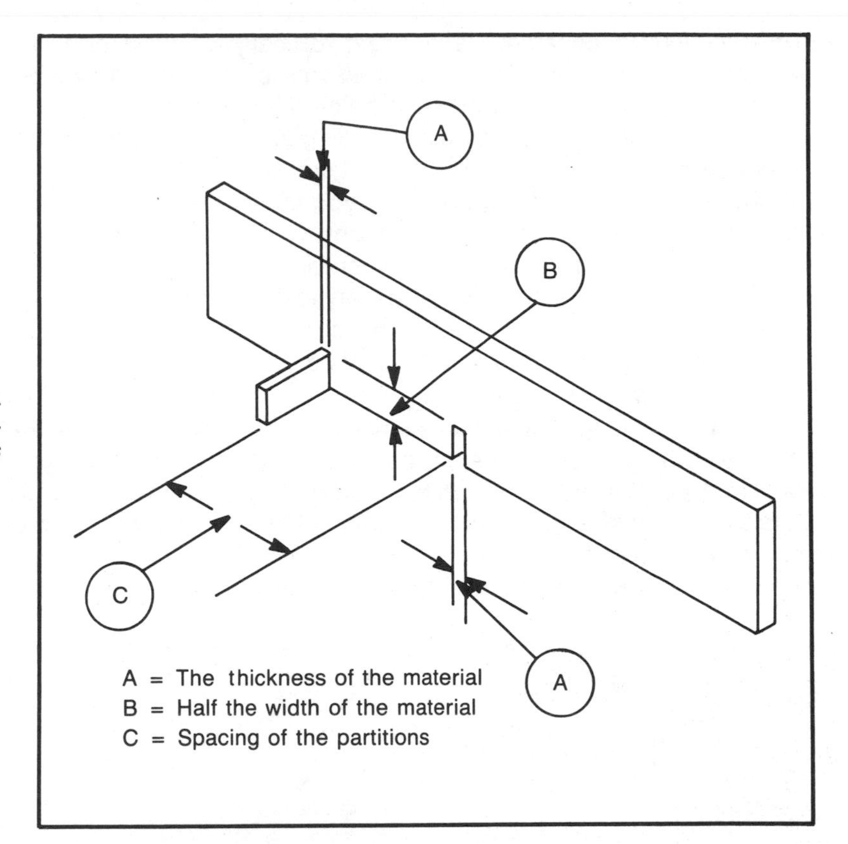

Fig. 10-44. The spacing of the partitions will be determined by the distance between the inside edge of the jig's guide block and the facing side of the cutting tool.

the stock flat on the table. Many times, you can work more accurately and get cleaner cuts by using the rip fence as a control and holding the part on end as you pass it across the cutting tool. Because the workpiece has insufficient surface to bear safely on the table, however, you must never follow the procedure using only your hands to secure the stock. To do such operations accurately and safely, it is necessary to employ a tenoning jig that will hold the work so it can't tilt or twist and that will let you place your hands well away from the cutting tool.

Tenoning jigs, like those shown in Figs. 10-45 and 10-46, are available commercially. Both of these jigs are designed to be guided by bars that ride in the table slots and have adjustable clamps to keep the work secure as the cut is made. The upper part of the jigs can be moved toward or away from the saw blade so that material of various thicknesses can be accommodated.

The tenoning jig in Fig. 10-47 is one that you can make to suit the machine you own. In order to allow adjustments for stock thicknesses, the jig is designed to straddle the rip fence. Special guides are used to position miter-cut pieces so that grooves for feathers or splines can be cut without difficulty. The type of rip fence you work with might affect the mounting arrangement. The jig I use is designed to suit the narrow, top edge of the Delta Unifence. The plan offered in Fig. 10-48 is a basic one that should be suitable for most conventional rip fences. An alternate method for mounting the jig is given in Fig. 10-49.

Whatever mounting arrangement you use,

Fig. 10-45. Tenoning jigs contribute to accuracy and safety when you need to make cuts on the end of narrow workpieces. This Delta product is a heavy-duty version. A sliding clamp can be situated appropriately for the size of the stock.

be sure to be precise so the jig won't wobble and so it can be moved without needing to be forced. Polishing the rip fence and the contact areas of the jig with a hard, paste wax will help the accessory to move more easily. Be sure that the angle between the face of the jig and the saw table is 90 degrees (Fig. 10-50). When you drill the holes in the right-angle guide for the screws that will secure it, make them a bit oversize. This will allow some adjustment each time you use the jig so you can be sure that the angle between its bearing edge and the saw table is 90 degrees (Fig. 10-51).

To form slots in the end of narrow workpieces, position and clamp the work as demonstrated in Fig. 10-52. To be sure the slot is centered, make a second pass after you have repositioned the workpiece so its outboard surface is against the face of the jig. You can make repeat passes to widen the slot, or you can work with a dadoing tool so you can make the slot you need in a single pass.

You can form tenons with a saw blade by first making the shoulder cuts with the stock flat on the table and moved with a miter gauge, and then using the tenoning jig for the cheek cuts.

Fig. 10-46. The face of this Craftsman tenoning jig is drilled in various places so the security clamp can be aptly located. The jigs are designed to be used with various thicknesses of stock.

Fig. 10-47. This homemade tenoning jig straddles the rip fence so it is adjustable for the distance between it and the cutting tool. The design includes miter guides, which are identified so they can always be used in the same position.

(10-53). If you work with a dadoing tool, the whole job can be done by working only with the jig. Whether you work with a saw blade or a dadoing tool, be sure that the angle between the side of the cutter and the saw table is 90 degrees.

As for the right-angle guide, drill the holes for the screws to secure the mitering guides slightly oversize so you can adjust for a precise angle between them whenever you put them to use (Fig. 10-54).

Figure 10-55 shows how the jig is used to cut a groove for a feather spline to reinforce a miter joint. It's essential, of course, that the miter cuts on the components be accurate at the beginning. Note also in Fig. 10-55 that a strip of wood that is secured to the jig with a screw serves nicely as a hold-down. The groove does not need to be centered exactly, if you take care to place similar surfaces of the workpieces against the jig for each of the cuts you make.

The grooves for straight splines that are used to reinforce miter joints are cut on the pieces individually. The arrangement is shown in Fig. 10-56. After you cut the first groove, turn the stock end for end and place it against the second miter guide for the groove in the opposite end. Be sure that the same surface of the work is against the face of the jig. It's often necessary, because of the width of the parts, to use the miter guides one at a time. You can cut the grooves for feathers or splines with a saw blade. If you need wider grooves, as for thick material, you can do the cutting with a dadoing tool.

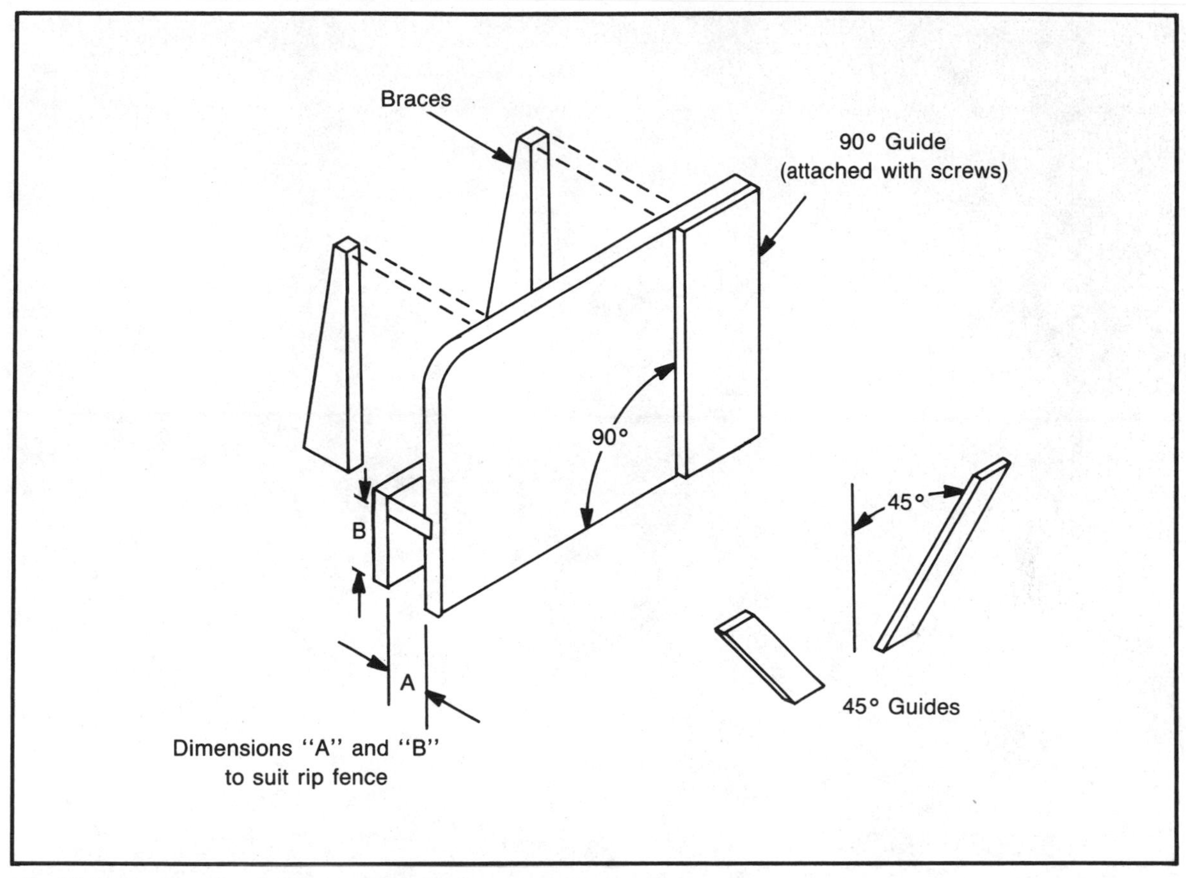

Fig. 10-48. Construction details to custom-design a tenoning jig.

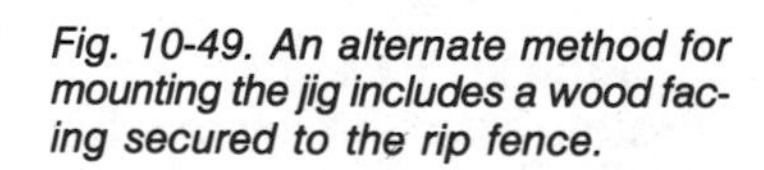

Fig. 10-49. An alternate method for mounting the jig includes a wood facing secured to the rip fence.

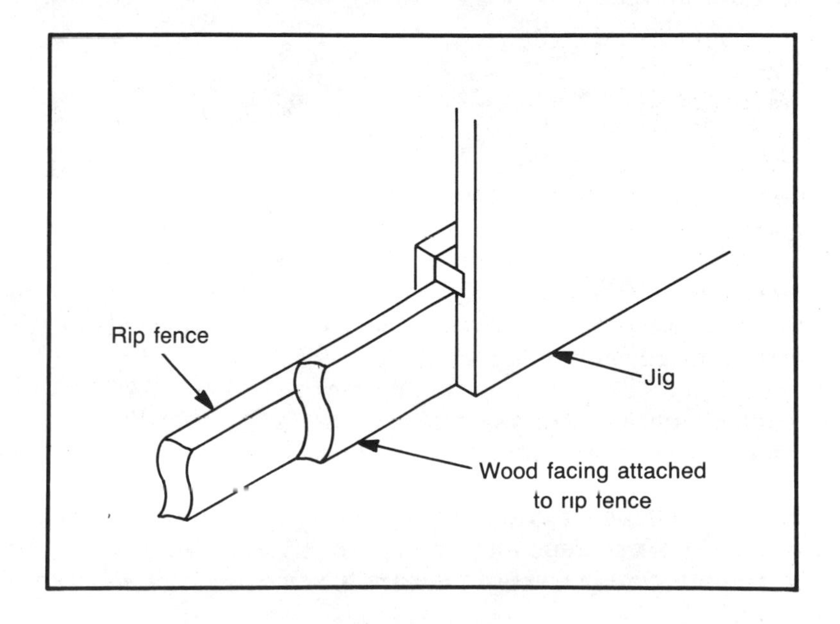

Fig. 10-50. Be sure the angle between the face of the jig and the table is 90 degrees before you assemble the components permanently.

AN ALTERNATE TENONING JIG DESIGN

If you are up to a bit more work, you can put together a tenoning jig that will produce accurate results without reliance on the rip fence. The concept involves two assemblies (Fig. 10-57). One is a base guided by a bar that rides in the table slot. The other is a longitudinally adjustable work-support system that slides between guides and is secured with a clamp bar, both of which are components of the base assemb' Construction details for the jig are in Fig. 10-58; Fig. 10-59 demonstrates how the clamp bar works.

When you assemble the base, be sure that the inboard edge of the platform will be parallel to the saw blade and that the angle between the guides and that same edge will be 90 degrees. You can form the slot in the handgrip by first boring end holes and then cleaning out the waste between them with a jigsaw or coping saw. If you

Fig. 10-51. The bearing edge of the right-angle guide must be 90 degrees to the table. The holes for the screws that attach the guide are oversize, so a degree of adjustment is available when the guide is used.

want to be fancier, you can use the handle of a handsaw as a pattern for the jig's handgrip. Be sure that the base and face of the work-support assembly are square to each other and that the face is at an angle of 90 degrees to the saw table before you permanently attach the brace. Also, to avoid any movement that will result in inaccurate cuts, be sure that the width of the base matches the distance between the guides very closely.

In action, the jig is handled pretty much like the fence-straddling concept. Typical applications are demonstrated in Figs. 10-60 through 10-63.

MAKING AND USING SPLINES AND FEATHERS

The basic purpose of splines and feathers is to reinforce a joint, but they also make it easier to do assembly work because they position the components for correct alignment. This feature is an advantage since the parts of miter joints, and even edge-to-edge joints that are simply abutted to each other, can slip away from a correct relationship when they are under clamp pressure.

A *spline*, when used in a miter joint, runs the full length of the connection. It doesn't matter what the joint angle is (Figs. 10-64 and 10-65). A *feather*, often called a *key*, fits a groove that is cut across the outside corner of the joint (Figs. 10-66 and 10-67).

Most times, the thickness of the reinforcement piece is arbitrary although, for the sake of convenience, it pays to choose a material whose thickness matches a standard-sized groove. For example, if the frame material is ¾ inch thick, I form the grooves with a saw blade that cuts a ⅛-inch kerf and then use a material of similar thickness for the spline or feather. For joint components of thicker material, I will set a dadoing tool to cut a ¼-inch-wide groove since plywood and hardboard, both of which are excellent materials for splines or feathers, are available in exactly those thicknesses. An important point is that the reinforcement components are strongest when the grain direction runs across the narrow dimension. Obviously, this does not apply when you use a grainless material like hardboard.

Other factors, of course, also must be considered. On utility projects, it might not matter what material you use for the reinforcement pieces. On some projects, you might wish to use a contrasting material to add a decorative detail. Other times, you may want the spline or feather to be the same material you use for the project so that the method of connection will not be obvious.

You must custom-make the reinforcement pieces when you wish them to contrast or blend. One way to work is to first cut grooves into the end of the material. Use a tenoning jig for the cuts if the parent material is narrow. Space the grooves so the remaining material will have the thickness required for the splines. Next, use a crosscutting operation that will separate the splines (Fig. 10-68). The width of the cutoff pieces must equal the total depth of the grooves

Fig. 10-52. Workpieces are held against the guide and moved for the cut by sliding the tenoning jig. You can use a saw blade to form slots, or you can work with a dadoing tool. Use a clamp to hold the work for extra security.

that are cut into the joint members. When you work this way, the grain direction of the splines will be across the small dimension, which is how it should be.

The use of splines to reinforce miter joints is not limited to flat-frame projects. They can also serve when miter joints are the connection between components of a box-type structure. The difference in the procedure is in how you cut the spline grooves in long pieces of work. You can set up by tilting the saw blade to 45 degrees and then using the rip fence to guide the work, or you can make a guide jig (Fig. 10-69) so the cut can be accomplished without tilting the saw blade. In either case, the cut should favor what will be the inside corner of the joint, and the angle between the groove and the plane of the miter cut must be 90 degrees.

Using Feathers for Cross or Rip Miters

Triangular feathers, sometimes called keys, can be used as reinforcements when components of box-type projects are connected with miter joints. When the feathers are made from a contrasting material, they also serve as decorative details (Fig. 10-70).

Fig. 10-53. Make the cheek cuts for tenons after you have formed the shoulder cuts. Work with a miter gauge and with the work flat on the table. You can skip the preliminary shoulder cut if you work with a dadoing tool.

Form the grooves for the feathers as shown in Fig. 10-71. You can make the cuts with a saw blade or a dadoing tool, depending on the thickness of the material you will use for the feathers. Use the rip fence to control the edge distance of the cuts. Have a stop block on the fence to

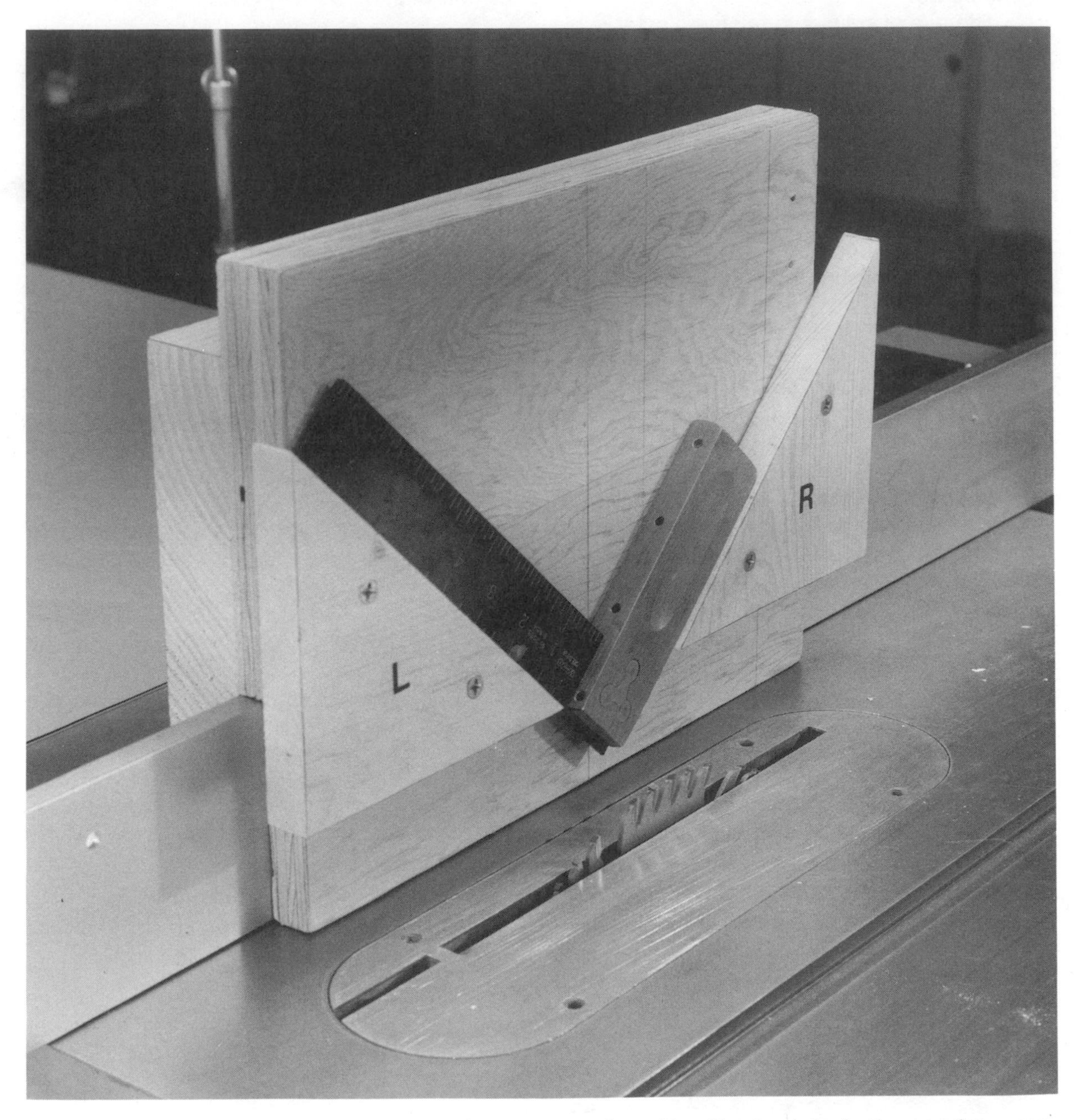

Fig. 10-54. Use a square to check alignment when you mount the miter guides. Here too, holes for the mounting screws are oversize to allow for adjustment.

control how far you can advance the work. Make the cuts so they stop short of the inside surface of the part by about ⅛ inch. Be sure to make the same cut on each component before you change the position of the fence for other cuts.

Make the feather pieces a bit longer and wider than necessary (Fig. 10-72). In this way, you can trim and sand them perfectly flush after the glue dries. You must shape the inside edge of the feathers somewhat like the example in Fig. 10-73 because the groove cut ends in an arc. The relief area does not need to be V-shaped. If you wish, you can form it to fit the arc in the groove more precisely.

Fig. 10-55. To form grooves for feathers in miter joints, mount both components in the jig. The cut does not need to be centered perfectly if you place all pieces so similar surfaces are against the jig. The same surfaces must be up or down at assembly time.

Fig. 10-56. Spline grooves are cut in the pieces individually. You can still use the hold-down, either by spanning across the miter guides or by using a spacer under its free end. The width of the workpieces might make it necessary to use one guide at a time.

Splines in Edge-to-Edge Joints

Splines are often used to join narrow pieces of stock to form large slabs (Fig. 10-74). The grooves are formed by moving the stock on edge, guided by the rip fence. The thickness of the spline material will determine whether you should cut the grooves with a saw blade or a dadoing tool.

Mark the same surface of each piece and have that surface against the fence when you are making the cuts. If you work so in this way, you will be sure of matching cuts even though they might not be perfectly centered in the edges of the workpieces. Those same surfaces must face either up or down when you assemble the parts. Splines that run the full length of the slab pieces will be exposed at each end of the project (Fig. 10-75). This is not a problem if the slab will be framed. Often, when framing is not part of the design, the splines are made from a contrasting material to supply a decorative detail.

When exposed splines are not acceptable, you can form the grooves by using the stop-cut technique explained in Chapter 8. Cut and shape the splines to fit the stopped groove (Figs. 10-76 and 10-77). The splines will be concealed, but will still do the job.

Be sure the width of splines is not greater,

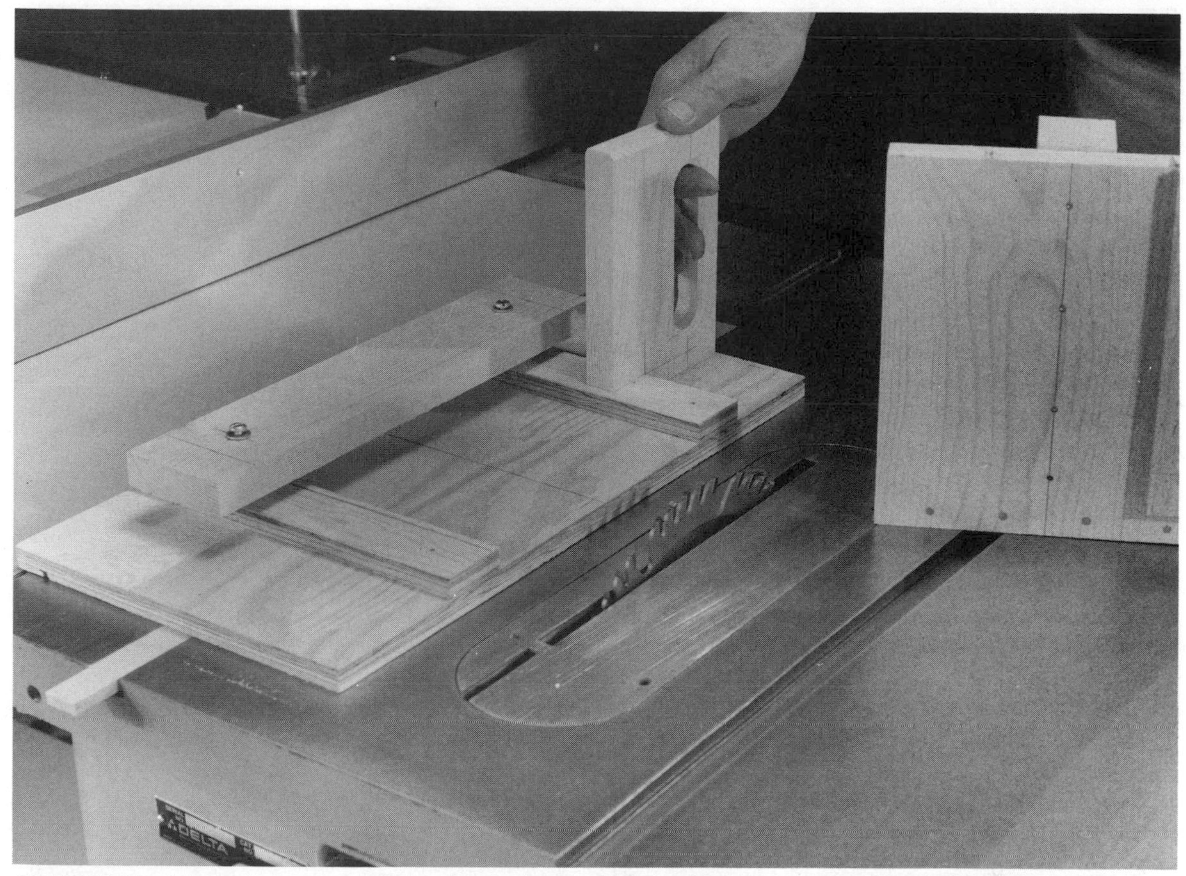

Fig. 10-57. The independent tenoning jig has its own base, which is guided by a bar that rides in the table slot. The work-support component slides between the base's guide bars so it can be set for the correct distance from the cutting tool.

by even a smidgen, than the combined depth of the grooves. If it is, there will be gaps between components that can't, or shouldn't be, closed with clamp pressure. Parts that are forced together under these conditions will be stressed at joint areas and might split apart at some future time. Some workers take the opposite approach, using a spline width that is a fraction less than the total groove depth. This method ensures that mating edges will come together firmly.

Splines as Decoration

A unique spline system is displayed in Fig. 10-78. Here, the cross splines supply a suitable connection between components, but they are deliberately a visible part of the project. The idea is most effective when the splines are cut from a contrasting material.

The grooves in the slab pieces are formed by conventional methods, but the cross splines require particular attention. Cutting them individually would require very persnickety setups. A better way is to form a host of them by cutting matching grooves in a large piece of stock by working with a dadoing tool or a molding head that is fitted with blank knives (Fig. 10-79). You then can separate the splines by making rip cuts. Very careful cutting and close attention to setups for each cut are necessary for the project to be successful.

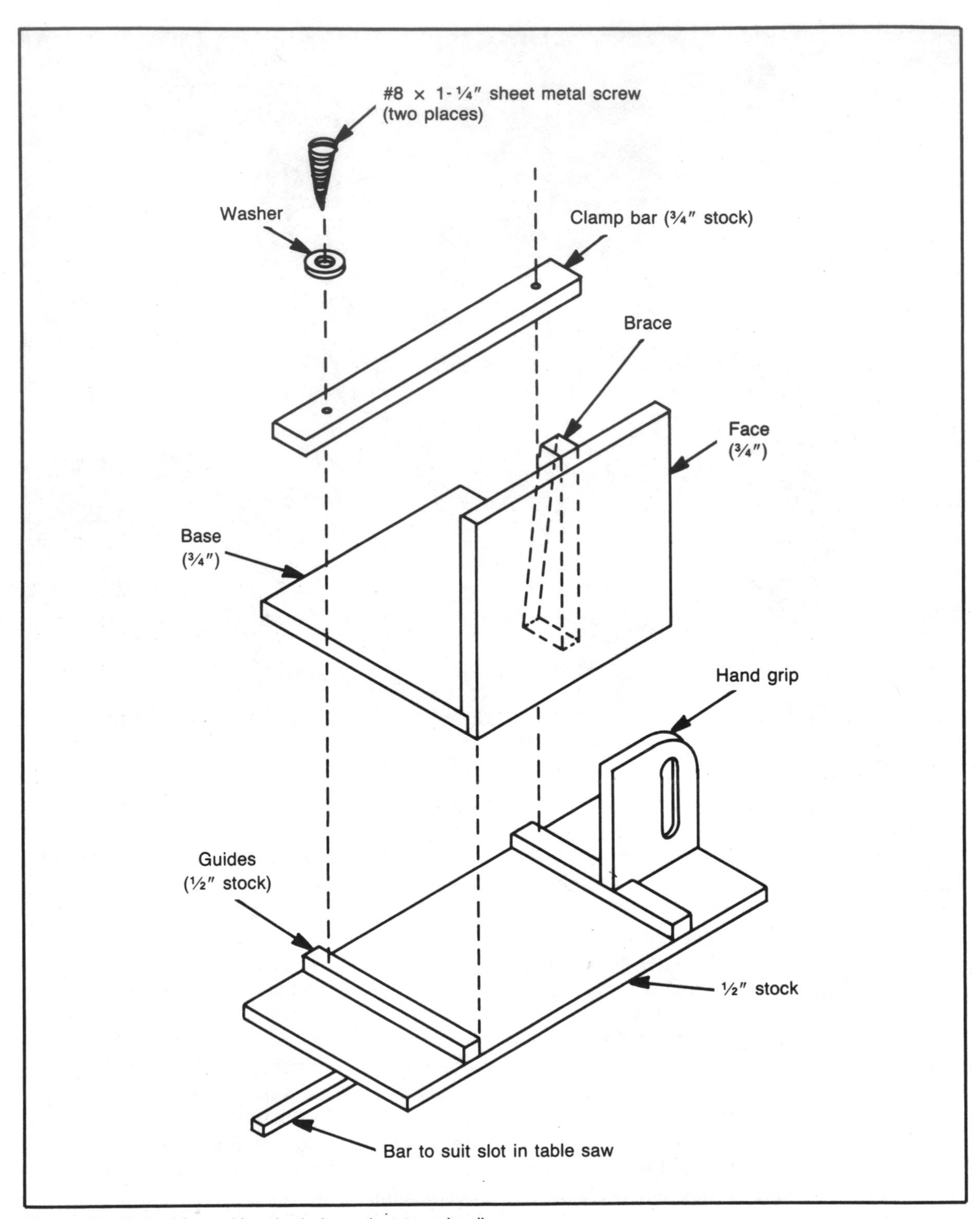

Fig. 10-58. Method for making the independent tenoning jig.

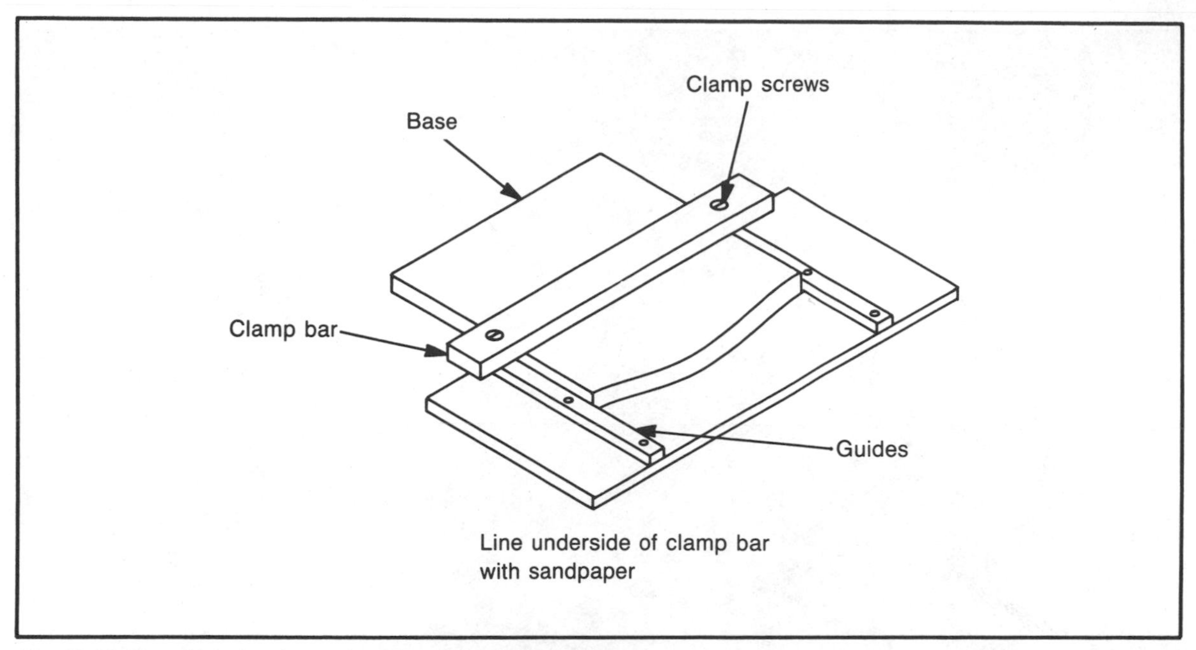

Fig. 10-59. The clamp bar is a strip of hardwood that bears down on the base of the work-support assembly when the screws are tightened. The base must slide between the guide bars, but without lateral play.

Fig. 10-60. The work support is adjusted to suit the cut and the thickness of the workpiece. Here, the outside blades of a dado assembly are used to form a ¼-inch-wide slot for an open mortise-tenon joint.

Fig. 10-61. To form tenons, first make cheek cuts. Then use the miter gauge with the work flat on the table for the cuts that form the shoulders.

Fig. 10-62. You can form tenons more quickly with a dadoing tool. The thickness of the tenon is determined by the distance between the work support and the inside face of the cutter.

Fig. 10-63. Forming a slot in dowels or rounds is a unique application for a tenoning jig. Set the work support so the cut will be on the diameter of the workpiece. Place the work firmly against the right angle guide and secure it with a clamp.

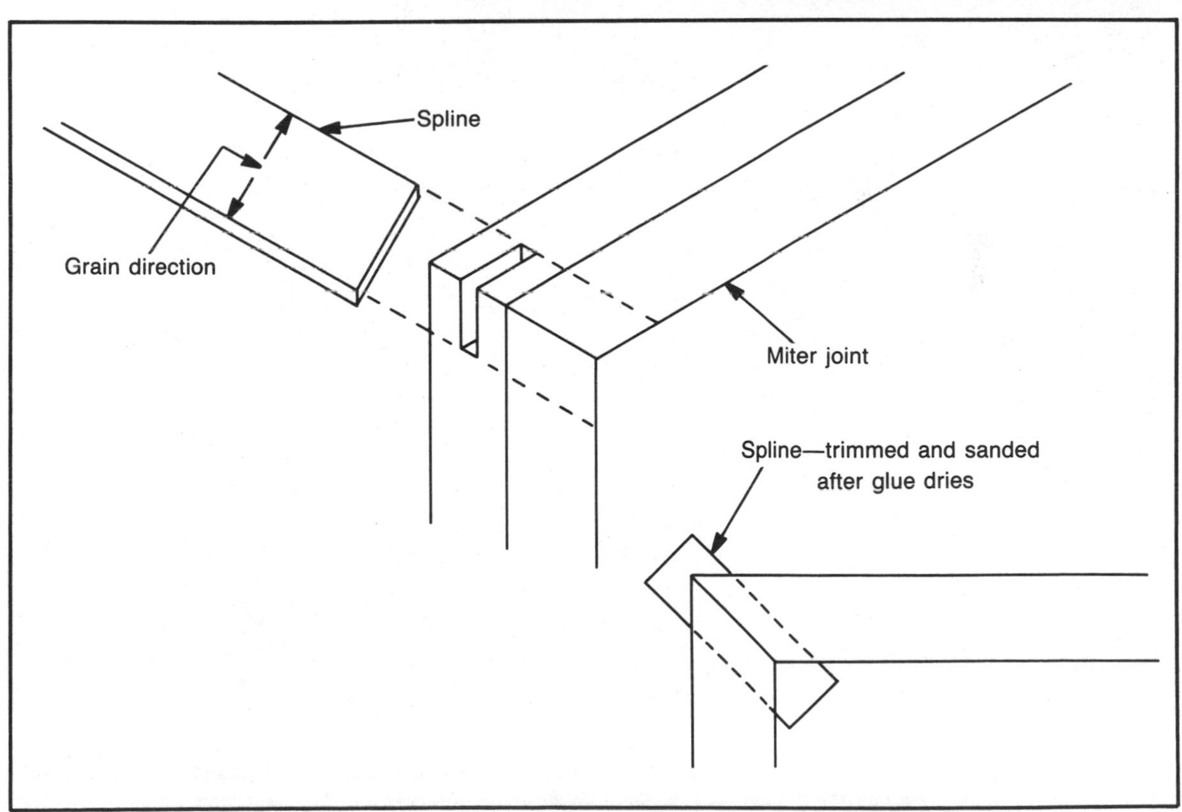

Fig. 10-64. Splines are strips of material that fit matching grooves formed in the edges of miter cuts.

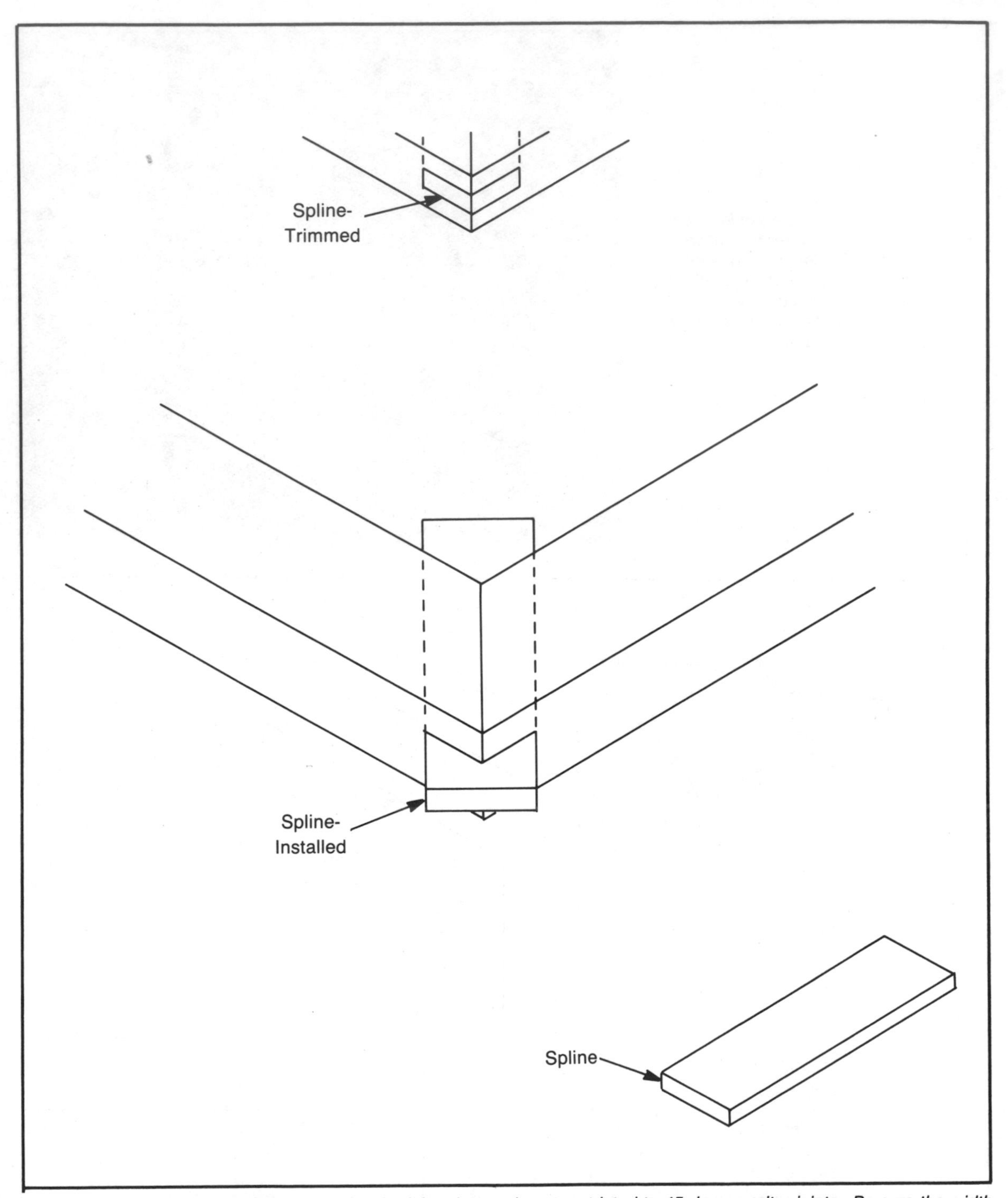

Fig. 10-65. A spline runs completely across the joint. Its use is not restricted to 45-degree miter joints. Be sure the width of the spline is not the least bit greater than the combined depth of the grooves.

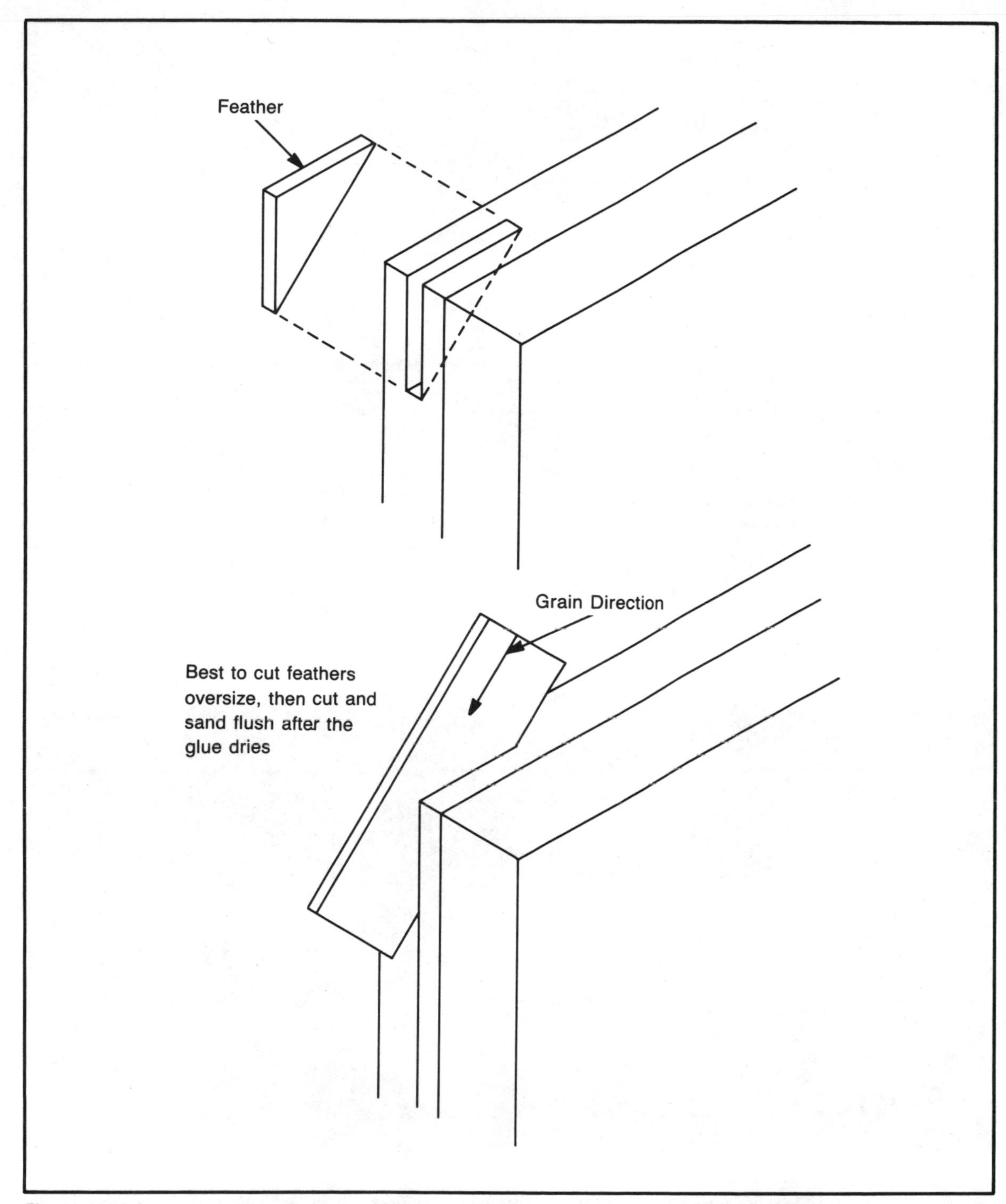

Fig. 10-66. A feather runs through the outside corner of the joint. Note the suggested grain direction.

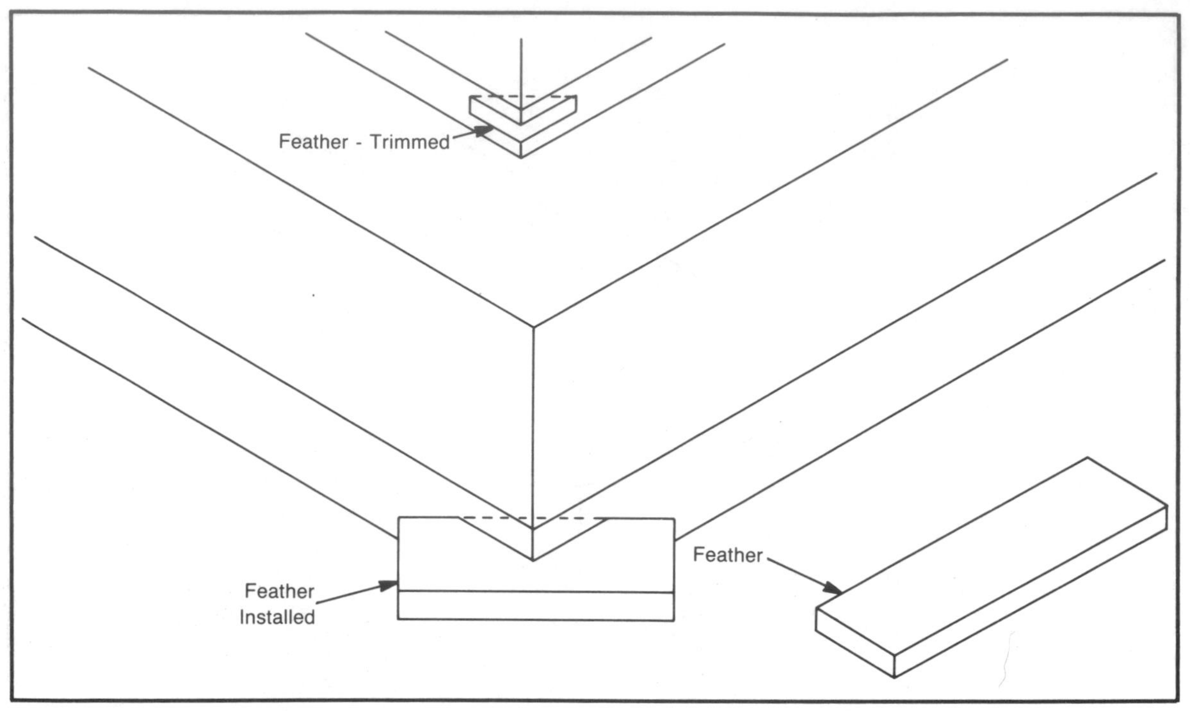

Fig. 10-67. The maximum depth of the groove for feathers is arbitrary, but should not be greater than one-half to three-quarters the length of the joint.

Fig. 10-68. Cut splines off from parent stock after edge grooving has been done by working with a tenoning jig. Use this technique when the splines must be a special material.

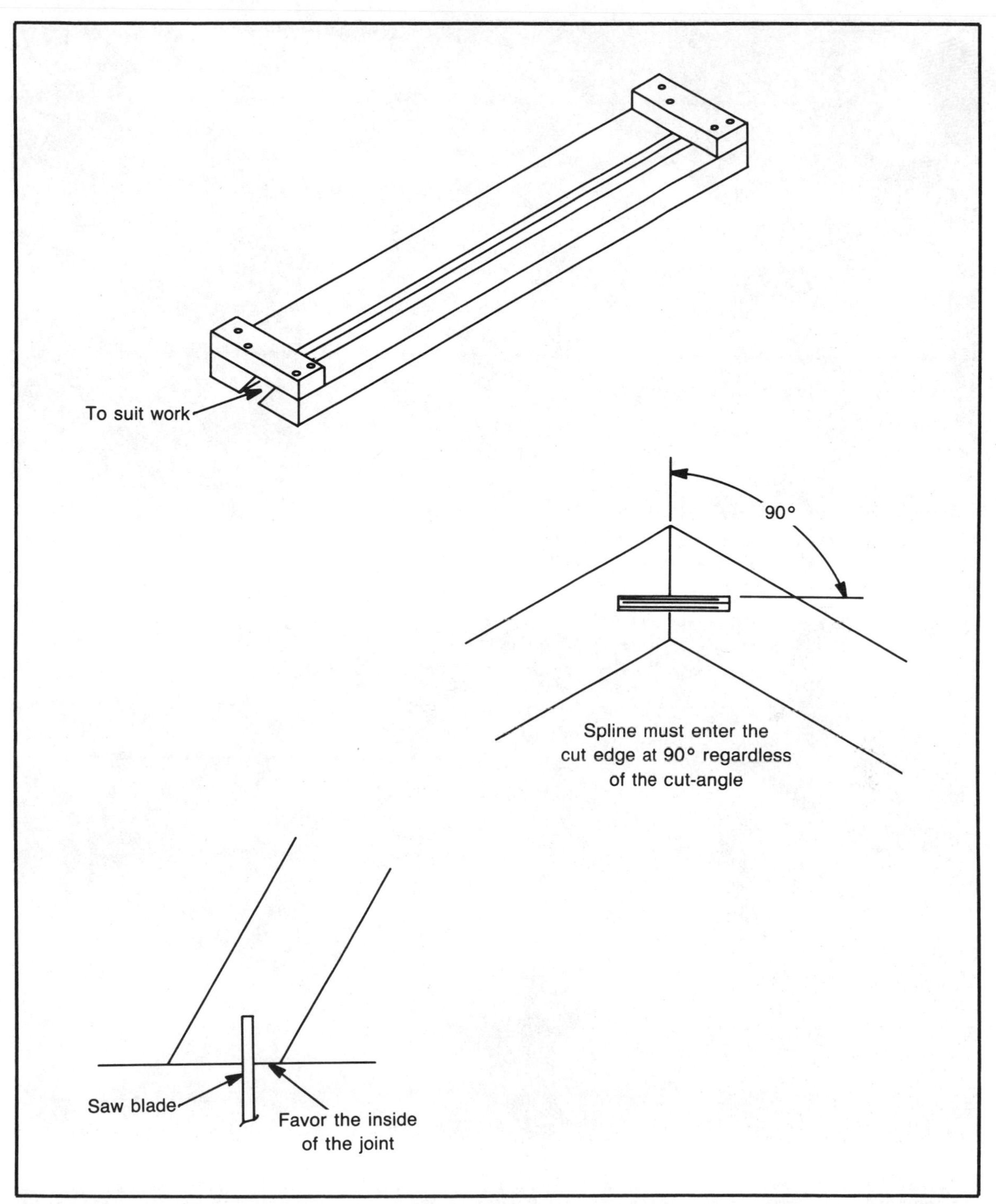

Fig. 10-69. You can make accurate grooves for splines that will be used in long, angular cut edges by making a special guide jig. The saw blade must be at right angles to the plane of the angular cut on the workpiece.

Fig. 10-70. Feathers can be used across the corner of cross-miter or rip-miter joints. They provide for reinforcement, as well as a decorative detail if they are made from a contrasting material.

Fig. 10-71. Form the grooves for the feathers this way. The inside surface of the workpiece is up; the stop block controls the length of the cut, which should stop short of the work surface by about ⅛ inch. Make the same cut in each piece before changing the position of the rip fence.

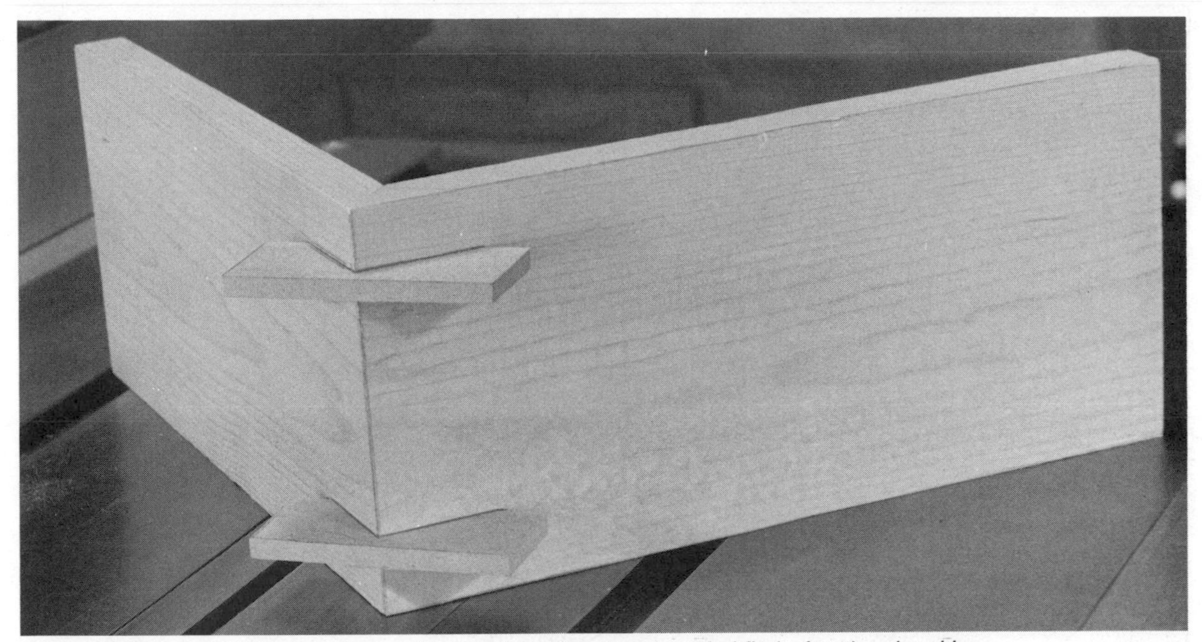

Fig. 10-72. Cut the feathers oversize so they can be trimmed and sanded flush after the glue dries.

Fig. 10-73. The inside edge of the feather must be shaped to clear the arc left by the cutting tool. A V-cut will do it, or you can shape the feather to conform with the arc.

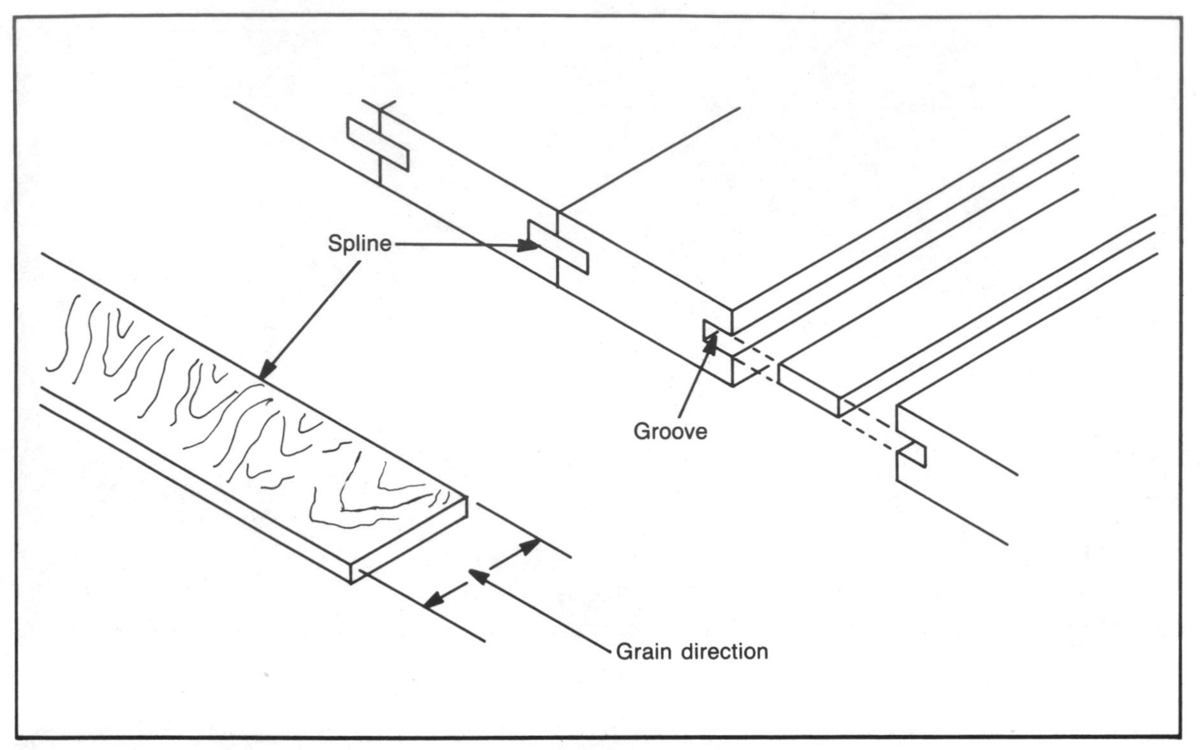

Fig. 10-74. Narrow pieces of stock, joined with splines, are often used to form large slabs. The system guards against the possibility of warpage, which can occur when solid stock is used.

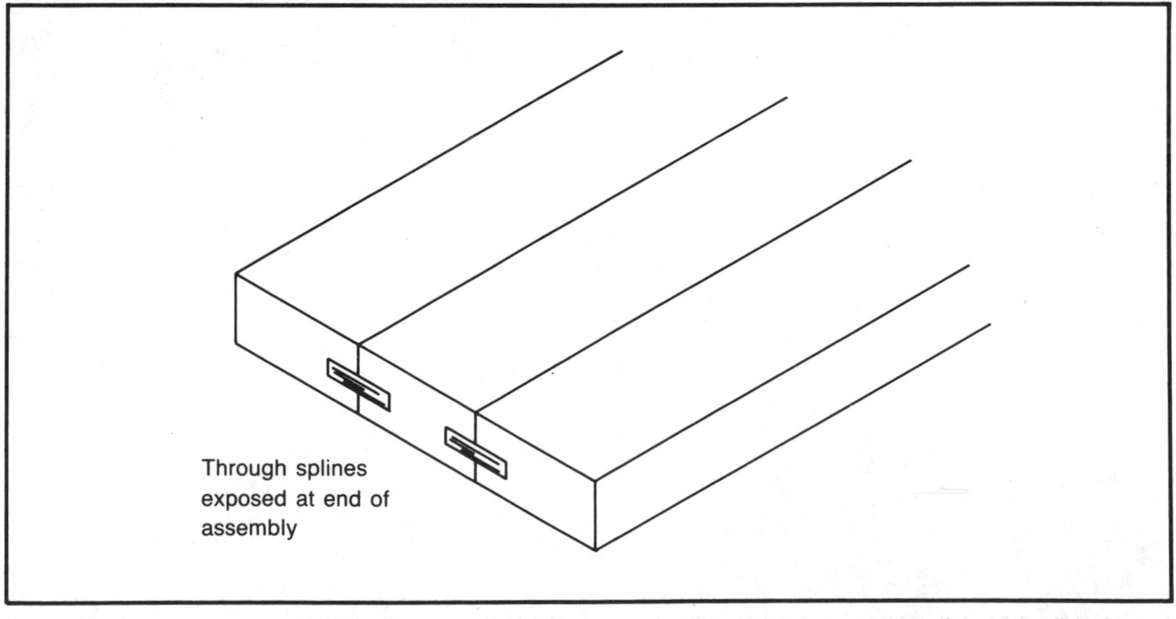

Fig. 10-75. Splines are exposed when they run completely across the joint. This is not a problem if the slab will be framed, or if you choose to make them from a contrasting material and deliberately leave them exposed as decorative details.

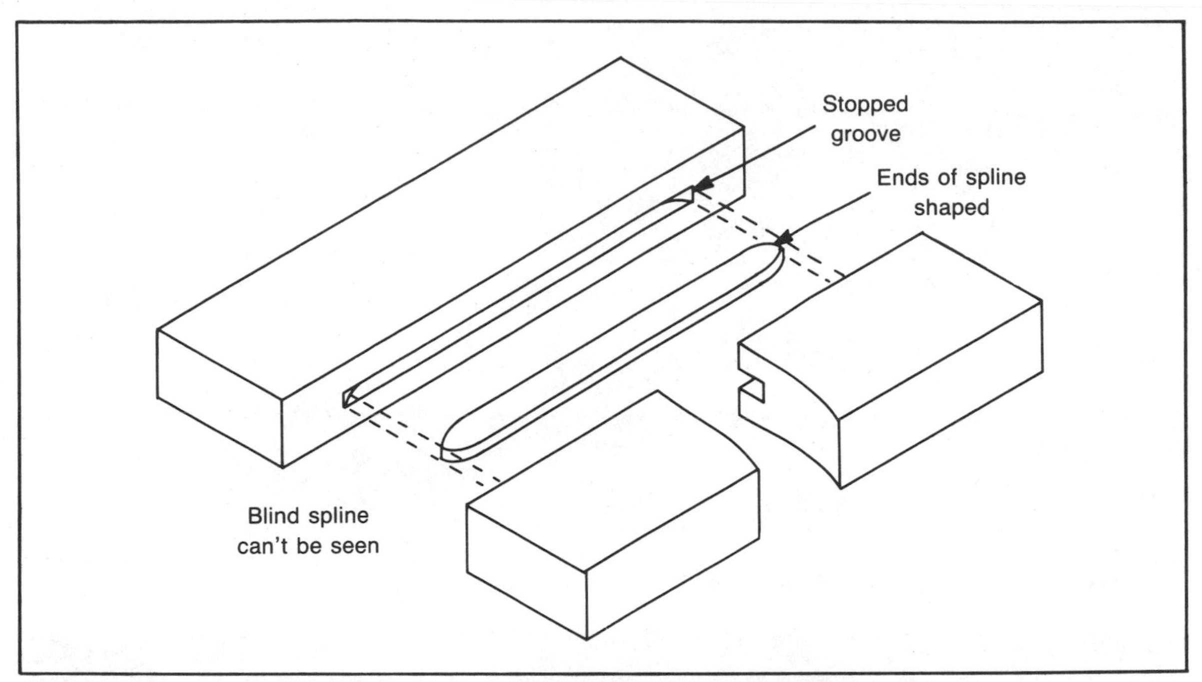

Fig. 10-76. You can conceal splines by forming stopped grooves and shaping the splines to fit.

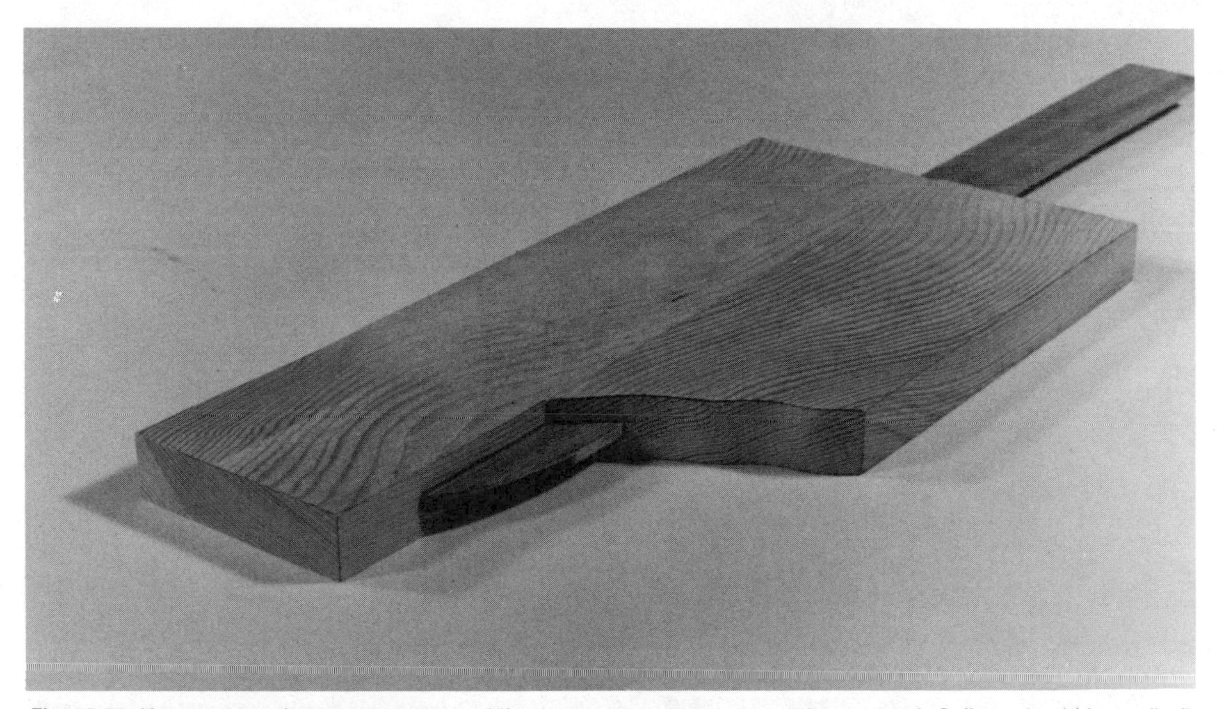

Fig. 10-77. You can stop the groove at one end if only one edge of the slab will be exposed. Splines should be a slip fit in the grooves. You will have a lot of trouble at assembly time if the splines must be forced into position. Forcing will also stress the joint, so that splitting can occur.

Fig. 10-78. Cross splines can be used to supply the necessary connection between parts while contributing an inlaid effect. Very precise cutting is required for the results to be successful.

Fig. 10-79. Don't try to shape the cross splines from individual pieces. Instead, make cuts across stock that is wide enough for safe handling. The grooving can be done with a dado assembly or with a molding head fitted with blank knives. The width of the grooves should be enough to allow for the kerf that will separate the parts. Have the bulk of the stock between the fence and the saw blade when ripping so the cutoffs can fall free.

Chapter 11

Special Applications

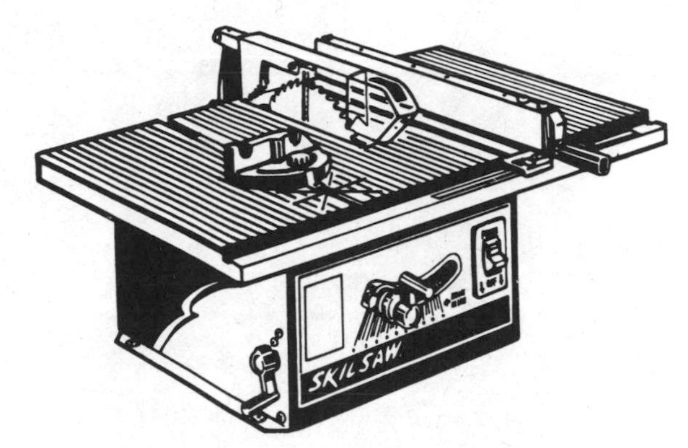

ACTUALLY, ANYTHING YOU CAN DO WITH A TABLE saw is special. The time and effort that you save and the automatic accuracy the tool provides in basic chores like crosscutting, ripping, and mitering are impressive, especially when compared with hand tools. Many of the jigs that have been demonstrated are special, but this chapter will introduce you to what might be referred to as *off-beat* methods, but methods that don't shun practicality. You might never want to work with two saw blades or two miter gauges, or use a conventional saw blade to cut circles, but the ideas are available when you need them to solve a problem, work more accurately, or produce multiple, similar pieces faster.

WORKING WITH TWO MITER GAUGES

Having two miter gauges for a table saw is not an indulgence. Any setup that helps you work more conveniently and that lessens the chance of human error is a good stride toward faster, more accurate sawing. The miter gauges do not need to be twins, as long as the bars fit the table slots as they should (Fig. 11-1).

Using an extension to span across two miter gauges (Fig. 11-2) can supply good support when you are crosscutting long pieces of stock. It's also an idea to use when you need to trim a piece that is too short for safe cutting with a single miter gauge. When setting up, check for the correct 90-degree angle between one miter gauge and the saw blade before attaching the extension. Be sure the head of the second miter gauge is flush against the extension when you make the connection at that point.

Figure 11-3 demonstrates how to use an extension with the gauges set at 45 degrees. Working this way does much to guard against the pivoting and creep action that can spoil miter cuts. Attach the extension after you have checked for the correct angle between one of the gauges and the saw blade. Since the miter gauges can be set at any angle, this idea is not limited to 45-degree miters.

Fig. 11-1. The miter gauges you use don't need to be identical. They will work well together as long as the bars fit the table slots correctly. It's a good idea, though, for the extra one to be as good as the one that came with the saw.

The accuracy problem of mitering shaped pieces of stock was discussed in Chapter 7. Working with a single gauge means having to use the tool in each of the table slots often after resetting the head of the gauge. You can preset two miter gauges so the included angle between them is 90 degrees. After adjusting one gauge to 45 degrees, use a square to determine the correct setting for the second gauge (Fig. 11-4). Then you can saw complementary miters with minimum fuss and little chance of human error.

PATTERN SAWING

Pattern sawing can solve the kind of problems that are encountered when the shape of the component make it hazardous to work conventionally. The system also has advantages when you must produce a certain number of similar, odd-shaped parts, or workpieces that are shaped, for example, as octagons or hexagons. It's a good way to work because it organizes a mechanical means of gauging cuts. Since a *pattern*—that is, a precise example of the parts you need—is used, all the pieces you cut will be exactly alike.

The jig required for pattern sawing doesn't need to be more elaborate than the assembly in Fig. 11-5. The horizontal member, which is the guide, is secured firmly to what is not more than a facing for the rip fence. The width of the guide should allow ample room between the saw blade and the fence, and its height above the table should be a bit more than the thickness of the stock that will be cut (Fig. 11-6). Be sure the rip fence is parallel to the saw blade and that the outboard edge of the guide is on the same plane as the outside of the saw blade when the jig is clamped in place (Fig. 11-7).

Fig. 11-2. Spanning across two miter gauges with an extension can provide extra support for crosscutting operations. Be sure that the head of the first miter gauge to which you attach the extension is square to the saw blade.

Fig. 11-3. Dual miter gauges set at 45 degrees, or other angles, and connected with an extension will make it easier to cut accurate miter joints, especially if the extension is faced with sandpaper.

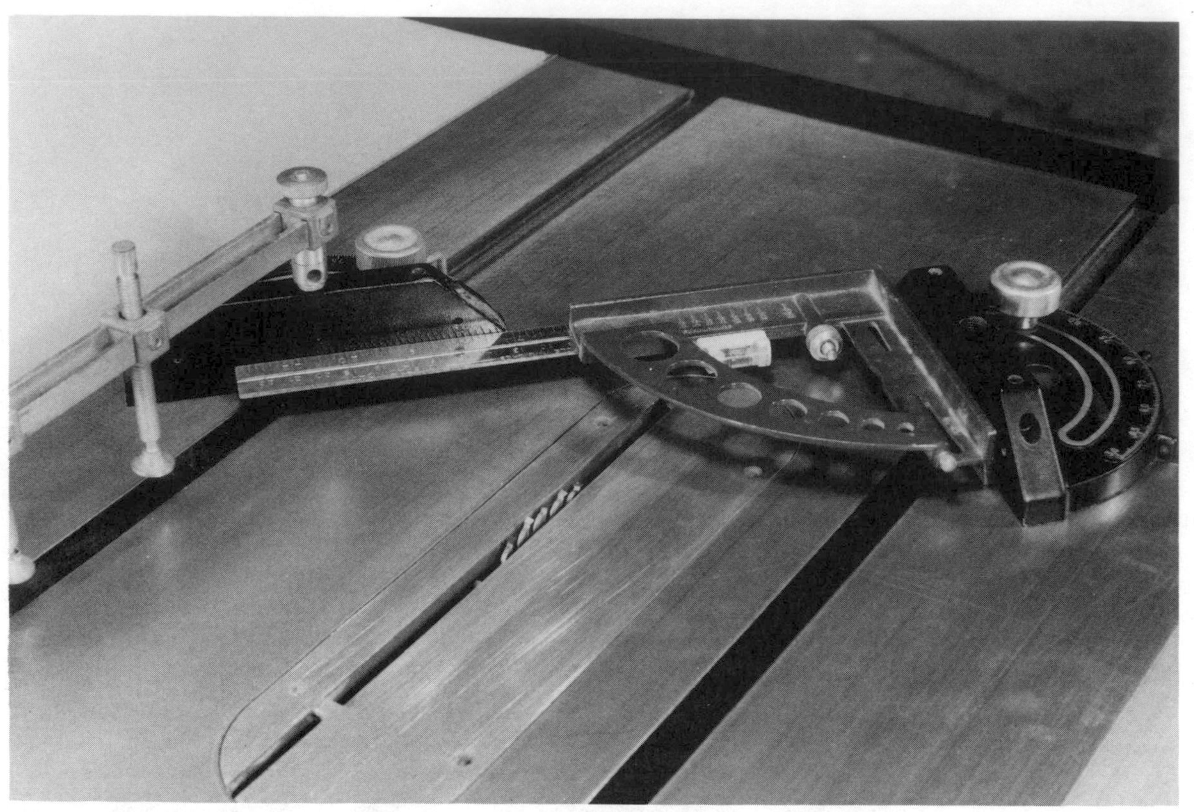

Fig. 11-4. Miter gauges can be set to opposing angles for miter cuts that can't be completed by working only on one side of the blade. Set one gauge to the correct angle and then use a square to organize the second one.

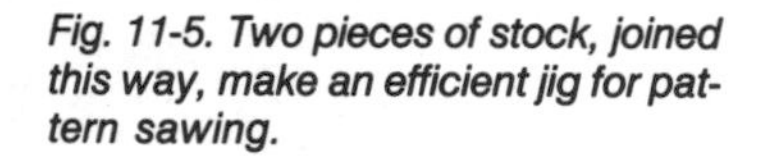

Fig. 11-5. Two pieces of stock, joined this way, make an efficient jig for pattern sawing.

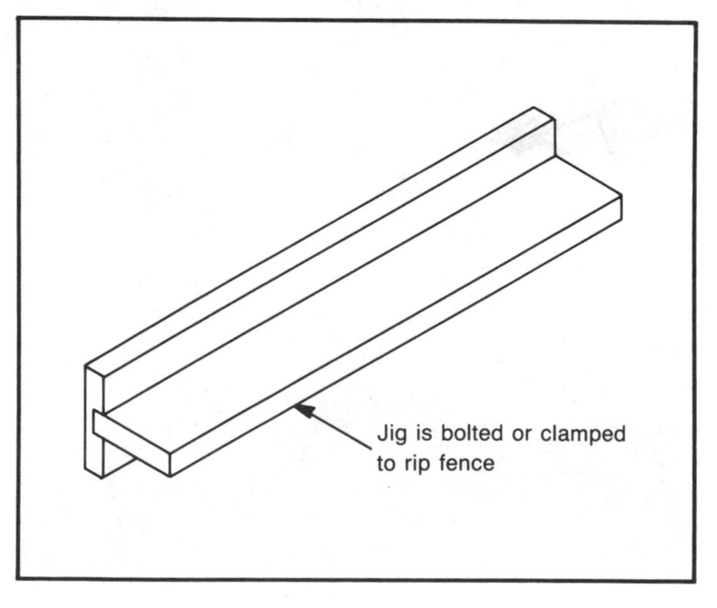

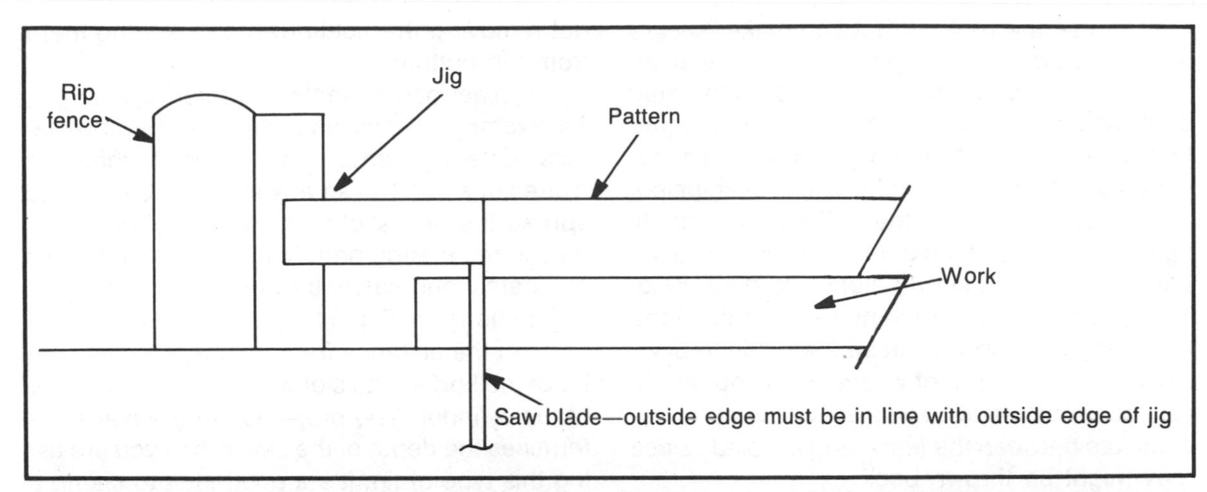

Fig. 11-6. Secure the pattern-sawing jig to the rip fence with clamps or screws. Lock the rip fence in place so the bearing edge of the jig will be correctly related to the saw blade.

Fig. 11-7. The outside of the saw blade and the bearing edge of the jig must be on the same vertical plane. Make the jig long enough to supply adequate support for the pattern you use.

To use the pattern that you make, secure it to workpieces that have been rough-cut to approximate shape. You can accomplish this step by driving slim nails or screws through the pattern so they project enough to become anchor points on which the workpieces can be impaled.

To cut, abut the edge of the pattern firmly against the edge of the guide and then move the pattern and work slowly along the guide (Fig. 11-8). Keep the pattern in place throughout the pass; any twisting will cause the blade to bind. Allow for a minimum of waste when you rough-cut the workpieces. Don't allow cutoffs to accumulate between the fence and the blade since they might be thrown back.

Results are best when the edges of the pattern and the edge of the guide are smooth, and sawing is done with a planer-type saw blade.

NOTCHED JIGS

Often, parts you need can't be sawed accurately or safely by working only with the rip fence or the miter gauge. The part might be too small or oddly shaped so a conventional procedure cannot, or should not, be used. A notched jig like the one in Fig. 11-9 can provide a solution.

A *notched jig* is a piece of stock that is wide enough for ample room between the fence and the blade, and that has a shape formed in its outboard edge which matches either the piece you need or the piece you wish to remove from the parent material. A good example is the one in Fig. 11-10, which was designed specifically for cutting the corners from a quantity of slim, square pieces of wood so they would have an octagonal shape.

Nestle each piece in the jig's cutout and feed it past the blade, as demonstrated in Fig. 11-11. The distance between the fence and the inside surface of the blade equals the width of the jig. Unless the pieces fit tightly in the notch, they can chatter, which will cause inaccuracies, and they might even be ejected by the action of the blade. Adding a simple hold-down to the jig will guard against these negative actions (Fig. 11-12). You can place workpieces in the jig without removing the hold-down by inserting them from the bottom.

Project components are often cylindrical; for example, rungs and rails for chairs or tables. Often, to ensure a tight joint, they are slotted at each end so a wedge can be used to spread the end slightly as it is inserted in the hole in the mating part. You can form the slots accurately and safely by working with the type of jig shown in Fig. 11-13.

Set the distance from the fence to the center of the kerf so the slot will be on the diameter of the cylinder. The projection of the blade determines the depth of the slot. When you are using this type of jig, it's a good idea to clamp a stop block to the rip fence (Fig. 11-14). The stop block will help to keep the jig flat on the table as you are cutting, and it will prevent the jig from being moved too far forward.

The jig in Fig. 11-15 is the type of design to think about when you need slim components that must be equal in length. Cutting splines to length is another application. The dado, sized to suit the workpiece, runs across the jig so the rip fence acts as a stop. The width of the jig determines the length of the workpieces.

Figure 11-16 shows how a notched jig can be used to guide work for the cheek cuts of a tenon. The jig, which is moved along the rip fence, is just a suitably sized length of wood that is dadoed so the workpiece will fit rather snugly. The position of the work is reversed after the first pass so the result is two matching slots. The depth of the dado will determine the thickness of the tenon. Shoulder cuts that remove the waste are made in routine fashion with the work flat on the table and moved with the miter gauge.

BENDING WOOD WITHOUT STEAMING

Commercial furniture manufacturers have equipment so they can make any species of wood flexible enough, usually by steaming, to literally be bent into knots. Unfortunately, the engineering and the equipment involved is not usually suitable for small shops, yet curving a piece of wood to suit a particular project is often necessary.

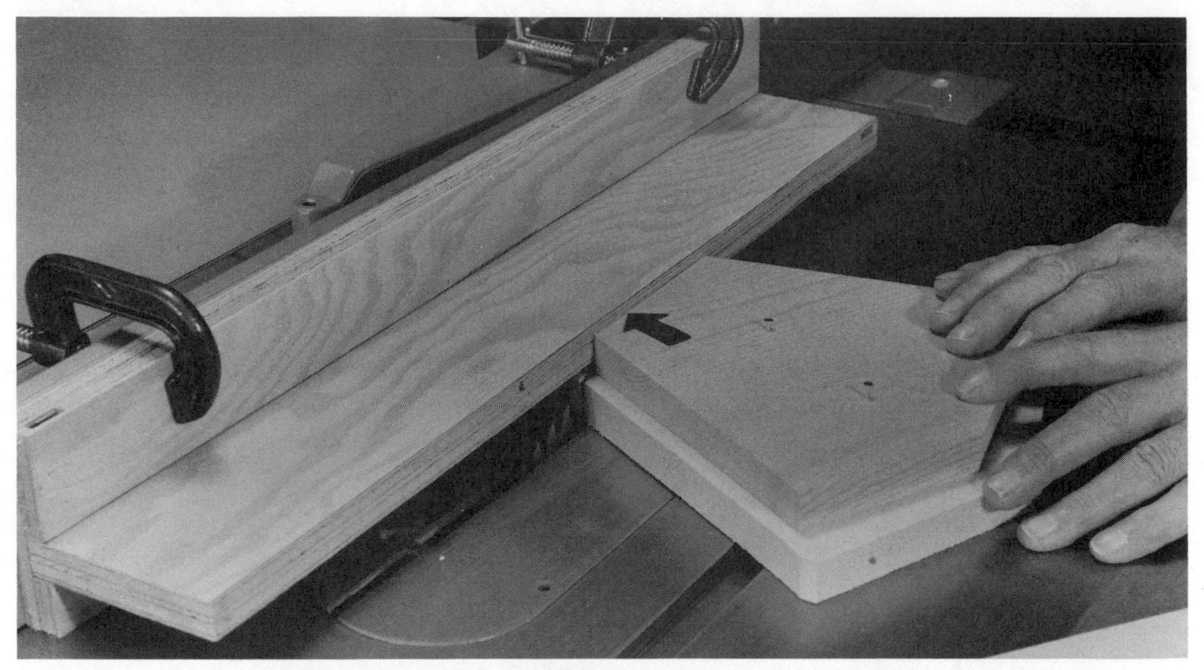

Fig. 11-8. Tack-nail the pattern to the rough-cut workpiece and make the passes by keeping the edge of the pattern firmly against the jig (arrow). Don't allow cutoffs to accumulate between the fence and the blade.

Fig. 11-9. A notched jig is just a wide piece of stock that has a cutout in its outboard edge shaped for the part you need or wish to cut off. This jig was used for cutting wedge-shaped pieces. Note the hold-down and the handle that is used to move the jig.

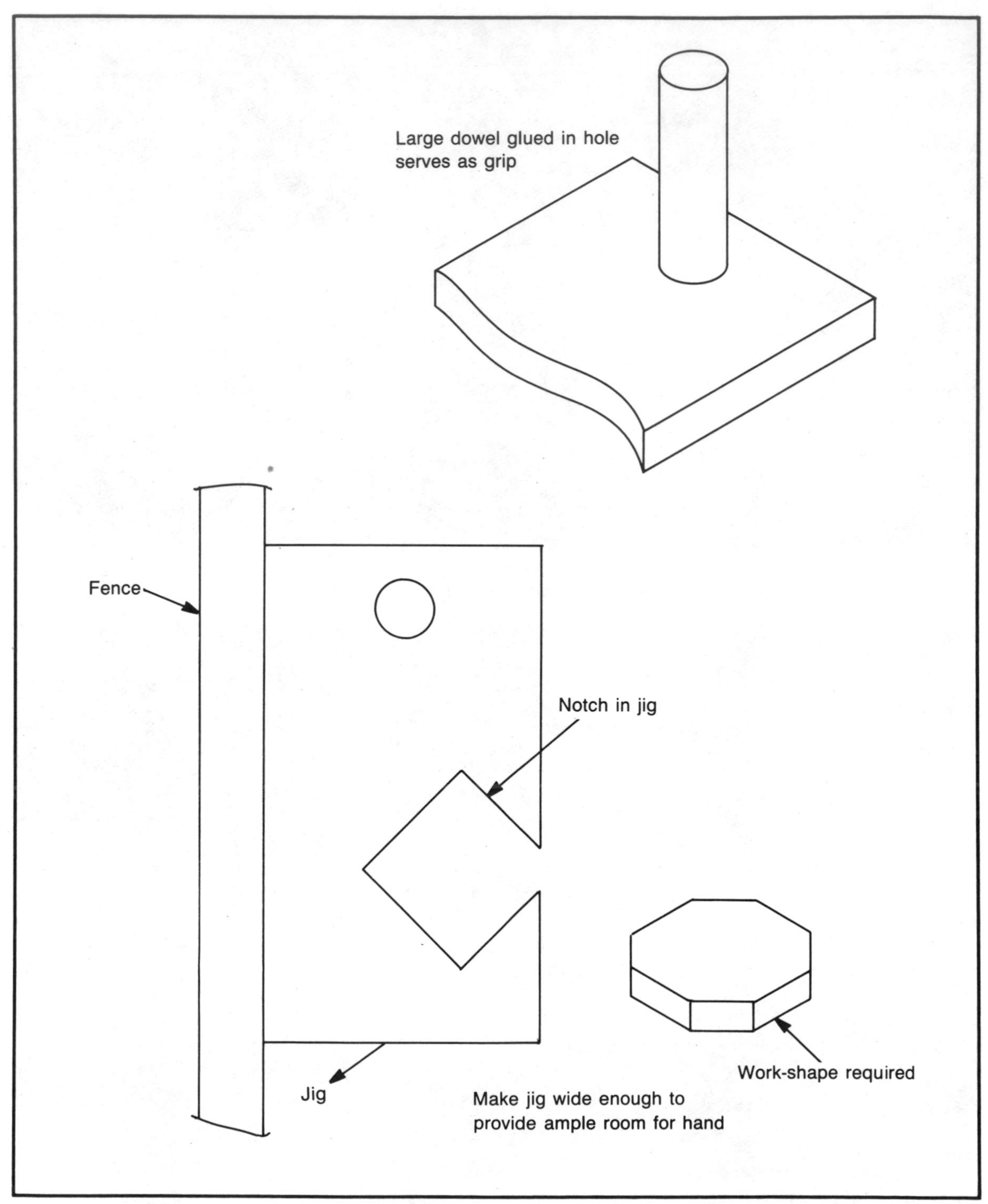

Fig. 11-10. This notched jig was designed so that small, square pieces could be shaped as octagons. The parts were needed for an inlaid table project.

Fig. 11-11. The width of the jig is optional, but should allow enough room between the fence and blade for safe cutting. Forming the notch to provide a tight fit for the workpieces contributes to accuracy.

Fig. 11-12. Incorporate a hold-down to keep parts flat on the table during all cuts. You can place workpieces in the notch from under the jig. Be sure to add a hand grip to any notched jig you make.

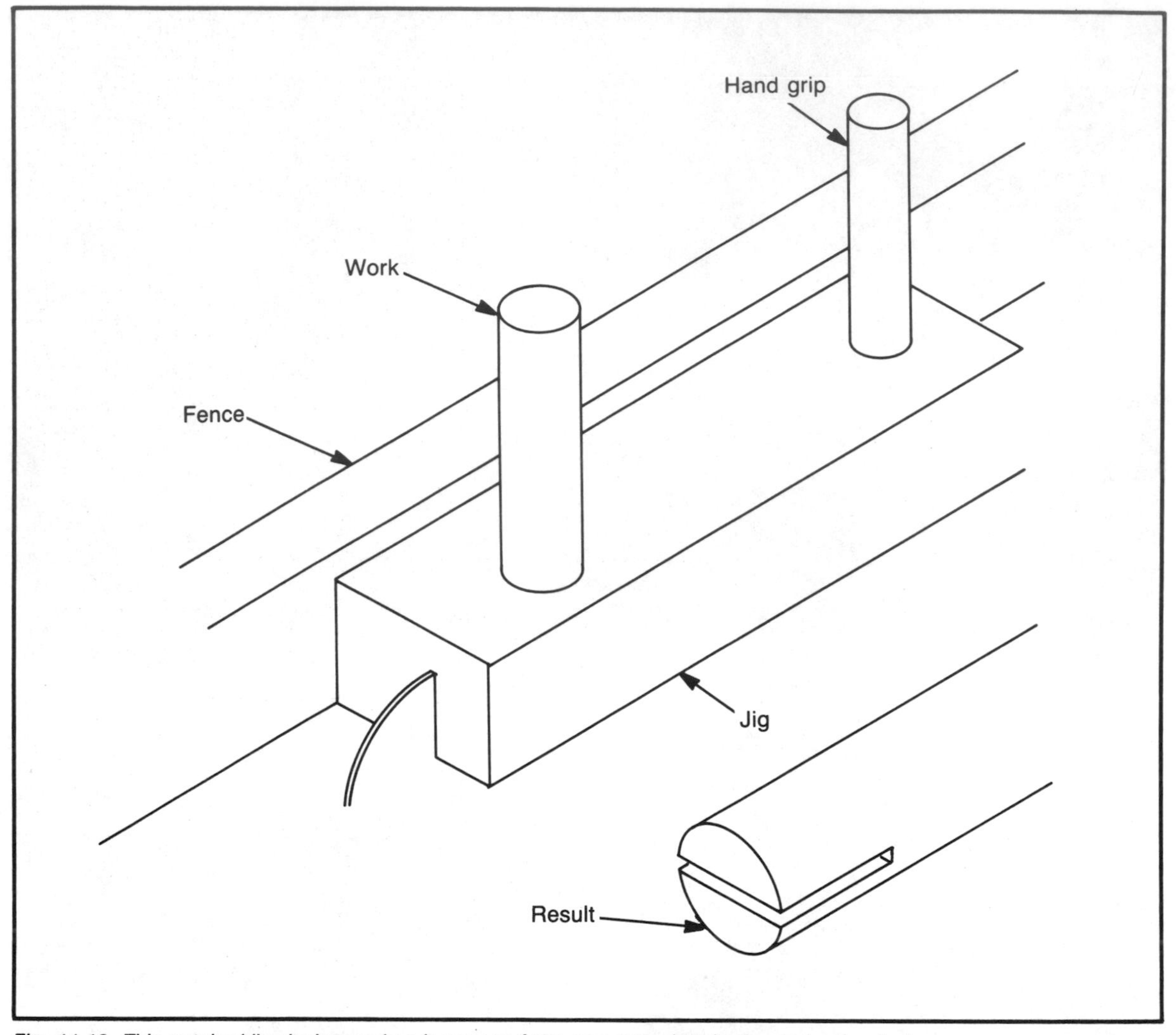

Fig. 11-13. This notched-jig design makes it easy to form accurate slots in the ends of cylinders.

Aprons on round stools, chairs, and drum tables, and the arched top for a garden arbor or a doorway are examples of when a wood-bending technique is required. The solution? If you can't steam it, kerf it. *Kerfing* involves making a series of equally spaced cuts into one surface of the material. Because of all the open areas provided by the kerfs, the solid surface, which is now something like a veneer, becomes flexible enough so the wood can conform to a curve or even a circle (Fig. 11-17).

The size and spacing of the kerfs is critical. The deeper they are and the closer they are spaced, the more flexible the wood will be. Deep and closely spaced kerfs also contribute to a smoother contour (Fig. 11-18). It isn't always wise to arbitrarily go about doing the kerfing, however. You might waste time by cutting too many, or weaken the project by cutting too deeply. It's best to arrive at least to an approximation of what you need by making a test kerf in a length of wood that is the same species and thickness as the material you plan to bend.

Fig. 11-14. The distance from the fence to the center of the kerf should be on the diameter of the cylinder. The clamped hold-down keeps the jig flat on the table and prevents you from moving it too far forward.

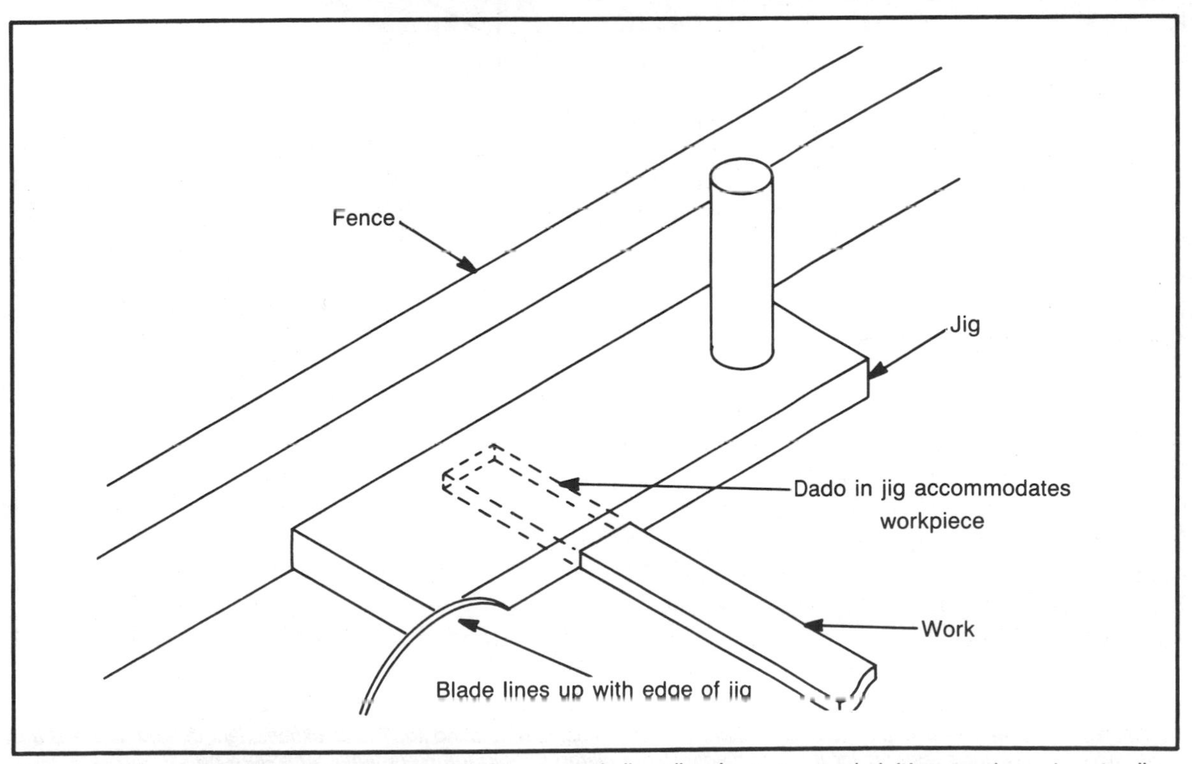

Fig. 11-15. This type of notched jig can be used when many similar, slim pieces are needed. It's a good way to cut splines to length. The width of the jig equals the length of the parts you need.

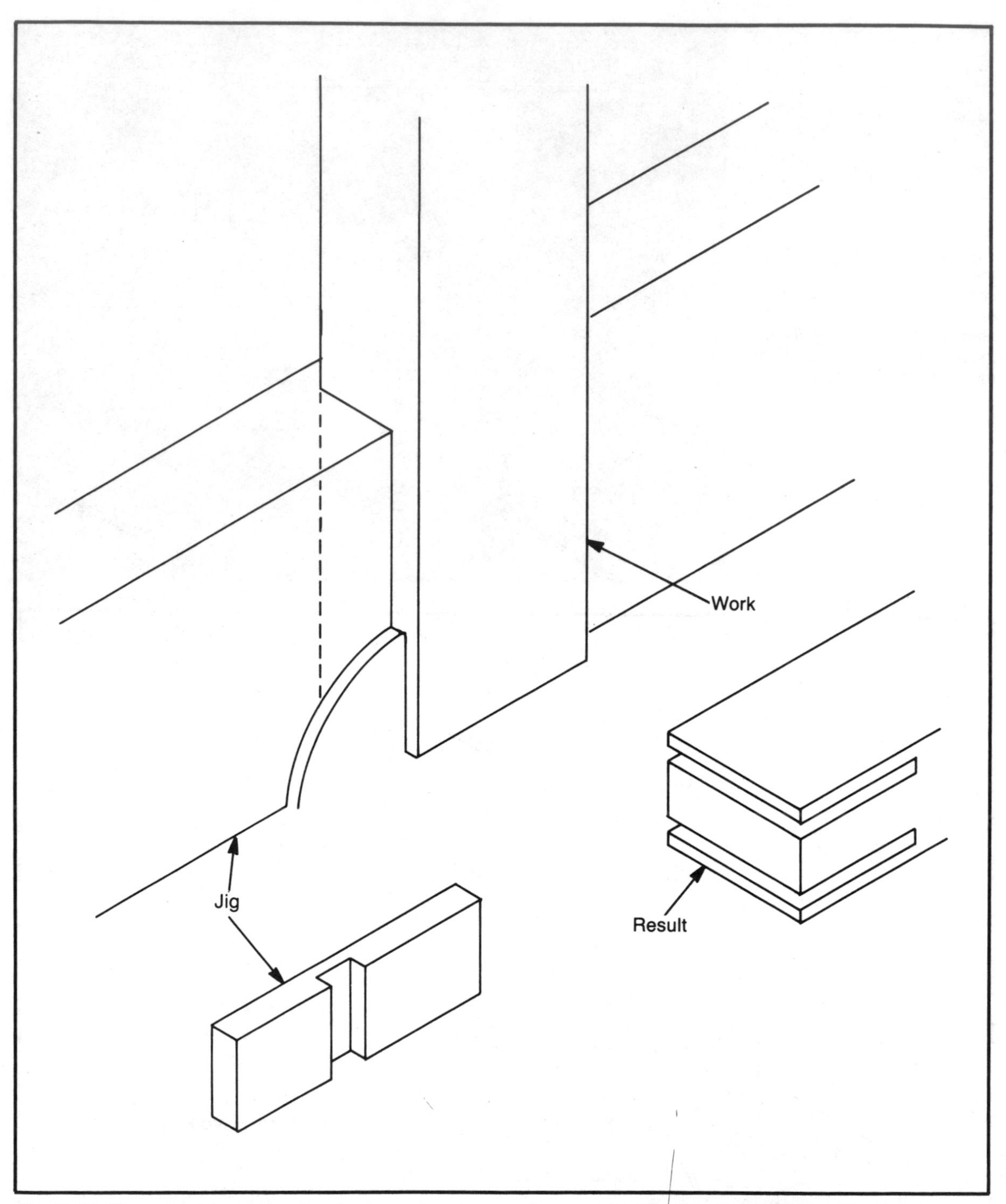

Fig. 11-16. The method for making the cheek cuts for a tenon using a notched jig. The thickness, length, and width of the jig should provide good support for the workpiece. Use a clamp to hold the work firmly while sawing.

Fig. 11-17. Kerfing is done so wood can be bent without having to steam it. The technique leaves a flexible veneer on one surface of the stock. The solid wood between the kerfs provides strength.

Fig. 11-18. The way you cut the kerfs will depend primarily on the wood species and whether the bend will be made with or across the grain. Deep kerfs, closely spaced, result in smoother contours, but overdoing it can result in weak components.

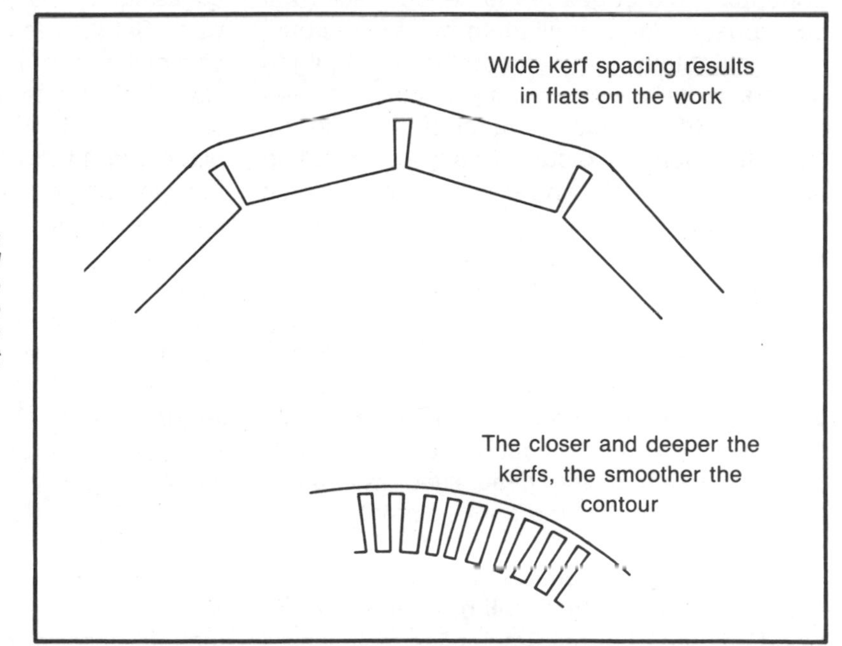

Make the kerf in the test piece so it leaves about 3/32 to 1/8 inch of solid material, and then organize it on a flat surface (Fig. 11-19). Lift the free end of the test piece until the kerf closes and then take the measurement indicated in the drawing. This will be the kerf spacing with which you should start.

Making the cuts to layout lines marked on the work is not the correct procedure. The job is more professional and much easier when you make a miter-gauge extension like the one in Fig. 11-20. The guide pin in the extension doesn't need to be more than a finishing or headless nail that is driven into the extension so the distance from it to the adjacent side of the blade will equal the spacing of the kerfs. Be sure the pin is placed low enough so you can place the kerf cuts over it. Place each cut you make over the pin to position the work correctly for the next cut (Fig. 11-21).

Do the actual bending slowly and, when necessary, after wetting the solid surface with hot water. You can hide the kerfs by gluing on a thin veneer, or you can leave them exposed if they will be concealed by other project components. If the edges of the kerfed pieces will be exposed, you can fill them with wood putty.

Incidentally, you can use the kerfing jig to make a dentil type molding if you work as shown in Fig. 11-22. It's just a matter of inverting the stock for each of the cuts. This type of molding is cut into thin strips and surface-mounted on a backing that becomes part of the total design. It's not unusual for workers to cut a groove in the backing piece so the dentil material can be inlaid.

Another way to make wood bendable is to thin out the bend area so what is left is a thin veneer (Fig. 11-23.) You can reduce the stock's thickness by making repeat passes with a dadoing tool or a molding head fitted with blank knives. Be sure that the thinned area is some inches longer than the actual length of the bend. The extra length is needed so the part will bend smoothly and the stress will be less where the wood changes from a straight line to a curved one. This wood-bending technique is more appropriate when the thinned section can be reinforced with glue blocks or solid components.

COVE CUTTING

Cove cutting is a table-saw technique that can be used with a conventional saw blade to produce the type of arched shapes shown in Fig. 11-24. Actually, it is not a true sawing procedure. Workpieces are moved obliquely across the saw blade many times with the blade raised about 1/16 inch for each pass. The cutting action is as much scraping as it is sawing.

Essentially, the procedure calls for clamping a guide that is just a straight strip of wood to the table at an angle to the saw blade. The angle sets the path along which the work should be moved. The angle can't be arbitrary, not if you are seeking a cove of particular shape and size, so the first step is to make the parallel rule detailed in Fig. 11-25. The parallel rule is used to establish the pass angle and the diameter of the cove.

When organizing it for the setup, adjust the long members so the distance between inside edges equals the diameter of the arch shape you want. Set the blade's elevation to equal the radius of the cove. Place the parallel rule over the saw blade so its opposite inside edges make contact with teeth at the front and rear of the blade (Fig. 11-26). With the parallel rule leading the way, clamp a guide strip to the table on the left side of the blade (Fig. 11-27). You might need to adjust the guide strip since it determines whether the cove will be on or off the longitudinal centerline of the workpiece. You can clamp an optional guide strip—one whose use is highly recommended because of the added security it will provide—parallel to the first one by using the workpiece to establish the correct spacing (Fig. 11-28).

Set the blade so it barely pokes above the table and make the first cut by moving the work very slowly across the blade. Make all other cuts the same way after raising the blade about 1/16 inch for each of them (Figs. 11-29 and 11-30).

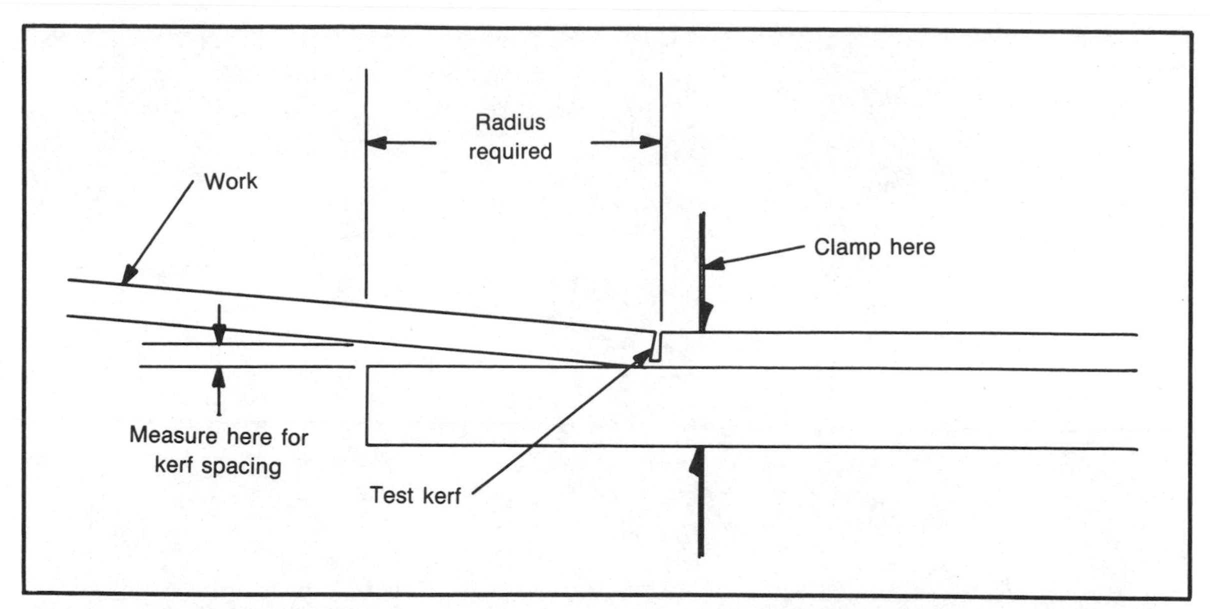

Fig. 11-19. Make the test on a piece of the stock you wish to bend. The kerf spacing that the test suggests might not be exact, but it will be close enough so a slight change in kerf depth or spacing will get the job done in good order.

Fig. 11-20. Use a miter-gauge extension to make accurate kerfs. The distance from the guide pin (arrow) to the facing side of the blade establishes the kerf spacing. Blade height equals kerf depth.

Fig. 11-21. Place each cut over the guide pin to position the workpiece for the following cut. Don't like this. You work more accurately and more safely when you don't rush.

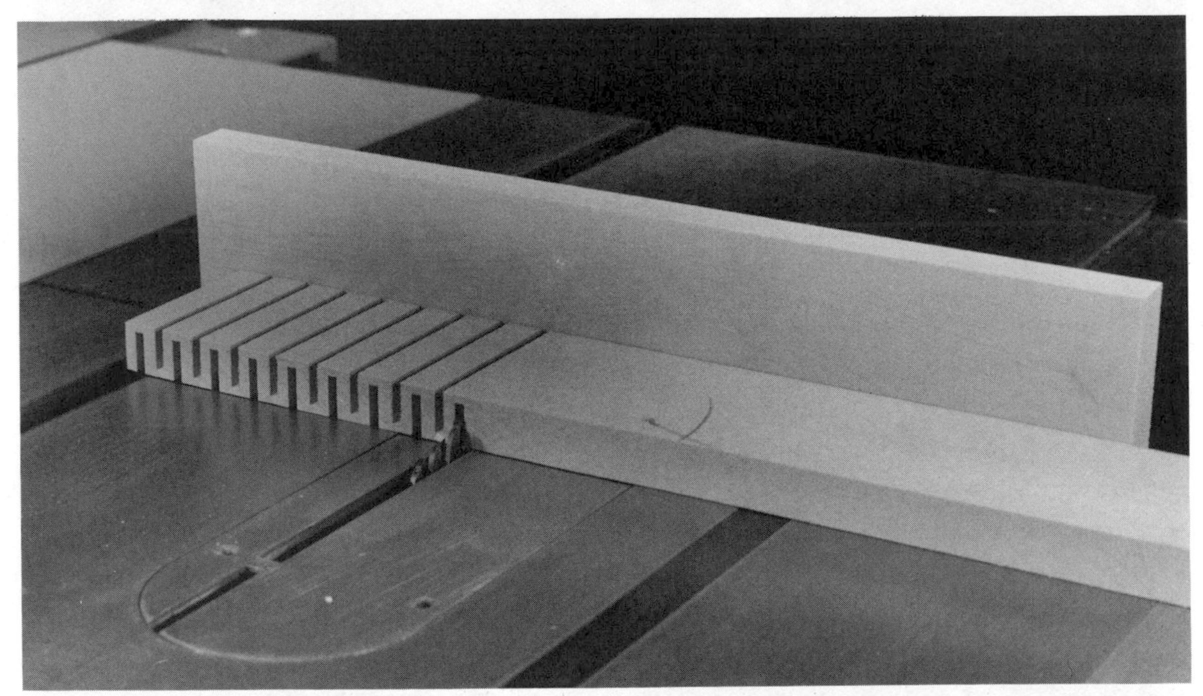

Fig. 11-22. The same jig can be used for the cuts required for dentil-type moldings. Turn the stock over for each cut. Work with a smooth-cutting blade. If you want more space between kerfs, just change the position of the guide pin.

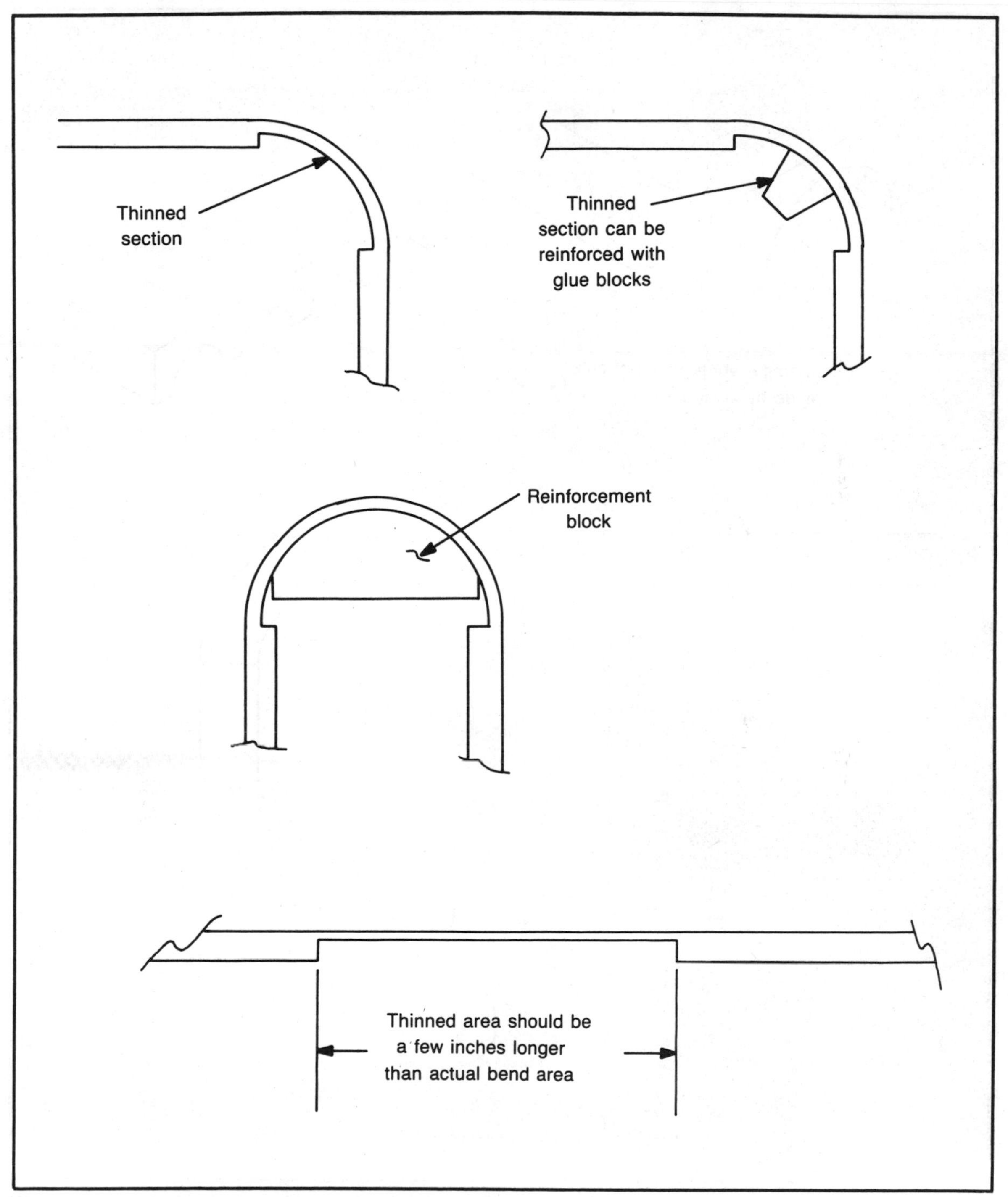

Fig. 11-23. You can also bend wood by reducing its thickness to leave a heavy veneer where the bend must occur. The thickness of the veneer will depend mostly on the radius and the length of the bend.

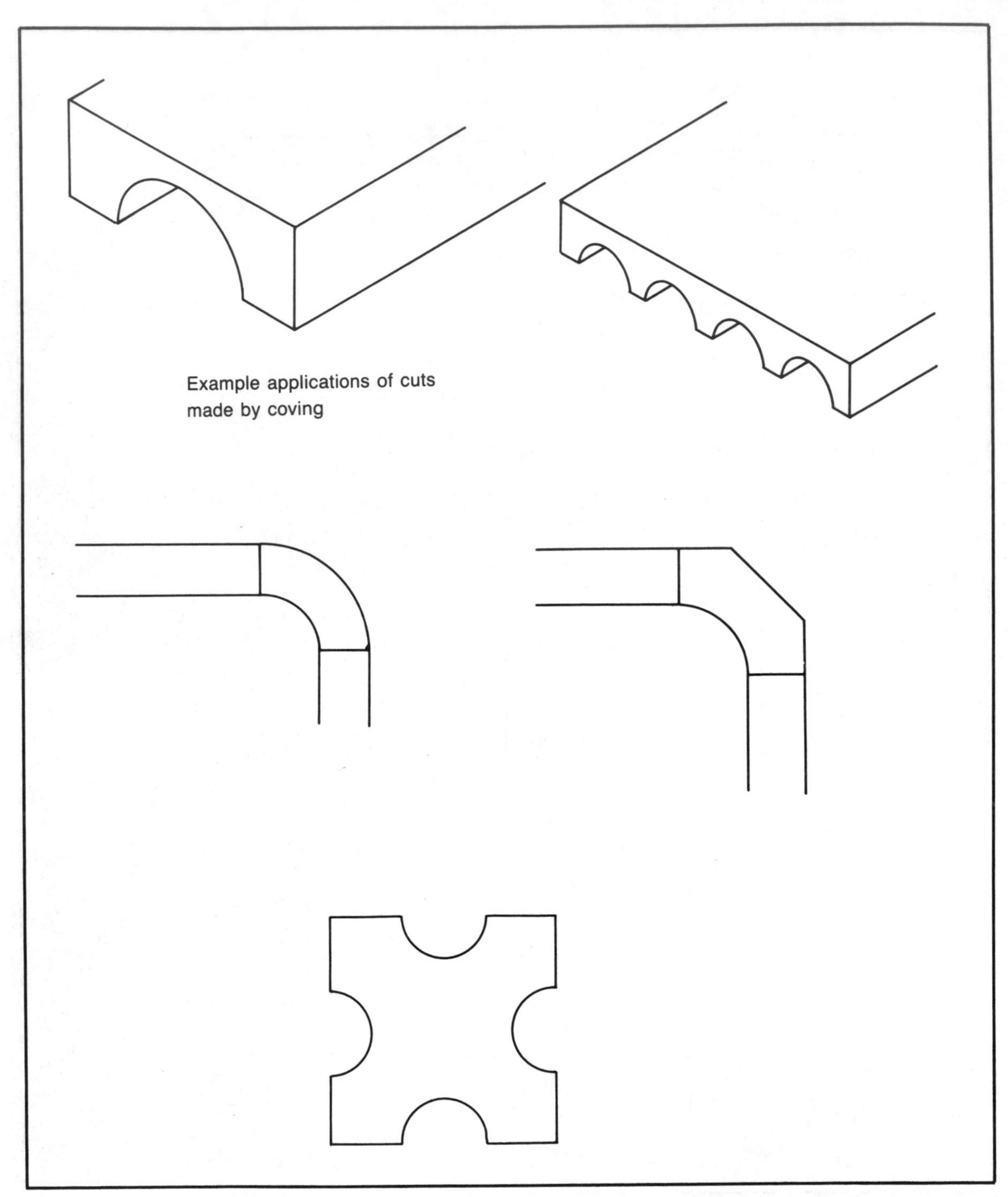

Fig. 11-24. To make cove cuts, pass work obliquely across a saw blade. The more oblique the angle, the closer you get to a true arc.

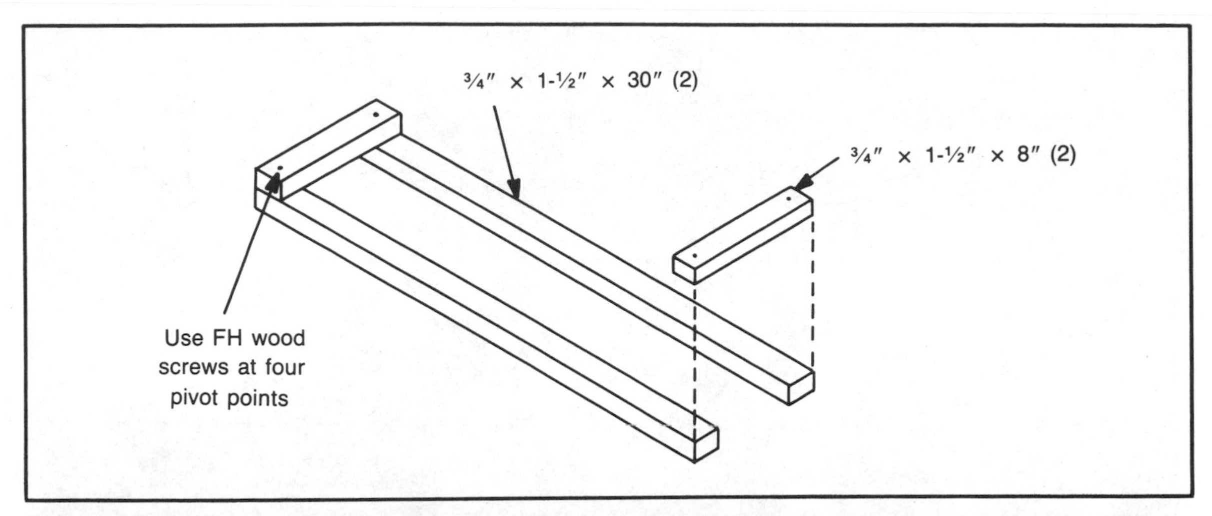

Fig. 11-25. The parallel rule is needed for coving to establish the pass angle for the work and the correct height for the saw blade.

Fig. 11-26. Set the legs of the parallel rule to equal the diameter of the cove cut, and set the blade height to equal its radius. Place the parallel rule so legs touch front and rear saw teeth to establish the pass angle for the work.

Fig. 11-27. Clamp a guide strip to the table on the left side of the blade. You might have to change the position of the guide strip, not its angle, depending on the width of the workpiece and whether you want the cove to be on or off the work's centerline.

Fig. 11-28. Take the precaution of adding a second guide strip so you will have rails to keep the work from moving "off track." Use the workpiece to establish the distance between the guides.

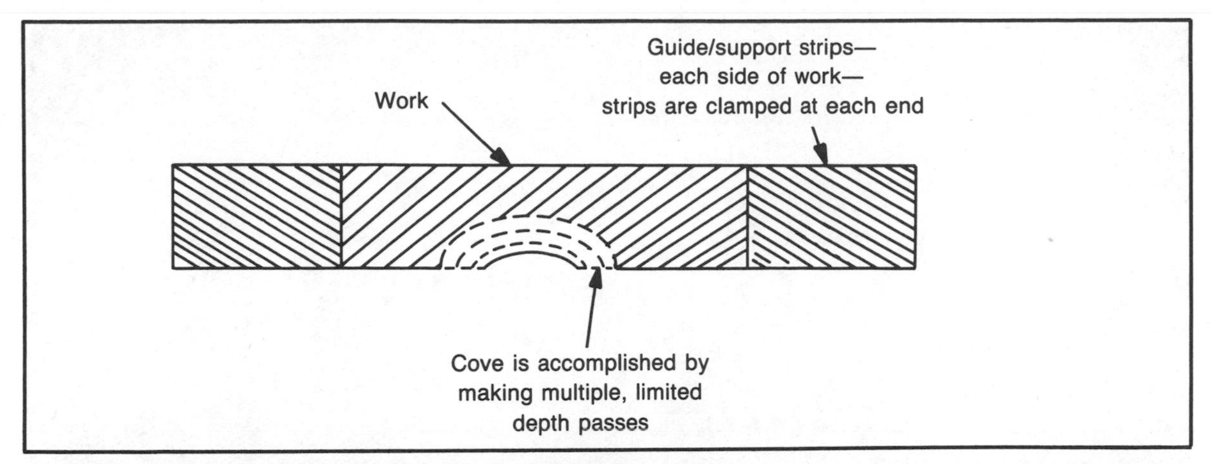

Fig. 11-29. Cutting is as much a scraping action as a cutting one. Raise the blade about 1/16 inch for each of the passes you must make before arriving at the cove depth you want.

Fig. 11-30. Move the work slowly. Check the cove depth frequently. You don't want to cut through the workpiece. Be sure the guides are parallel.

Fig. 11-31. To be certain of safe hand position, move the work by using a pusher/hold-down. Make final passes with the saw barely touching the wood so the cove will be as smooth as possible. Blades with set teeth will cut faster, but leave rougher surfaces.

You may be able to make cuts deeper than 1/16 inch, depending on the density of the wood and the pass angle, but cut depth must never be extreme. If you must force to move the wood or if the wood chatters, vibrates, or is lifted by the blade, you are almost certainly trying to remove too much wood in a single pass. It's not a bad idea to use a pusher/hold-down to move the work (Fig. 11-31).

The coving cuts will be easier to accomplish if you first make straight cuts to remove the bulk of the waste (Fig. 11-32). Use a saw blade or dadoing tool for this step. When you are following this procedure, it isn't necessary to try to get too close to the ultimate shape of the cove.

You can form edge coves by following the same procedure (Fig. 11-33). Be sure, though, that there will be enough of a flat edge left on the workpiece for safe bearing surface on the table.

Coves produced by following this technique will not be true arcs, but they are close enough so that if a perfect one is essential you can finish the job with a drum sander or sandpaper wrapped around a cylinder. Remember that the more oblique the pass angle, the closer you get to a true arc.

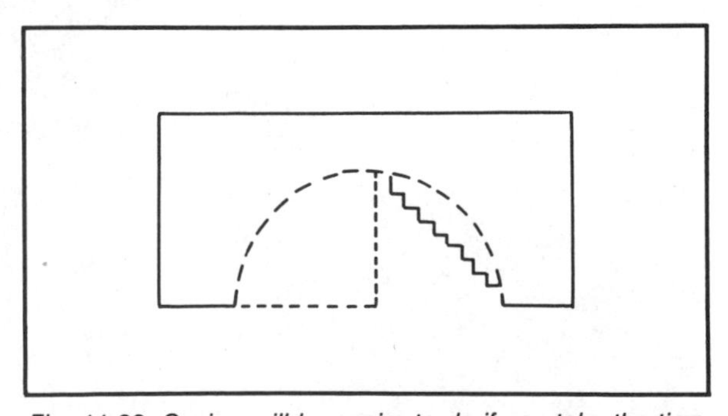

Fig. 11-32. Coving will be easier to do if you take the time to remove waste by making straight cuts with a saw blade or a dadoing tool.

Fig. 11-33. You can cove edges if you make the passes this way. The saw blade cuts into one of the guides, as well as the work. Be sure the guides are thick enough to provide good work support and that enough edge is left on the work so it can bear solidly on the table.

WHEN TWO SAW BLADES ARE BETTER THAN ONE

The idea is simple. If you mount two saw blades on the arbor, you can make two cuts simultaneously. Industry does it routinely and to extreme, spewing out components with multiple cuts in a single pass. Small shops can't compete, but can profit from the idea to a limited degree. A sampling of what can be accomplished with a double-blade setup, or with the outside blades from two dado assemblies if the machine allows it, is shown in Fig. 11-34.

The basic idea is to mount a second blade on the arbor, separating it from the first one using large, heavy washers (Fig. 11-35). Before getting involved, you will need to remember some very important rules:

☐ Never mount an assembly of components on the arbor that does not leave enough threads for the locknut to be tightened securely.

☐ Be sure that the maximum rpm of any extra saw blade you buy is not less than the rpm of the machine. Actually, you'll get better results if you work with twin blades.

☐ Be sure that the tool has the power to handle the extra load. It will probably be suitable if it can handle cutters like dado assemblies and molding heads.

Using two saw blades negates the use of the standard insert. It's possible that you might be able to work with a dado insert, but it's good practice to make a special one that will provide maximum support around the cutting area (Figs. 11-36 and 11-37). This is especially important when you are doing multiple-blade work that creates pieces which can be pulled down beneath the table by the action of the blades.

You can't judge the spacing of the cuts the blades will make solely by the thickness of the washers, especially if the blades have set teeth. Measure across the blades themselves and, when in doubt, make a trial cut to test the setup. Don't make setups so wide that the saw guard

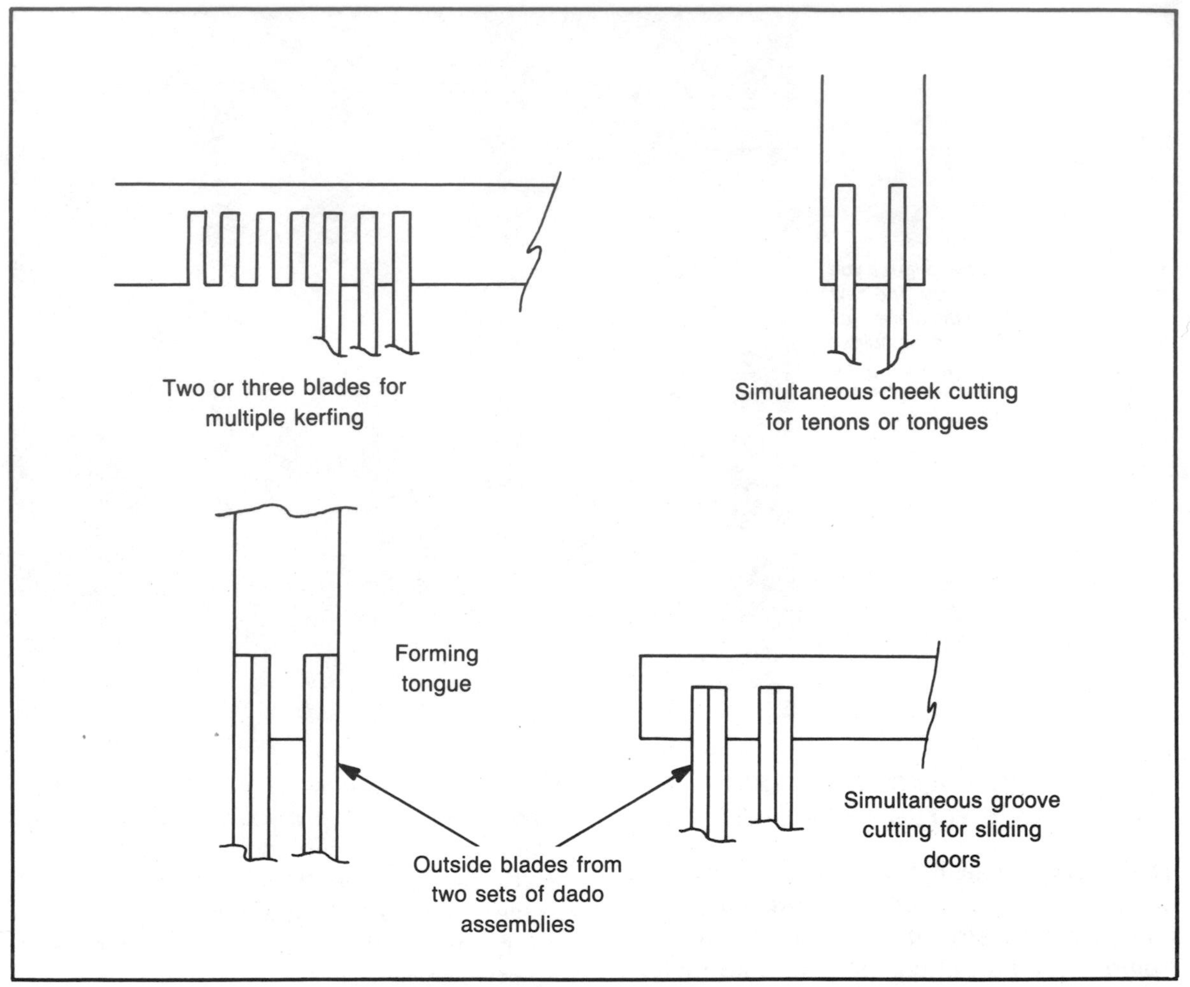

Fig. 11-34. Cutting with more than one saw blade can make many operations go faster. It's also possible to create setups for some cuts so they can be done accurately with minimum fuss.

can't be used. When necessary, provide for hold-downs and special pushers so you can work safely.

The practicality of double-blade work becomes obvious on jobs like sawing the kerfs for wood bending (Fig. 11-38). The miter-gauge extension that is required is made like that for a single blade. The only difference is that the extension for double blades will have twin kerfs. The spacing of the blades must equal the distance from the guide pin to the side of the nearest blade (Fig. 11-39).

You can halve the time required for sawing multiple strips if you work with two saw blades (Fig. 11-40). On operations like this, it is critical to make a special pusher/hold-down to get both cutoffs safely past the blades. The width of the accessory's base should equal the distance from the rip fence to *almost* the inside face of the outboard blade. Adjust the projection of the blades so they barely poke through the surface of the stock.

You can make the cuts required for slots simultaneously by working as shown in Fig. 11-41. Control the length of the cuts by using two stop blocks. You can start the cut by lowering

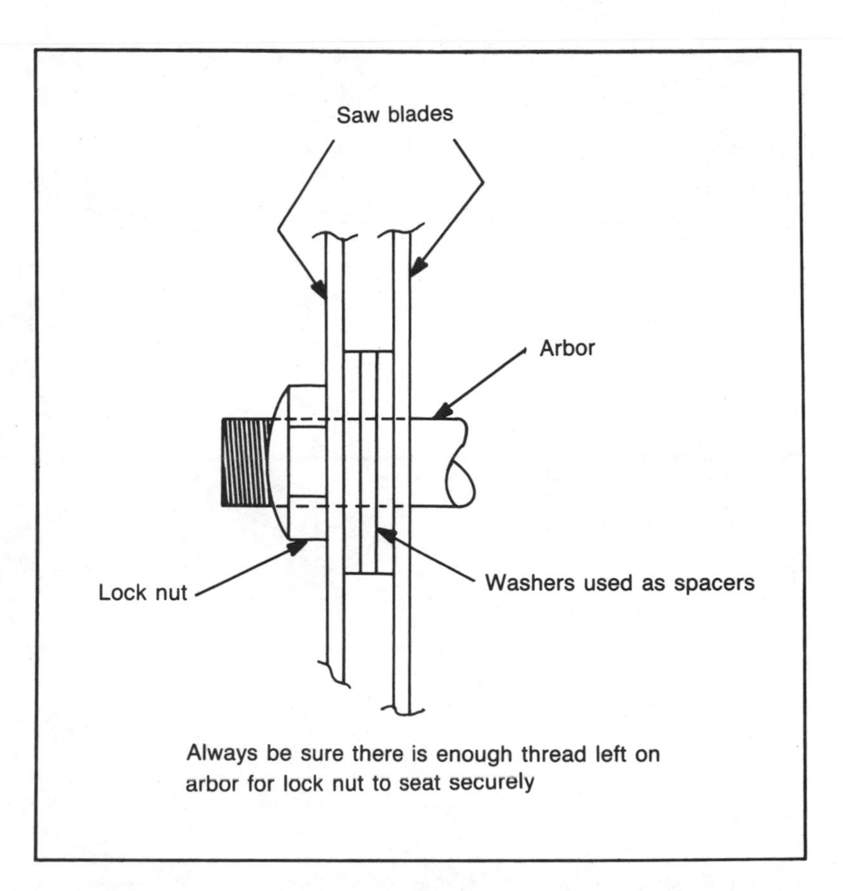

Fig. 11-35. Large washers used on the arbor establish the spacing between the blades.

the workpiece over the blades while they are turning, or you can have the work in place and get the blades working by raising them from beneath the table. The procedure using a single blade was described in detail in Chapter 6.

Simultaneous sawing with two saw blades, or with three as shown in Fig. 11-42, speeds production and can even spark original ideas for decorative screens and panels. The spacers used between the blades can be similar, or they can vary in thickness. It all depends on how you envision the result. You can "weave" panels that are cut in this way with wide or narrow strips of contrasting wood veneer or plastic (Fig. 11-43). The thinner the panel, the tighter the weave can be. You can frame the panels and use them as privacy screens or room dividers, for example.

Figure 11-44 demonstrates an arrangement I made with the parts of a dado assembly to cut twin, 1/4-inch-wide dadoes simultaneously. The setup involved using outside blades and 1/8-inch chippers in tandem, with washers between them that determined the spacing of the cuts. I used the same idea for the twin grooves being formed in Fig. 11-45, which is a nice way to work when you need twin grooves for sliding doors, for example.

An idea that suggests other practical applications for the twin-blade technique is shown in Fig. 11-46. Here, both of the cheek cuts for a tenon are cut in a single pass. You must use a tenoning jig and a special insert to do the work. The blade projection determines the length of the tenon; its thickness depends on the spacer used between the blades. The tenon can favor one surface of the stock or be centered, depending on how you adjust the jig. For a true tenon, make a second pass after placing an adjacent side of the workpiece against the face of the jig.

Fig. 11-36. You can't use the standard insert when doing multiple-blade work, but a dado insert might do. Check for freedom of the blades in the slot before you plug in the machine.

Fig. 11-37. Make special inserts whenever they are needed. Their purpose is to leave zero space around the cutters.

Fig. 11-38. Forming kerfs for wood bending is an example of how practical double-blade work can be.

Fig. 11-39. The distance from the guide pin to the adjacent blade and the spacing of the blades should be equal. Results are best when you use matching blades. Note the special insert.

Fig. 11-40. Using two blades to cut multiple, similar strips. You must work with a special pusher that will get both strips safely past the blade. Blades should barely poke through the stock.

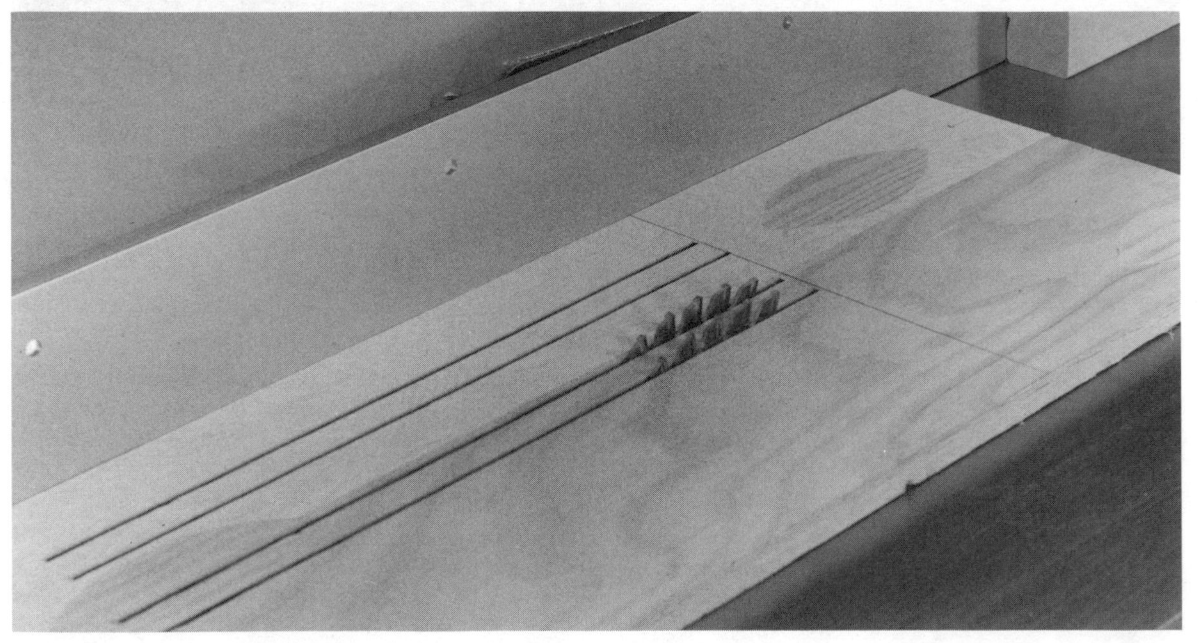

Fig. 11-41. Two blades contribute to fast, accurate cutting of slots. Work as you would when using a single blade. Check Chapter 6 for the proper procedure.

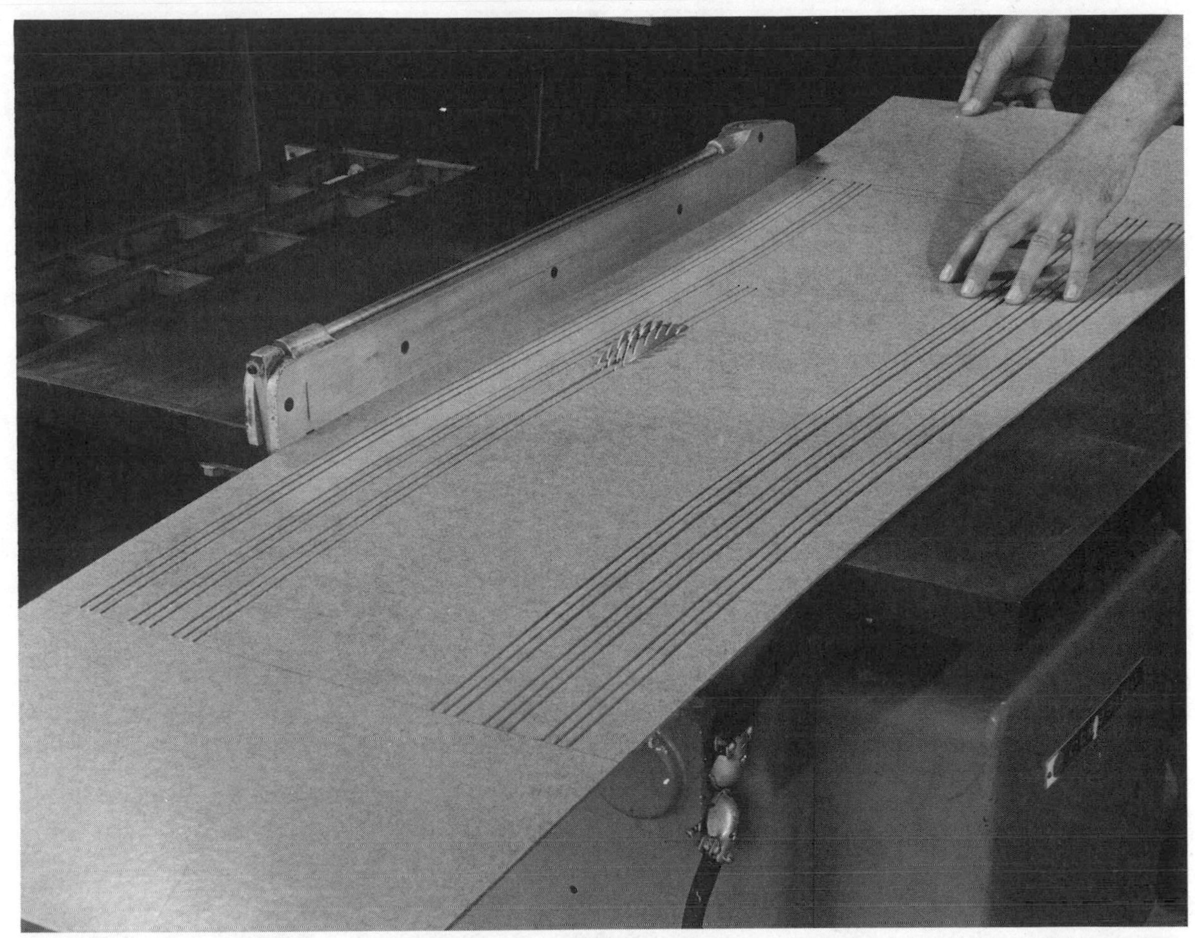

Fig. 11-42. I used a three-blade setup for the kerfs I needed to produce a decorative panel. Always be sure that the components you mount on the arbor leave enough room for the locknut to be tightened securely.

Remove waste by sawing with a single blade, with the work flat on the table and moved by the miter gauge.

CUTTING CIRCLES WITH A REGULAR SAW BLADE

It's puzzling to think that you can cut perfect circles with a saw blade designed for straight cutting, but the idea is as practical as it is simple. The first step is to make the sliding platform shown in Fig. 11-47. Its size is optional, but being generous will provide more support for workpieces. The bar the platform is attached to should fit precisely in the table slot. Don't use material thicker than ¾ inch. Actually, a thinner platform—say, ¼ or ⅜ inch—might be better since it will allow you to work with less blade projection.

Mount the workpiece by tapping a nail through its center into the platform. The distance from the nail, which is used as a pivot point, to the saw blade equals the radius of the circle. If a hole through the work is objectionable, drive the nail up from the bottom of the platform so it penetrates the workpiece enough to keep it secure.

Make the first cuts tangent to the target circle by holding the work firmly on the platform

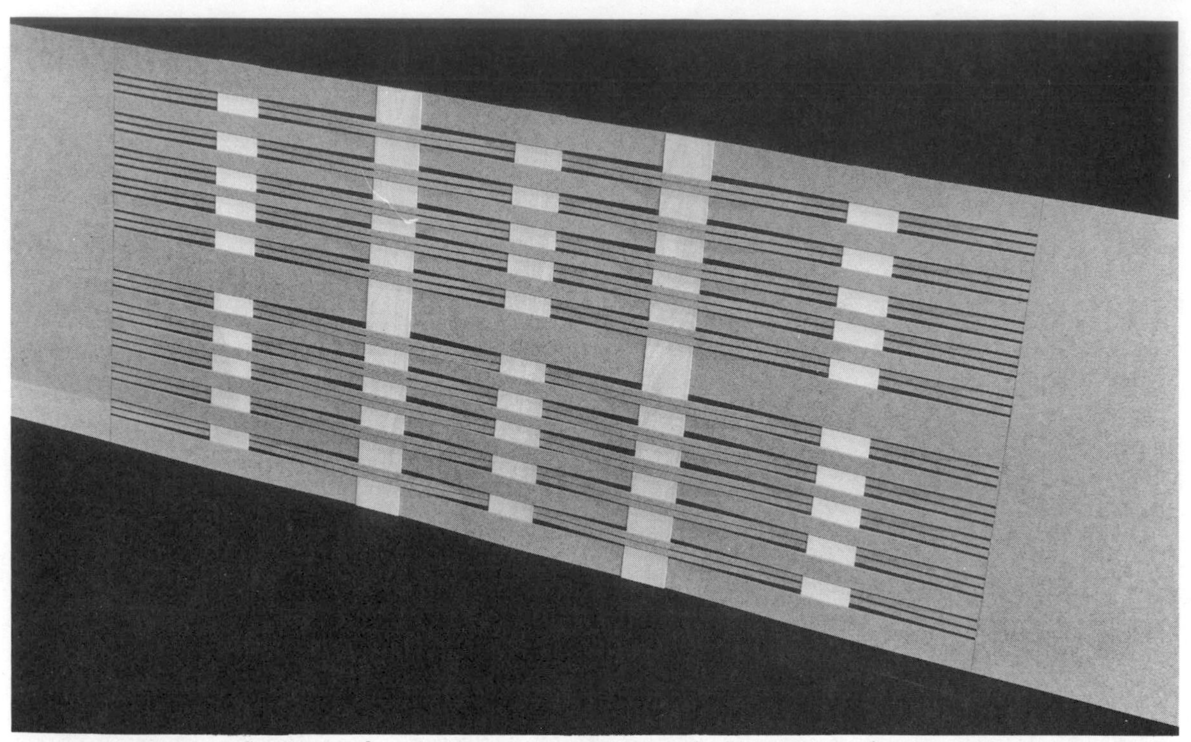

Fig. 11-43. The kerfed panel was woven with strips of thin veneer. The thinner the panel and the strips, the tighter the weave.

Fig. 11-44. The components of a dado assembly were organized to cut twin dadoes simultaneously. Outside blades and ⅛-inch chippers, in tandem, are separated by a heavy washer that establishes the spacing between the cuts.

Fig. 11-45. The same idea can be used to form matching grooves when, for example, grooves are needed for bypassing, sliding doors.

Fig. 11-46. Double blades allow for cutting of both tenon cheek cuts in a single pass. You can center the tenon or position it to favor one surface of the stock. This kind of work must be done with a tenoning jig.

Fig. 11-47. The jig you need for circle-cutting is a simple platform guided by a bar that rides in the table slot. Make the platform oversize, then true its right edge by making a pass across the blade.

and moving both past the saw blade. Make repeat passes, rotating the work anywhere between 10 and 20 degrees for each of the cuts (Fig. 11-48). The purpose of the straight cuts is to remove the bulk of the waste stock. The more straight cuts you make while rotating the workpiece a minimum amount, the easier the final step will be.

After the bulk of the waste is removed, clamp the platform to the table so the pivot point is on line with the teeth at the front edge of the blade. Then rotate the work very slowly *against* the blade's direction of rotation (Fig. 11-49). You can cut just about any size circle so long as the platform is wide enough to allow significant radius distance from the pivot point to the saw blade. Remember that the final step is done with the platform clamped in position.

You can follow the same procedure if the circular piece requires a beveled edge. For the tangent cuts and the final cut made by rotating the work, simply tilt the saw blade to the angle you need (Fig. 11-50).

It is possible to form circular pieces directly, skipping the tangent-cutting step, but it is best to use this idea on thin material. Raise the saw blade to penetrate the workpiece after it is mounted on the platform. Then rotate the work, as usual, in a counterclockwise direction (Fig. 11-51). If you try this idea on thick material, you must get through the stock by making repeat passes, raising the saw blade just a fraction for each of the cuts. If you examine the edge of the waste, you will see that the procedure is some-

Fig. 11-48. Mount workpieces on the platform with a nail (arrow), which serves as a pivot. Blade-to-nail distance equals the radius of the target circle. Make first tangent cuts, to remove waste, by moving jig and work together.

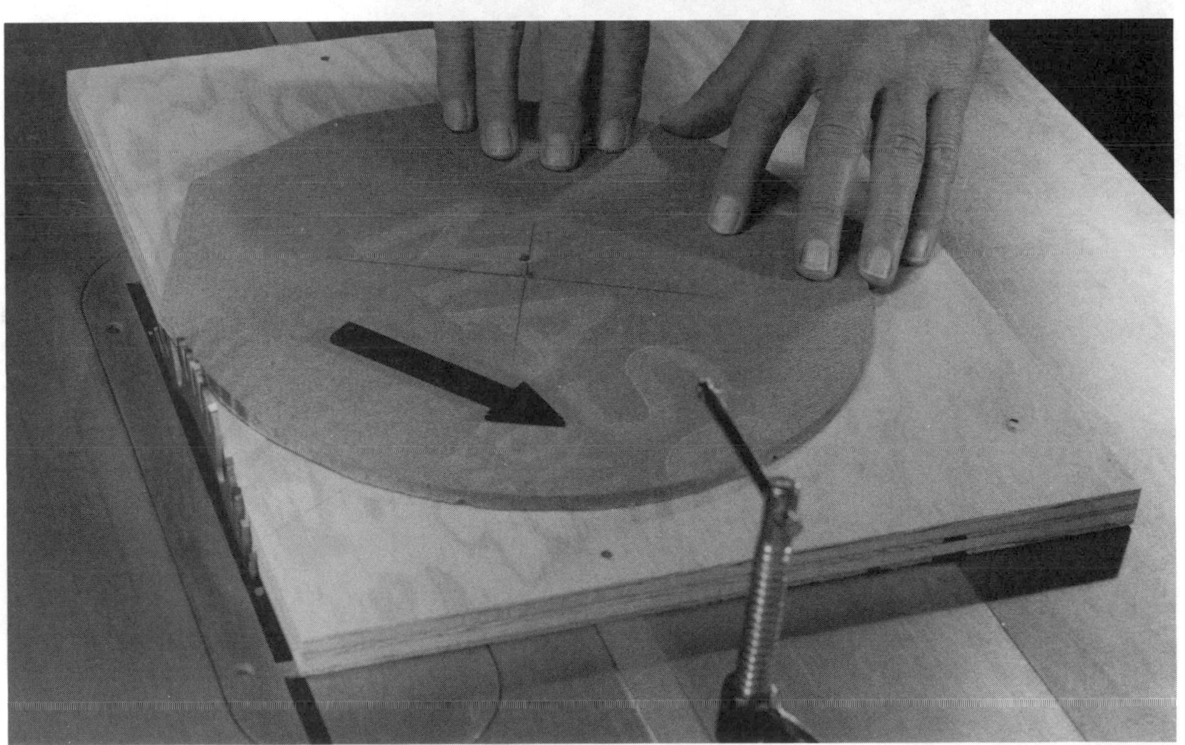

Fig. 11-49. The next step is to position and clamp the jig so the pivot point is in line with the front of the blade. Hold the work firmly and turn it very slowly against the blade's direction of rotation.

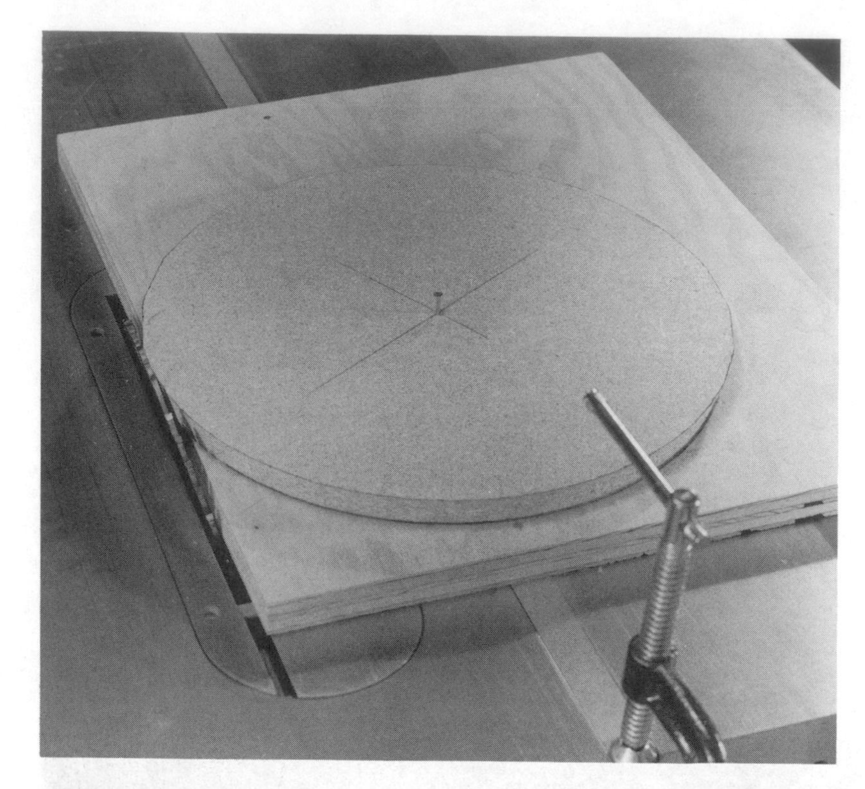

Fig. 11-50. If the disk you need must have a beveled edge, follow the same procedure, tangent cuts included, with the blade tilted to the angle you want.

Fig. 11-51. You can form circles without the tangent-cutting step, but this idea is most applicable when the workpiece is thin. Raise the blade after you position the work. Allow the blade to barely poke through the stock.

Fig. 11-52. You can rabbet circular pieces on their perimeters by using the pivot-cutting technique. Cutting starts with the dadoing tool under the table. Raise it a bit for each revolution of the workpiece.

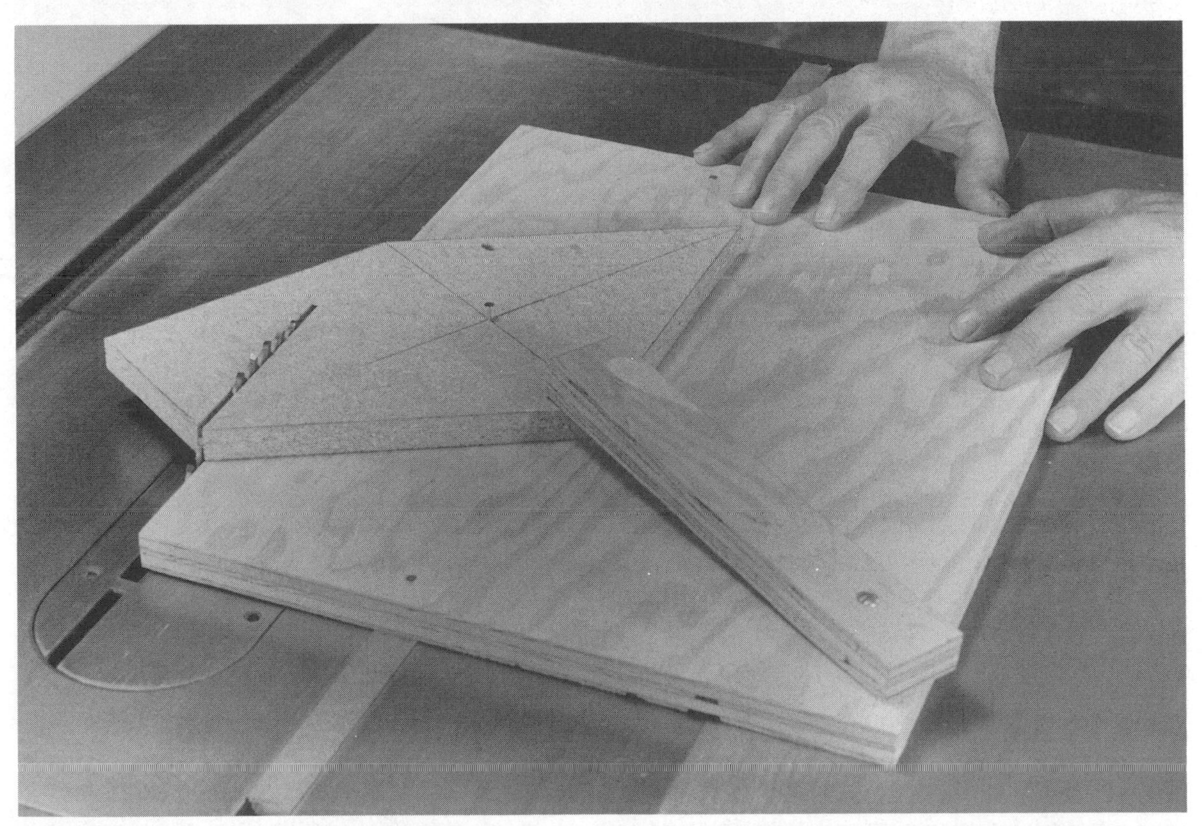

Fig. 11-53. You can use the platform for odd-angle cuts or to shape figures like hexagons and octagons. Note the use of a simple hold-down.

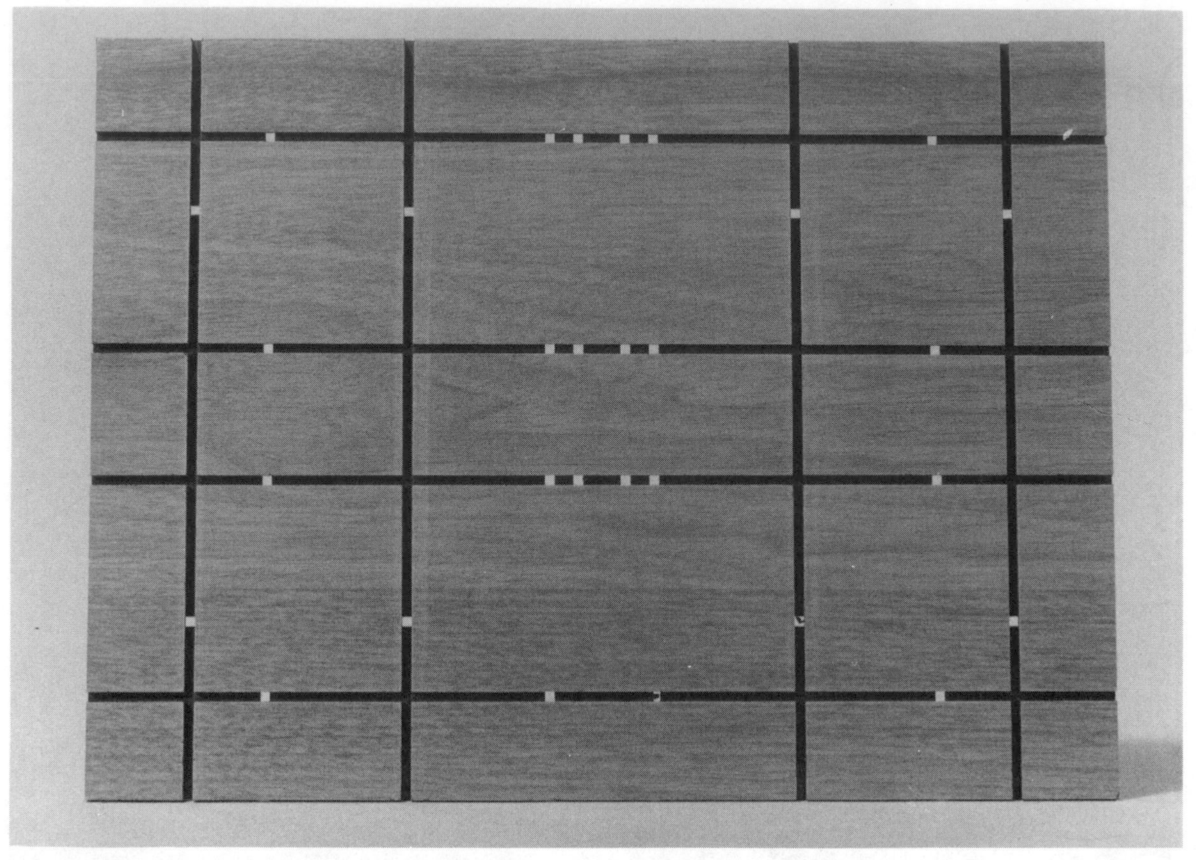

Fig. 11-54. Pierced panels result when intersecting cuts are made on opposite surfaces of the stock.

thing like a coving operation. The outside edge (the waste) will be arc-shaped, even though the edge of the workpiece will be square.

Workpieces that have been brought to true round by using the pivot-cutting technique or by other means, can be rabbeted on the perimeter with a dadoing tool (Fig. 11-52). Set the cutting tool under the table when positioning the work for the cut. Then raise it a small amount for each of the passes required to get to the cut depth you need.

You can use the same platform for odd-angle cuts or to produce multiple-sided figures (Fig. 11-53). Draw the shape you need on the workpiece to establish cut paths that can be lined up with the blade side of the platform. Rotate the work about its pivot point for each of the required cuts. For shapes like hexagons and octagons, you can draw perpendicular diameters on the platform and the workpiece. To position the piece for the cuts, align the mark on the work with a line on the platform. Add a simple hold-down to the jig, like the one shown in Fig. 11-53, to keep the work secure for each of the cuts you must make.

PIERCING

Piercing is a procedure you can use to make decorative panels (Figs. 11-54 through 11-56). When kerfs are cut in one surface of the workpiece to a depth that is a bit more than half the stock's thickness and then repeated on the opposite surface but following a different pattern (Fig. 11-57), you end up with openings at all points where the cuts cross. It's that simple. The

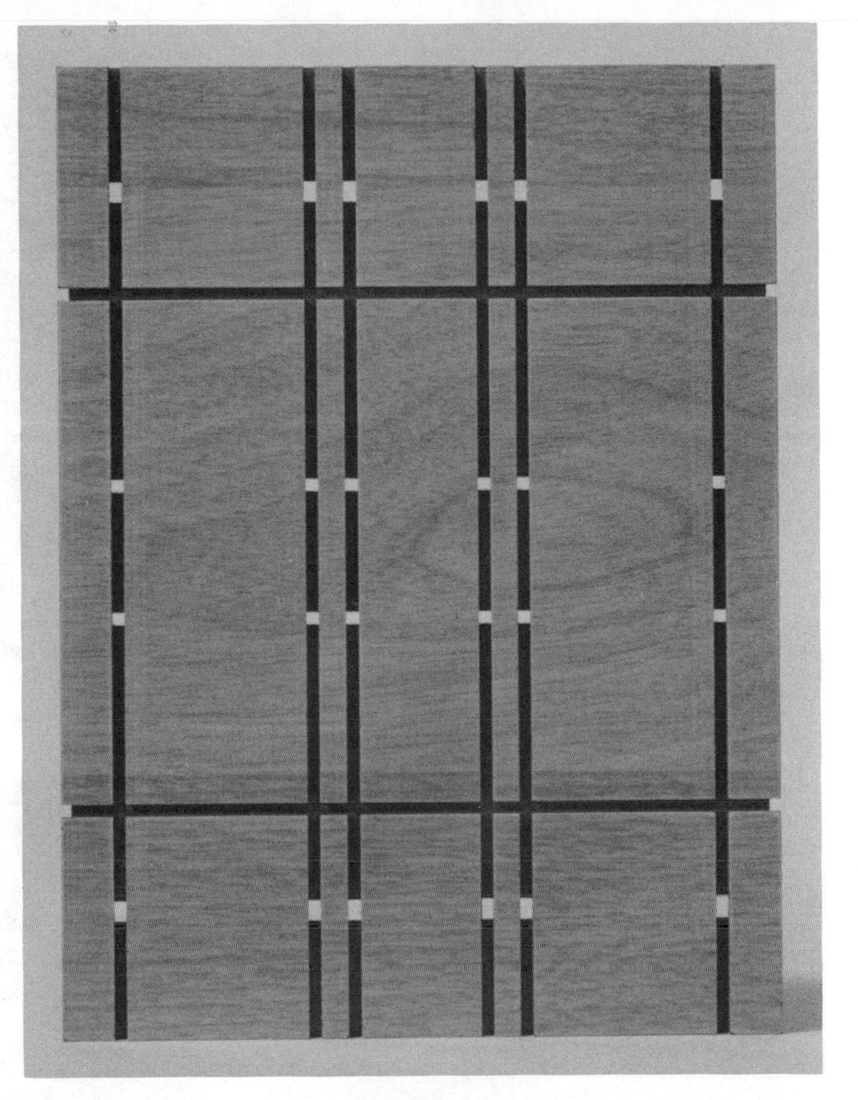

Fig. 11-55. The shape of the openings depend on how you make the cuts. They can be square or oblong, even diamond shaped. The panels shown here were done with a saw blade, but you can use a dadoing tool for wider cuts.

openings can be square or oblong, or shaped like diamonds; some can be wider or longer than others. The variations are limitless, and depend solely on how you do the cutting. The openings will be square when the kerfs are equal in width and at right angles to each other. Diamond shapes occur when kerfs are cut at an angle (Fig. 11-58).

Making the cuts is simple, but previewing the results might not be. It is best to "see" the cuts you can make on one surface of the stock by drawing them on tracing paper and then placing the paper over an opaque sheet so you can judge what cuts to make on the opposite surface for the most pleasing results. Remember that the visible kerfs contribute to the design, so you must include them when you are judging results.

You can add decorative details to panels or to any project component simply by cutting intersecting kerfs on only one surface (Fig. 11-59). Piercing is often done on doors, case sides, and drawer fronts to break up drab areas and to add visual interest.

DOODLING

Doodling (my own name for the idea) is a kind

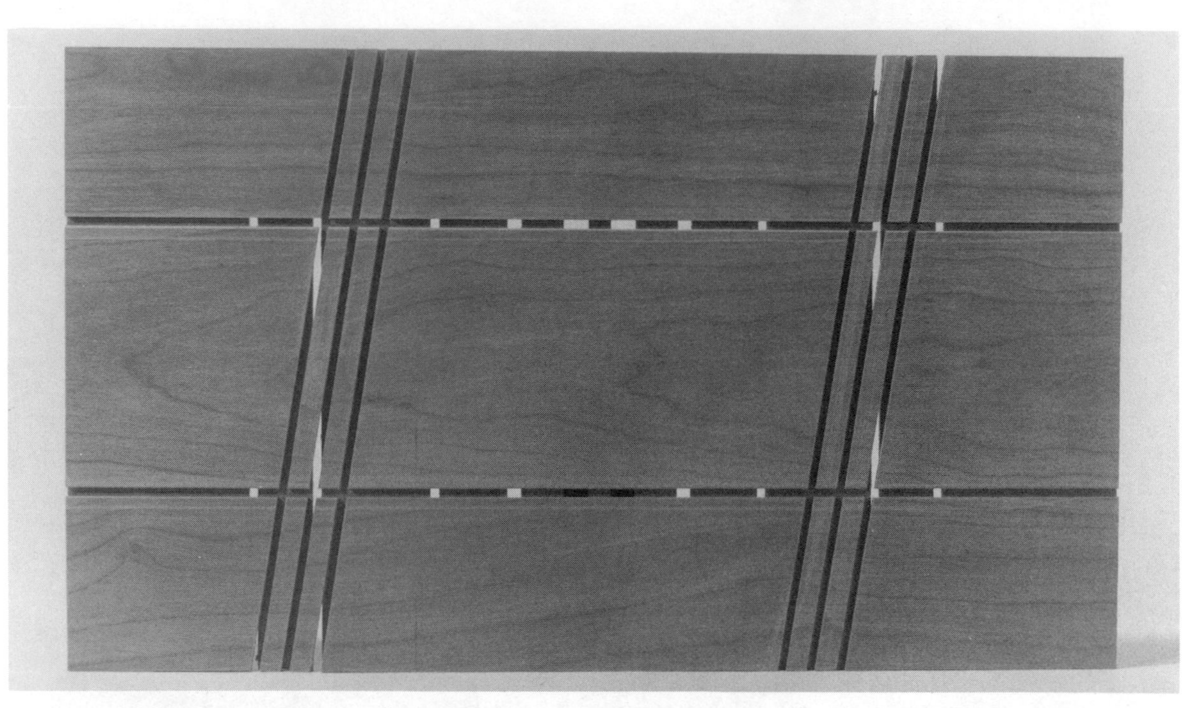

Fig. 11-56. Combining angular cuts with straight ones can provide many variations. Be careful with the spacing of close cuts to avoid weak areas.

Fig. 11-57. The secret of piercing with a saw blade is to make all cuts so they are a bit deeper than half the stock's thickness. Openings occur wherever the cuts cross.

Fig. 11-58. Make angular cuts by using a tapering jig. Remember that the visible kerfs contribute to the design, so take them into account when you plan the project.

of origami done with wood. You don't, of course, fold the wood as you would paper, but results are similar to what occurs when you fold a sheet of paper lengthwise several times and then cut into an edge with scissors. Unfold the paper and you have a decorative "panel." The technique with a table saw calls for making matching cuts across one or both surfaces of solid stock, and then ripping the prepared material into thin or thick strips that can be joined edge to edge as a panel (Fig. 11-60).

It is a fun chore, even though it requires a little time and careful cutting, but it has practical applications. Doodling in wood can result in original molding designs when the strips are used individually, or they can be assembled as filigree panels for grids, screens, doors, and so on. You can create unique tambour doors by mounting the strips on a flexible material like canvas, which is traditional for tambour doors, or on a colorful cloth, which might be more appropriate since the assembled strips will have see-through areas.

Results depend entirely on how you plan the saw cuts, something you can do on paper. The cuts can be simple kerfs of varying depths, like those in Fig. 11-61. When the wood is strip-cut and the pieces are assembled edge to edge, the final product will look like the sample in Fig. 11-62. Figure 11-63 demonstrates how V-cuts can be combined with straight kerfs for effects like those in Fig. 11-64.

Limit the depth of closely spaced cuts made on opposite surfaces of the stock to less than half the stock's thickness to avoid weak areas. You can doodle with any species of wood, but it should be kiln dried and free of blemishes. When you strip-cut the pieces, be sure to use a pusher that will let you get the parts safely past the saw blade. Make the initial cuts and the ripping with a planer-type saw blade.

You can use contact cement or conventional glue to hold parts together when you need a solid panel. Sand the panel after the glue has dried. If you have worked carefully and used a smooth-cutting blade, a light touch with a pad sander will finish the job. If the strips are thick and the project—say, a door or screen—calls for a frame, you might be able to get by without the gluing chore.

The dimensions of the panel you end up with depends on the size of the wood you start with, but remember to allow for the kerfs made when ripping. Panels do not need to be small. For example, if you start with wood that is 2 inches thick, 6 inches wide, and 6 feet long, and the blade you use cuts a ⅛-inch kerf, you can produce a panel that is 4 feet wide and 6 feet long.

HOW TO MAKE A VERTICAL TABLE

When you look at the homemade accessory in Fig. 11-65, you might think "tenoning jig." Actually, the project is used for many such applications, but the concept is a step up from a fence-straddling design because it incorporates a "miter slide" that is used on a plane positioned at right angles to the saw's table. Also, the jig is solidly secured to the rip fence so it can't rock or tilt (Fig. 11-66). Jig-to-blade settings are established by the position of the rip fence. The miter slide incorporates a clamping device,

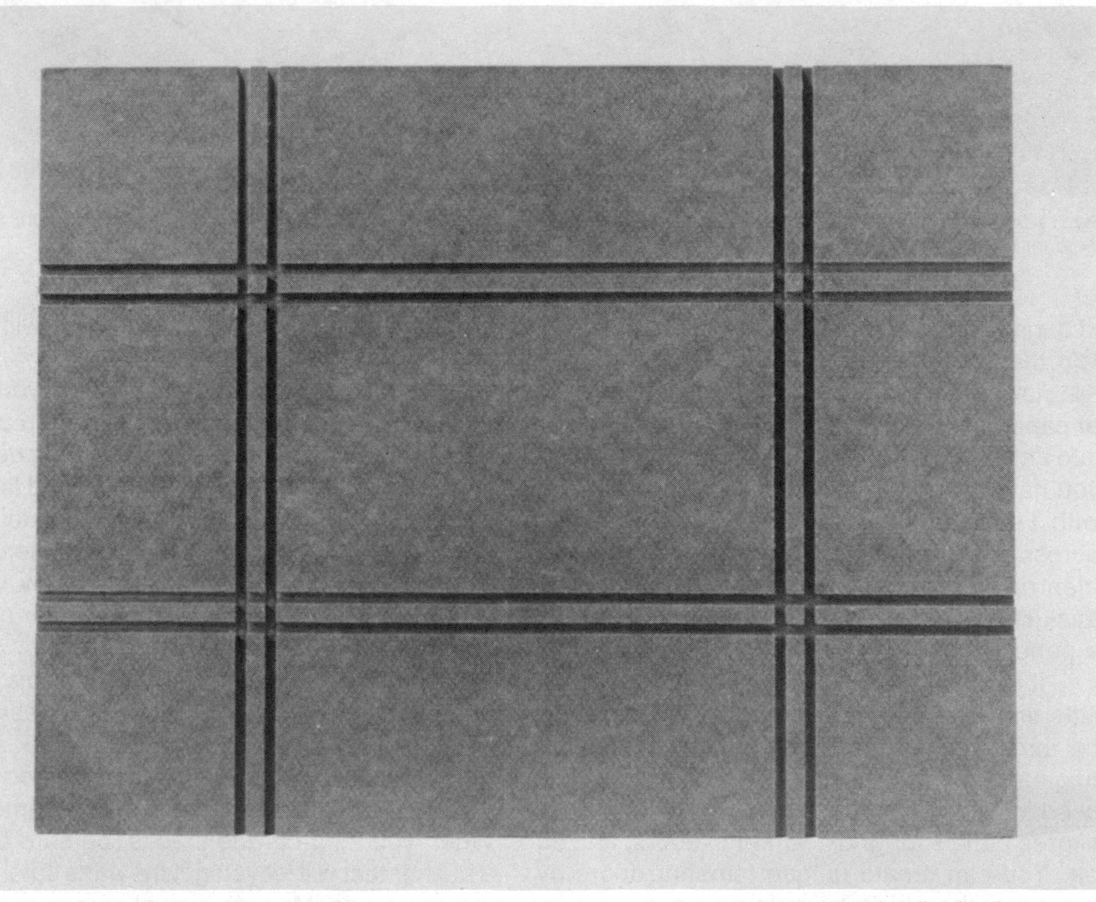

Fig. 11-59. Intersecting kerfs made on only one side of the panel can produce interesting and visually pleasing designs. Use a smooth-cutting blade for this kind of work and for piercing.

which is just an eyebolt, so extra clamps to hold workpieces securely are not needed (Fig. 11-67).

Figures 11-68 through 11-70 demonstrate one of the basic functions for the jig. An asset of the concept is that the body of the jig remains solidly in position while workpieces are moved past the blade with the miter slide.

Construction details for the jig and for an extra slide that will allow you to position mitered pieces for spline or feather grooves are shown in Figs. 11-71 and 11-72.

You can use plywood for jig projects, but it is wise to choose a maple or birch cabinet-grade with a sound surface veneer. Be careful when you assemble the bottom slide guide to a backboard because the L shape that is formed should suit the rip fence nicely. The length of the L can be a bit less than the height of the fence so the jig won't scrape on the table when you position the fence.

Be sure that the top edge of the bottom slide guide is parallel to the table surface. The best way to assemble the parts, after the rabbets are cut, is to clamp the guide to the fence and then clamp the backboard in place while you drill holes for the screws that hold the parts together. Drive the screws through the backboard. Then, remove the assembled pieces, put the miter

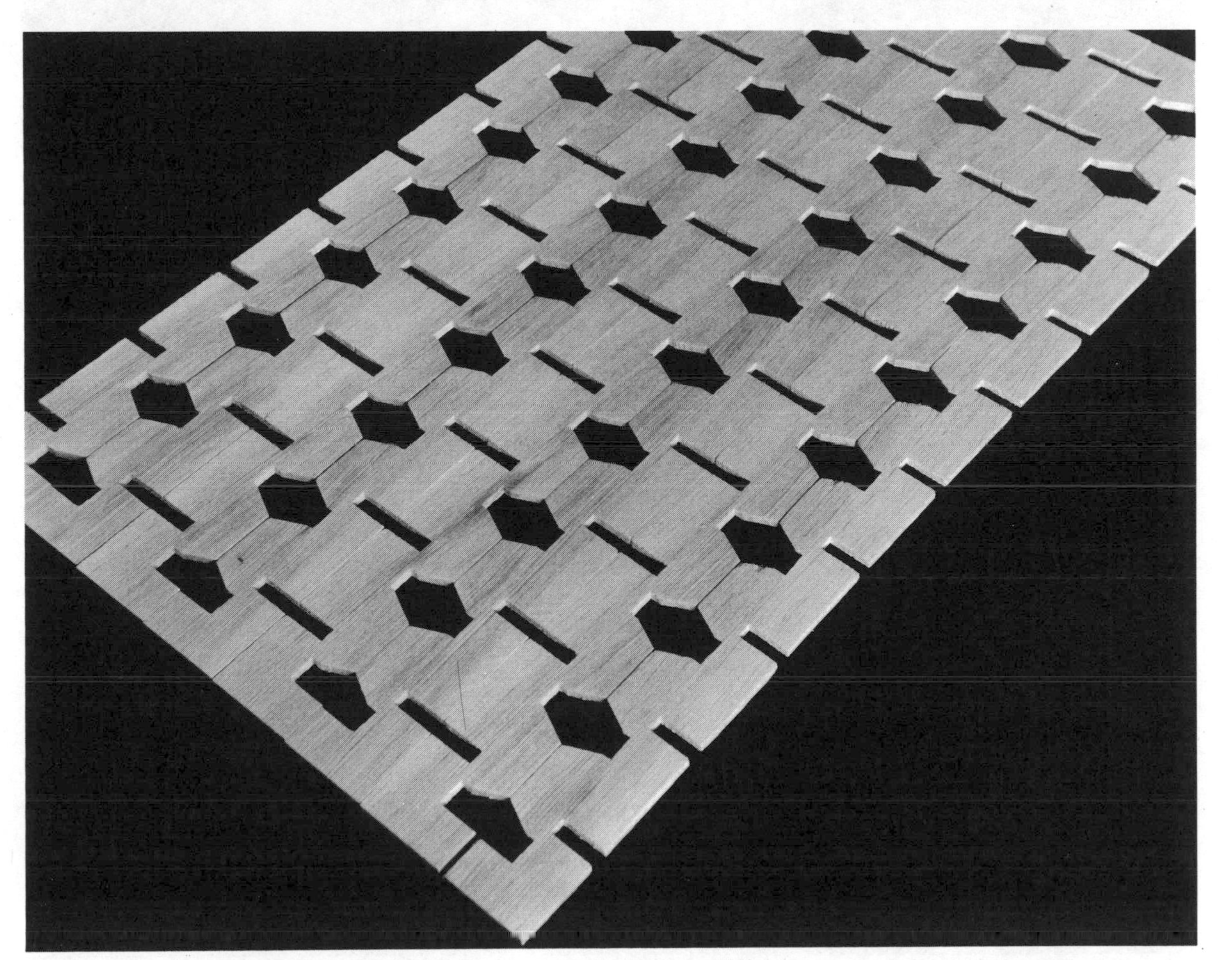

Fig. 11-60. Origami with wood? Well, not quite, but if you rip thin or thick pieces from a block that has been cut on one or both surfaces in a particular fashion and join them edge to edge, you can create panels that look like this.

Fig. 11-61. The starting cuts can be simple kerfs of varying depths. When you use the rip fence to gauge the spacing of the kerfs, be sure that the workpiece is wide enough to provide good contact surface.

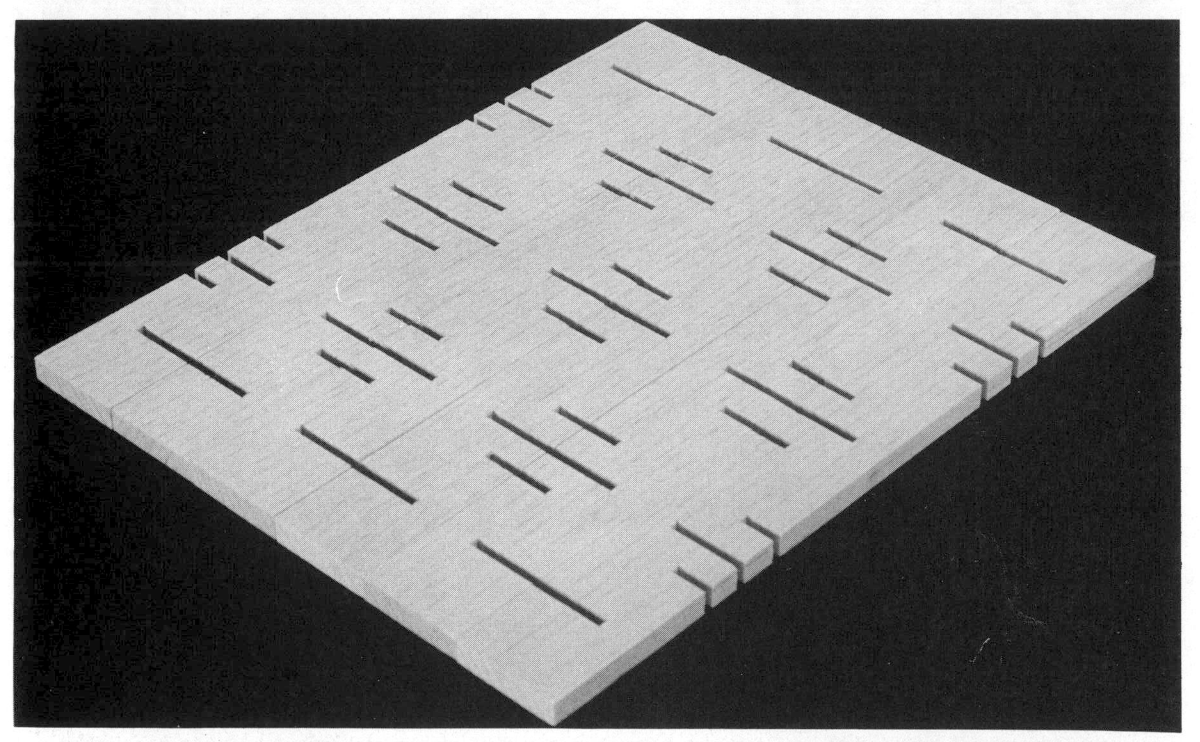

Fig. 11-62. Kerf-cut pieces that have been sliced from the parent stock will look like this when they are joined edge to edge. You can bond them with contact cement or plain glue. If they are thick enough and will be framed, you might be able to skip the gluing chore.

Fig. 11-63. Another example of how to plan the initial cuts. The Vs are formed by tilting the blade for matching cuts. Final cuts are straight kerfs at the center of the V.

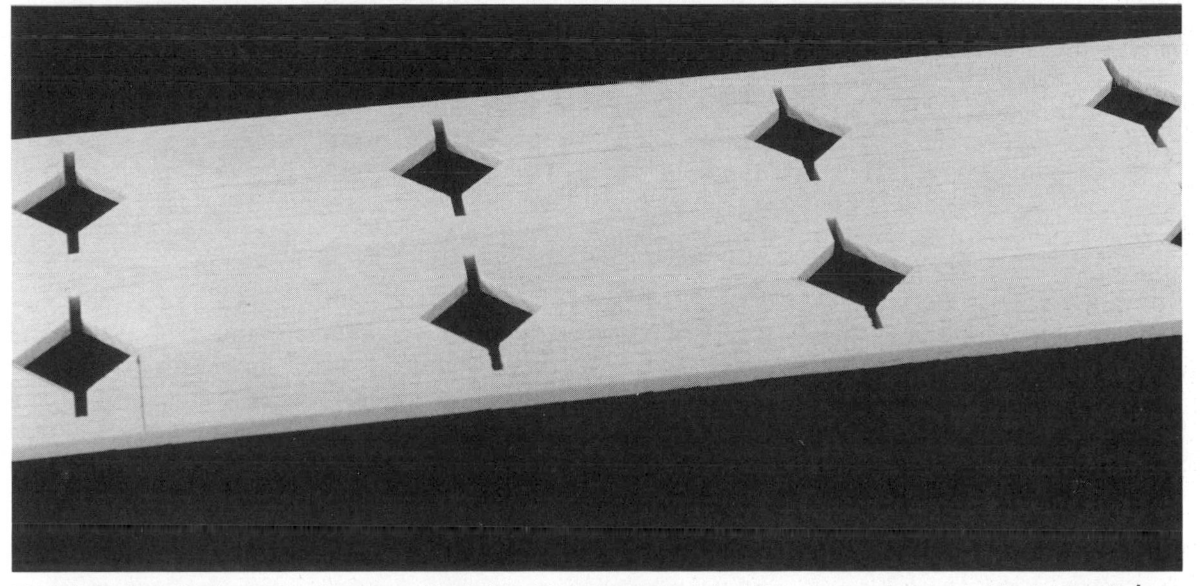

Fig. 11-64. The combination of V-cuts and kerfs has this appearance when the strip-cut pieces are joined. The designs that can be created are limitless. You can preview results by sketching on paper.

Fig. 11-65. The vertical table incorporates a miter slide that moves on a plane at right angles to the machine's table. The miter slide is rabbeted on each edge to match similar cuts made in the top and bottom slide guides.

slide in place, and add the top slide guide, again by driving screws through the backboard. The miter slide must be a fairly snug fit, but not so tight that you must use force to move it. When you add the head to the miter slide, be sure that the angle between its bearing edge and the saw's table is 90 degrees. You must install the 3/8-inch T-nut for the eyebolt, which serves as a clamp, before you attach the clamp block.

Sand all parts carefully before and after assembly. Take special care with edges and surfaces that must slide against each other. Apply a coat of sealer to all components and do a final smoothing with fine sandpaper. A well-rubbed application of paste wax on all surfaces is a good idea. Repeat the wax application frequently, especially over the surfaces and edges of the miter slide and the area of the jig in which it must slide.

Take good care when using the table saw or any power tool. Obey the rules. Read and reread instructions that come with the tool and with accessories. Remember that overconfidence and casualness are poor associates of expertise. A happy and productive shop is a safe shop.

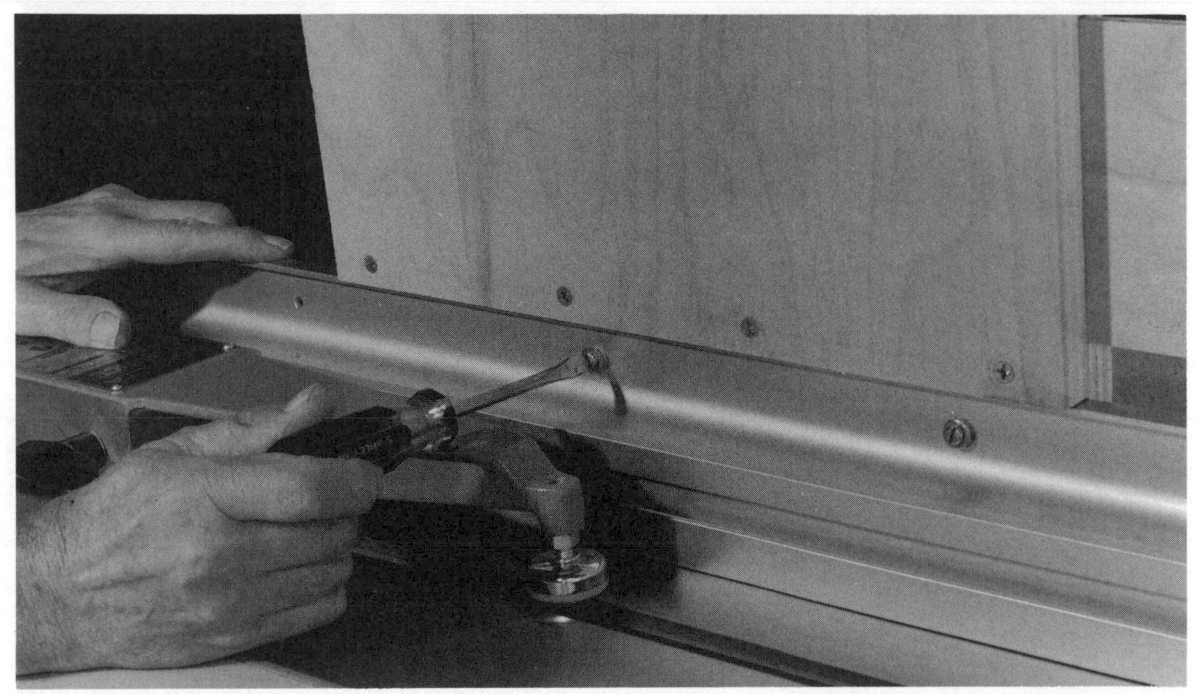

Fig. 11-66. The vertical table jig is attached firmly to the rip fence with screws. Most rip fences have holes for such purposes. If your fence lacks holes, you can drill a few to suit the jig.

Fig. 11-67. The clamping device is an eyebolt that threads through a T-nut set flush into the underside of the clamp block. Protect the workpiece by placing a thin piece of material between it and the shaft of the eyebolt.

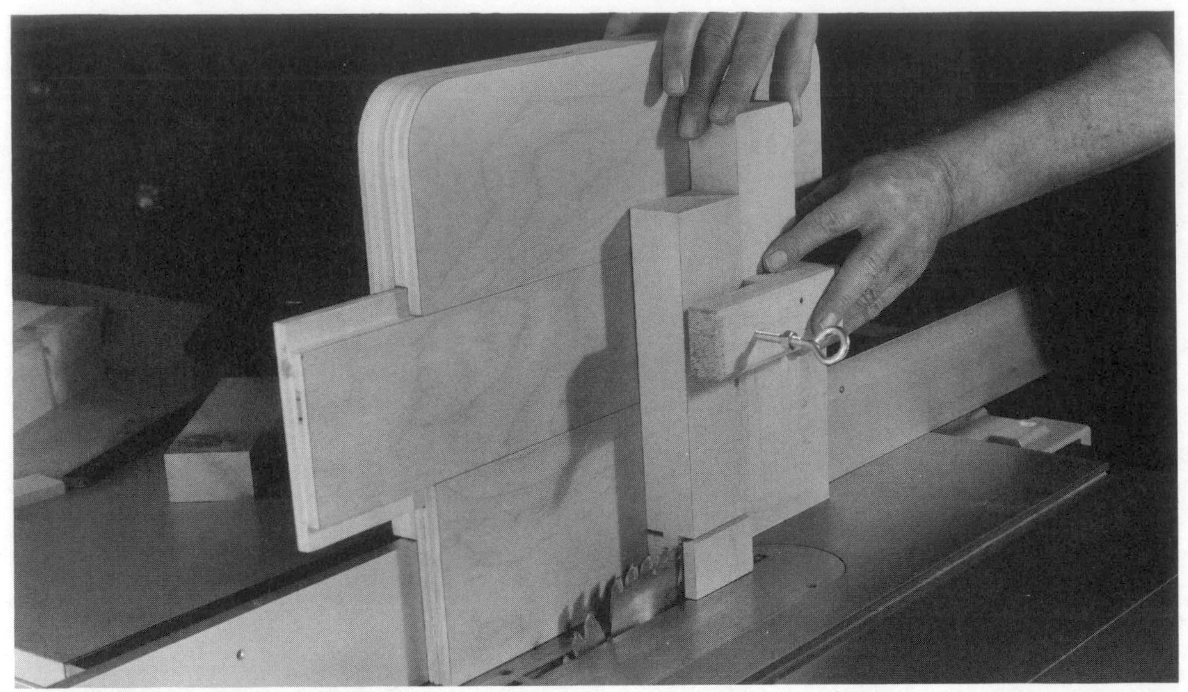

Fig. 11-68. Make the cheek cuts for tenons after you have formed the shoulder cuts. The position of the rip fence determines the distance between the face of the jig and the saw blade.

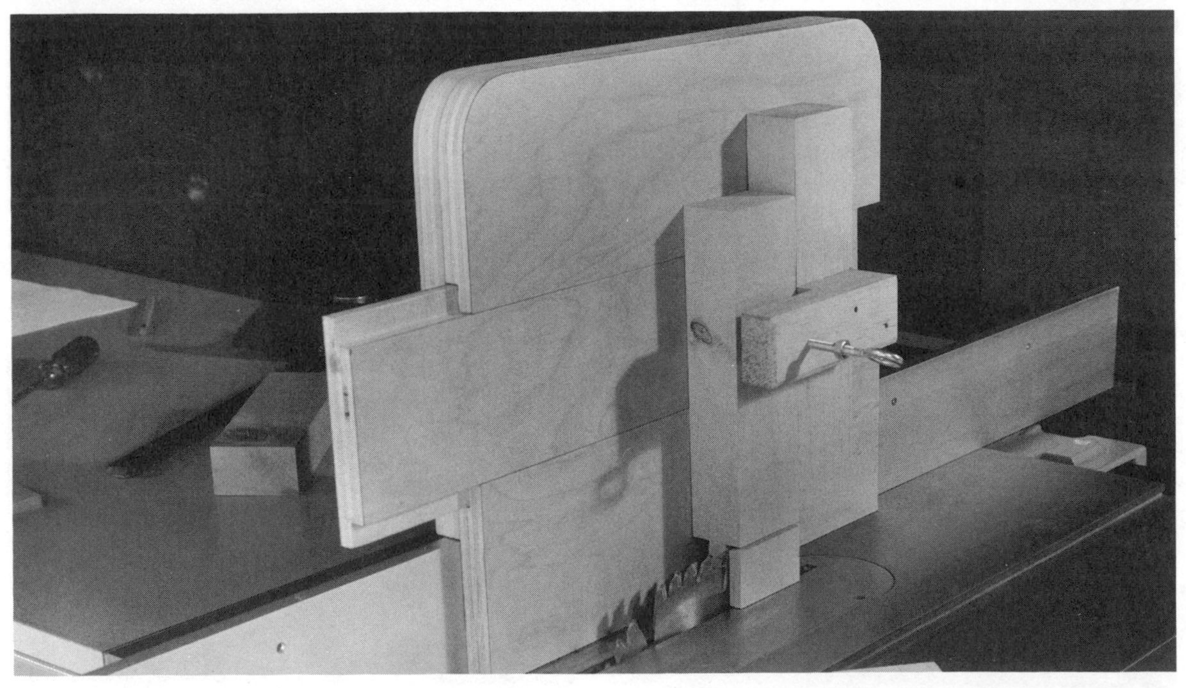

Fig. 11-69. Make the second cheek cut after you have placed the opposite surface of the stock against the jig. Work so waste will be on the outboard side of the blade. Keep mating surfaces waxed so the miter slide will move easily.

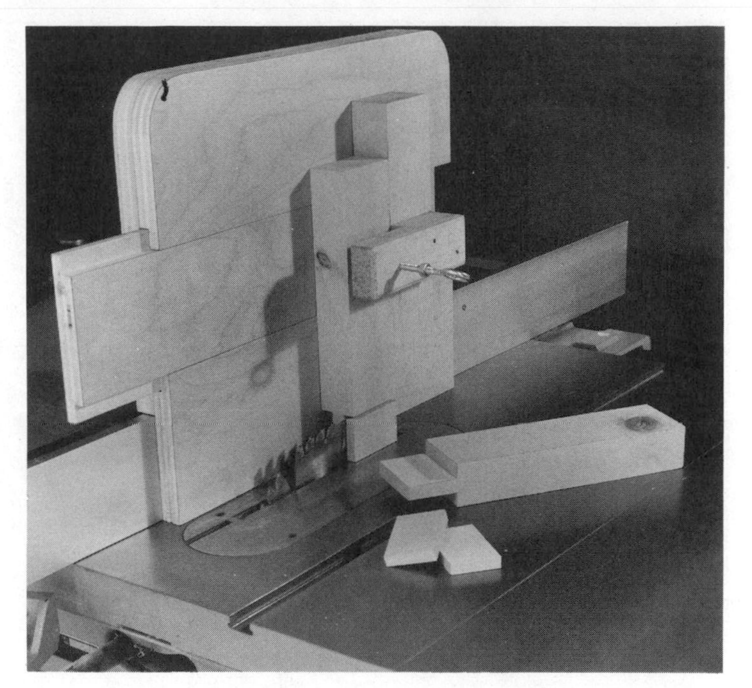

Fig. 11-70. The jig provides consistent accuracy so duplicate pieces will be alike. If you use a dadoing tool instead of a saw blade, the cheek and shoulder cuts required for a tenon would be formed in one pass.

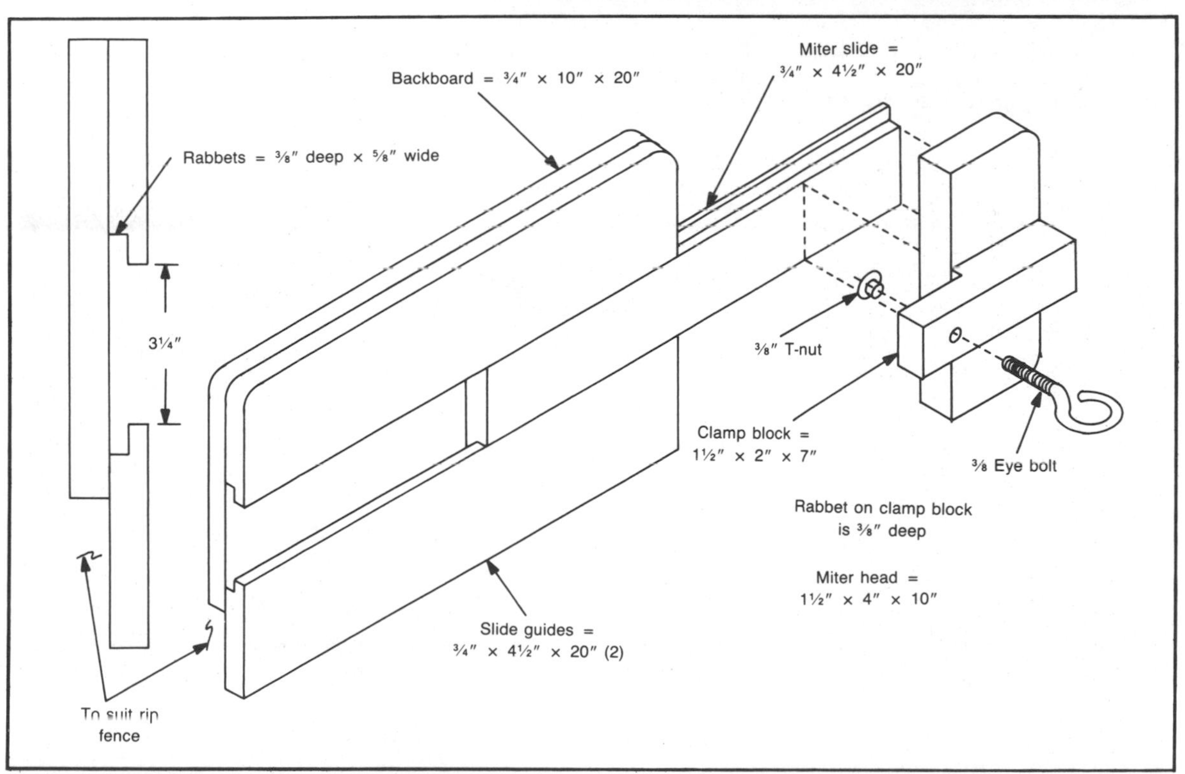

Fig. 11-71. Method for making the vertical table. Be certain that the miter slide will move parallel to the saw's table and that the angle between the head of the slide and the table is 90 degrees.

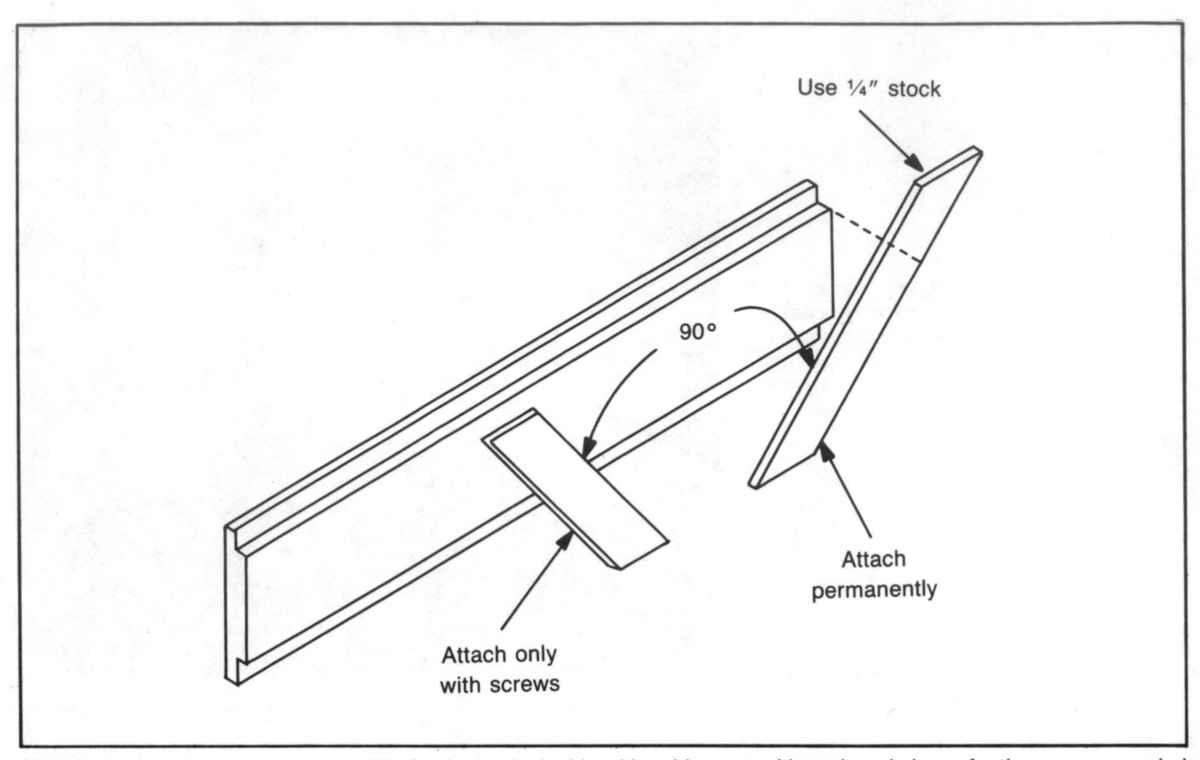

Fig. 11-72. You can make an extra slide for the vertical table with guides to position mitered pieces for the grooves needed to reinforce miter joints with splines or feathers.

Index

V

W

Edited by Suzanne L. Cheatle

Other Bestsellers From TAB

☐ **THE BAND SAW BOOK—with 20 Projects —R. J. De Cristoforo**

Considered by *Popular Science* magazine to be "the outstanding tool authority in the world," R. J. De Cristoforo offers a thorough and expert examination of the band saw, its operation, and its applications in this new book. Extended coverage on this multifaceted tool unavailable in other volumes is provided here. Projects to practice your skills include boxes, bird feeders, an end table, serving tray, and more. 320 pp., 367 illus.

Paper $16.95 **Hard $25.95**
Book No. 3189

☐ **CLOCKMAKING: 18 Antique Designs for the Woodworker—John A. Nelson**

Create timepieces of everlasting beauty and service using this illustrated, step-by-step guide to building antique clock reproductions. From the elegant long-case Grandfather clock to the one-of-a-kind Banjo clock, there is something here for woodworkers of all tastes and skill levels. Nelson's instructional savvy takes you through cutting the pieces, assembling the parts, and preparing, distressing, and finishing the wood to achieve an authentic look. 240 pp., 136 illus.

Paper $18.95 **Hard $26.95**
Book No. 3164

☐ **THE WOODWORKER'S SHOP: 100 Projects to Enhance Your Workspace—Percy W. Blandford**

Making tools, accessories, and equipment for for the home workshop, that will stay in the workshop, is the focus of this new book. More than 100 projects are presented in all, each with step-by-step instructions, detailed drawings, and materials lists. Ranging in complexity to suit any degree of skill, all are designed to help you do your work faster, more economically, and with greater accuracy. 270 pp.m 174 illus.

Paper $14.95 **Hard $22.95**
Book No. 3134

☐ **MAKING ANTIQUE FURNITURE—Edited by Vic Taylor**

A collection of some of the finest furniture ever made is found within the pages of this project book designed for the intermediate- to advanced-level craftsman. Reproducing European period furniture pieces such as a Winsdor chair, a Jacobean box stool, a Regency table, a Sheraton writing desk, a Lyre-end occasional table, and many traditional furnishings is sure to provide you with pleasure and satisfaction. Forty projects include materials lists and step-by-step instructions. 160 pp., Fully illustrated

Paper $15.95 **Hard $25.95**
Book No. 3056

☐ **THE WOODWORKER'S ILLUSTRATED BENCHTOP REFERENCE —William P. Spence and L. Duane Griffiths**

Whatever your woodworking question, you'll find the answer in this book, and what's more, you'll find it quickly and easily. Specifically aimed at the homecraftsman, *The Woodworkers Illustrated Benchtop Reference* covers every facet of woodworking, from choosing the right wood to bringing up a luster on your finished projects. Nearly 500 pages and more than 600 illustrations present the latest woodworking materials and technology, time-proven techniques, expert insight, and classic woodworking projects. 496 pp., 657 illus.

Paper $24.95 **Hard $34.95**
Book No. 3177

☐ **SMALL BUILDINGS—Percy W. Blandford**

This book is the do-it-yourselfer's solution to a diversity of homeowner's dilemmas. There is a small building project in this book to suit every need. Playhouses, treehouses, carports, workshops and studios, sun lounges, gazebos, greenhouses, stables, and plain old storage sheds are among the 37 structures included. Complete plans, step-by-step instructions, materials lists, and detailed illustrations are presented with each project. 300 pp., 190 illus.

Paper $16.95 **Hard $25.95**
Book No. 3144

☐ **101 OUTSTANDING WOODEN TOY AND CHILDREN'S FURNITURE PROJECTS—Wayne L. Kadar**

Turn inexpensive materials into fun and functional toys. Challenge and charm the youngsters in your life with building blocks, pull toys, shape puzzles, stilts, trains, trucks, boats, planes, dolls and more. This step-by-step guide is abundantly illustrated and provides complete materials lists. 304 pp., 329 illus.

Paper $15.95 **Hard $24.95**
Book No. 3058

☐ ***WOODWORKER'S* 30 BEST PROJECTS—Editors of *Woodworker* Magazine**

A collection of some of the finest furniture ever made can be found within the pages of this project book. Designed for the woodworker who has already mastered the basics, the projects presented in this book are for the intermediate- to advanced-level craftsman. Each furniture project comes complete with detailed instructions, a materials list, exploded views of working diagrams, a series of step-by-step, black and-white photos, and a photograph of the finished piece. 224 pp.m 300 illus.

Paper $14.95 **Hard $23.95**
Book No. 3021